Accelerated Optimization for Machine Learning

Zhouchen Lin • Huan Li • Cong Fang

Accelerated Optimization for Machine Learning

First-Order Algorithms

 Springer

Zhouchen Lin (iD)
Key Lab. of Machine Perception
School of EECS
Peking University
Beijing, Beijing, China

Huan Li
College of Computer Science
and Technology
Nanjing University of Aeronautics
and Astronautics
Nanjing, Jiangsu, China

Cong Fang
School of Engineering and Applied Science
Princeton University
Princeton, NJ, USA

ISBN 978-981-15-2909-2   ISBN 978-981-15-2910-8   (eBook)
https://doi.org/10.1007/978-981-15-2910-8

This Springer imprint is published by the registered company Springer Nature Singapore Pte Ltd.
The registered company address is: 152 Beach Road, #21-01/04 Gateway East, Singapore 189721, Singapore

*To our families. Without your great support
this book will not exist and even our careers
will be meaningless.*

Foreword by Michael I. Jordan

Optimization algorithms have been the engine that have powered the recent rise of machine learning. The needs of machine learning are different from those of other disciplines that have made use of the optimization toolbox; most notably, the parameter spaces are of high dimensionality, and the functions that are being optimized are often sums of millions of terms. In such settings, gradient-based methods are preferred over higher order methods, and given that the computation of a full gradient can be infeasible, stochastic gradient methods are the coin of the realm. Putting such specifications together with the need to solve nonconvex optimization problems, to control the variance induced by the stochastic sampling, and to develop algorithms that run on distributed platforms, one poses a new set of challenges for optimization. Surprisingly, many of these challenges have been addressed within the past decade.

The book by Lin, Li, and Fang is one of the first book-length treatments of this emerging field. The book covers gradient-based algorithms in detail, with a focus on the concept of acceleration. Acceleration is a key concept in modern optimization, supplying new algorithms and providing insight into achievable convergence rates. The book also covers stochastic methods, including variance control, and it includes material on asynchronous distributed implementations.

Any researcher wishing to work in the machine learning field should have a foundational understanding of the disciplines of statistics and optimization. The current book is an excellent place to obtain the latter and to begin one's adventure in machine learning.

University of California   Michael I. Jordan
Berkeley, CA, USA
October 2019

Foreword by Zongben Xu

Optimization is one of the core topics in machine learning. While benefiting from the advances in the native optimization community, optimization for machine learning has its own flavor. One remarkable phenomenon is that *first-order* algorithms more or less dominate the optimization methods in machine learning. While there have been some books or preprints that introduce major optimization algorithms used in machine learning, either partially or thoroughly, this book focuses on a notable stream in recent machine learning optimization, namely the *accelerated first-order* methods. Originating from Polyak's heavy-ball method and triggered by Nesterov's series of works, accelerated first-order methods have become a hot topic in both the optimization and the machine learning communities and have yielded fruitfully. The results have significantly extended beyond the traditional scope of unconstrained (and deterministic) convex optimization. New results include acceleration for constrained convex optimization and nonconvex optimization, stochastic algorithms, and general acceleration frameworks such as Katyusha and Catalyst. Some of them even have nearly optimal convergence rates. Unfortunately, existing literatures scatter across diverse and extensive publications. Mastering the basic techniques and having a global picture of this dynamic field thus becomes very difficult.

Fortunately, this monograph, coauthored by Zhouchen Lin, Huan Li, and Cong Fang, meets the need of quick education on accelerated first-order algorithms just in time. The book first gives an overview on the development of accelerated first-order algorithms, which is extremely informative, despite being sketchy. Then, it introduces the representative works in different categories, with detailed proofs that greatly facilitate the understanding of underlying ideas and the mastering of basic techniques. Without doubt, this book is a vital reference for those who want to learn the state of the art of machine learning optimization.

I have known Dr. Zhouchen Lin for a long time. He impresses me with solid work, deep insights, and careful analysis on the problems arising from his diverse research fields. With a lot of shared research interests, one of which is learning-based optimization, I am delighted to see this book finally published after elaborative writing.

Xi'an Jiaotong University   Zongben Xu
Xi'an, China
October 2019

Foreword by Zhi-Quan Luo

First-order optimization methods have been the main workhorse in the machine learning, signal processing, and artificial intelligence involving big data. These methods, while simple conceptually, require careful analysis and a good understanding of them to be effectively deployed. The issues such as acceleration, nonsmoothness, nonconvexity, parallel and distributed implementation are critical due to their great impact on the algorithm's convergence behavior and running time.

This research monograph gives an excellent introduction to the algorithmic aspects of first-order optimization methods, focusing on algorithm design and convergence analysis. It treats in depth the issues of acceleration, nonconvexity, constraints, and asynchronous implementation. The topics covered and the results given in the monograph are very timely and strongly relevant to both the researchers and practitioners of machine learning, signal processing, and artificial intelligence. The theoretical issues of lower bounds on complexity are purposely avoided to give way to algorithm design and convergence analysis. Overall, the treatment of the subject is quite balanced and many useful insights are provided throughout the monograph.

The authors of this monograph are experienced researchers at the interface of machine learning and optimization. The monograph is very well written and makes an excellent read. It should be an important reference book for everyone interested in the optimization aspects of machine learning.

The Chinese University of Hong Kong   Zhi-Quan Luo
Shenzhen, China
October 2019

Preface

While I was preparing advanced materials for the optimization course taught at Peking University, I found that accelerated algorithms is the most attractive and practical topic for students in engineering. Actually, this is also a hot topic of current machine learning conferences. While some books have introduced some accelerated algorithms, such as [1–3], they are nevertheless incomplete, unsystematic, and not up-to-date. Thus, in early 2018, I decided to write a monograph on accelerated algorithms. My goal was to produce a book that is organized and self-contained, with sufficient preliminary materials and detailed proofs, so that the readers need not consult scattered literatures, be plagued by inconsistent notations, and be carried away from the central ideas by non-essential contents. Luckily, my two Ph.D. students, Huan Li and Cong Fang, were happy to join this work.

This task turned out to be very hard, as we had to work among our busy schedules. Eventually, we managed to have the first complete yet crude draft right before Huan Li and Cong Fang graduated. Smoothing the book and correcting various errors further took us 4 months. Finally, we were truly honored to have forewords from Prof. Michael I. Jordan, Prof. Zongben Xu, and Prof. Zhi-Quan Luo. While this book deprived us of all our leisure time in the past nearly 2 years, we still feel that our endeavor pays when every part of the book is ready.

Hope this book is a valuable reference for the machine learning and the optimization communities. This will be the highest praise for our work.

Beijing, China   Zhouchen Lin
November 2019

References

1. A. Beck, *First-Order Methods in Optimization*, vol. 25 (SIAM, Philadelphia, 2017)
2. S. Bubeck, Convex optimization: algorithms and complexity. Found. Trends Mach. Learn. **8**(3–4), 231–357 (2015)
3. Y. Nesterov, *Lectures on Convex Optimization* (Springer, New York, 2018)

Acknowledgements

The authors would like to thank all our collaborators and friends, especially: Bingsheng He, Junchi Li, Qing Ling, Guangcan Liu, Risheng Liu, Yuanyuan Liu, Canyi Lu, Zhiquan Luo, Yi Ma, Fanhua Shang, Zaiwen Wen, Xingyu Xie, Chen Xu, Shuicheng Yan, Wotao Yin, Xiaoming Yuan, Yaxiang Yuan, and Tong Zhang. The authors also thank Yuqing Hou, Jia Li, Shiping Wang, Jianlong Wu, Hongyang Zhang, and Pan Zhou for careful proofreading. The authors also thank Celine Chang from Springer, who offered much assistance during the production of the book. This monograph is supported by National Natural Science Foundation of China under Grant Nos. 61625301 and 61731018 and Beijing Academy of Artificial Intelligence.

Contents

About the Authors

Zhouchen Lin is a leading expert in the fields of machine learning and computer vision. He is currently a Professor at the Key Laboratory of Machine Perception (Ministry of Education), School of EECS, Peking University. He served as an area chair for several prestigious conferences, including CVPR, ICCV, ICML, NIPS/NeurIPS, AAAI and IJCAI. He is an associate editor of the IEEE Transactions on Pattern Analysis and Machine Intelligence and the International Journal of Computer Vision. He is a Fellow of IAPR and IEEE.

Huan Li received his Ph.D. degree in machine learning from Peking University in 2019. He is currently an Assistant Professor at the College of Computer Science and Technology, Nanjing University of Aeronautics and Astronautics. His current research interests include optimization and machine learning.

Cong Fang received his Ph.D. degree from Peking University in 2019. He is currently a Postdoctoral Researcher at Princeton University. His research interests include machine learning and optimization.

Acronyms

AAAI	Association for the Advancement of Artificial Intelligence
AACD	Asynchronous Accelerated Coordinate Descent
AAGD	Asynchronous Accelerated Gradient Descent
AASCD	Asynchronous Accelerated Stochastic Coordinate Descent
AC-AGD	Almost Convex Accelerated Gradient Descent
Acc-ADMM	Accelerated Alternating Direction Method of Multiplier
Acc-SADMM	Accelerated Stochastic Alternating Direction Method of Multiplier
Acc-SDCA	Accelerated Stochastic Dual Coordinate Ascent
ADMM	Alternating Direction Method of Multiplier
AGD	Accelerated Gradient Descent
APG	Accelerated Proximal Gradient
ASCD	Accelerated Stochastic Coordinate Descent
ASGD	Asynchronous Stochastic Gradient Descent
ASVRG	Asynchronous Stochastic Variance Reduced Gradient
DSCAD	Distributed Stochastic Communication Accelerated Dual
ERM	Empirical Risk Minimization
EXTRA	EXact firsT-ordeR Algorithm
GIST	General Iterative Shrinkage and Thresholding
IC	Individually Convex
IFO	Incremental First-order Oracle
INC	Individually Nonconvex
iPiano	Inertial Proximal Algorithms for Nonconvex Optimization
IQC	Integral Quadratic Constraint
KKT	Karush–Kuhn–Tucker
KŁ	Kurdyka–Łojasiewicz
LASSO	Least Absolute Shrinkage and Selection Operator
LMI	Linear Matrix Inequality
MISO	Minimization by Incremental Surrogate Optimization
NC	Negative Curvature/Nonconvex
NCD	Negative Curvature Descent
PCA	Principal Component Analysis

PG	Proximal Gradient
SAG	Stochastic Average Gradient
SAGD	Stochastic Accelerated Gradient Descent
SCD	Stochastic Coordinate Descent
SDCA	Stochastic Dual Coordinate Ascent
SGD	Stochastic Gradient Descent
SPIDER	Stochastic Path-Integrated Differential Estimator
SVD	Singular Value Decomposition
SVM	Support Vector Machine
SVRG	Stochastic Variance Reduced Gradient
SVT	Singular Value Thresholding
VR	Variance Reduction

Chapter 1
Introduction

Optimization is a supporting technology in many numerical computation related research fields, such as machine learning, signal processing, industrial design, and operation research. In particular, P. Domingos, an AAAI Fellow and a Professor of University of Washington, proposed a celebrated formula [23]:

machine learning = representation + optimization + evaluation,

showing the importance of optimization in machine learning.

1.1 Examples of Optimization Problems in Machine Learning

Optimization problems arise throughout machine learning. We provide two representative examples here. The first one is classification/regression and the second one is low-rank learning.

Many classification/regression problems can be formulated as

$$\min_{\mathbf{w} \in \mathbb{R}^n} \frac{1}{m} \sum_{i=1}^{m} l(p(\mathbf{x}_i; \mathbf{w}), y_i) + \lambda R(\mathbf{w}), \qquad (1.1)$$

where $\mathbf{w}$ consists of the parameters of a classification/regression system, $p(\mathbf{x}; \mathbf{w})$ represents the prediction function of the learning model, l is the loss function to punish the inconformity between the system prediction and the truth value, $(\mathbf{x}_i, y_i)$ is the i-th data sample with $\mathbf{x}_i$ being the datum/feature vector and y_i the label for classification or the corresponding value for regression, R is a regularizer that

© Springer Nature Singapore Pte Ltd. 2020
Z. Lin et al., *Accelerated Optimization for Machine Learning*,
https://doi.org/10.1007/978-981-15-2910-8_1

enforces some special property in $\mathbf{w}$, and $\lambda \geq 0$ is a trade-off parameter. Typical examples of $l(p, y)$ include the squared loss $l(p, y) = \frac{1}{2}(p - y)^2$, the logistic loss $l(p, y) = \log(1 + \exp(-py))$, and the hinge loss $l(p, y) = \max\{0, 1 - py\}$. Examples of $p(\mathbf{x}; \mathbf{w})$ include $p(\mathbf{x}; \mathbf{w}) = \mathbf{w}^T \mathbf{x} - b$ for linear classification/regression and $p(\mathbf{x}; \mathbf{W}) = \phi(\mathbf{W}_n \phi(\mathbf{W}_{n-1} \cdots \phi(\mathbf{W}_1 \mathbf{x}) \cdots))$ for forward propagation widely used in deep neural networks, where $\mathbf{W}$ is a collection of the weight matrices $\mathbf{W}_k$, $k = 1, \cdots, n$, and ϕ is an activation function. Representative examples of $R(\mathbf{w})$ include the ℓ_2 regularizer $R(\mathbf{w}) = \frac{1}{2}\|\mathbf{w}\|^2$ and the ℓ_1 regularizer $R(\mathbf{w}) = \|\mathbf{w}\|_1$.

The combinations of different loss functions, prediction functions, and regularizers lead to different machine learning models. For example, hinge loss, linear classification function, and ℓ_2 regularizer give the support vector machine (SVM) problem [21]; logistic loss, linear regression function, and ℓ_2 regularizer give the regularized logistic regression problem [10]; square loss, forward propagation function, and $R(\mathbf{W}) = 0$ give the multi-layer perceptron [33]; and square loss, linear regression function, and ℓ_1 regularizer give the LASSO problem [68].

There are also many problems investigated by the machine learning community that are not of the form of (1.1). For example, the matrix completion problem, which has wide applications in signal and data processing, can be written as:

$$\min_{\mathbf{X} \in \mathbb{R}^{m \times n}} \|\mathbf{X}\|_*,$$

$$s.t. \quad \mathbf{X}_{ij} = \mathbf{D}_{ij}, \forall (i, j) \in \Omega,$$

where Ω is the locations of observed entries. The low-rank representation (LRR) problem [50], which is powerful in clustering data into subspaces, is cast as:

$$\min_{\mathbf{Z} \in \mathbb{R}^{n \times n}, \mathbf{E} \in \mathbb{R}^{m \times n}} \|\mathbf{Z}\|_* + \lambda \|\mathbf{E}\|_1,$$

$$s.t. \quad \mathbf{D} = \mathbf{D}\mathbf{Z} + \mathbf{E}.$$

To reduce the computational cost as well as the storage space, people observe that a low-rank matrix can be factorized as a product of two much smaller matrices, i.e., $\mathbf{X} = \mathbf{U}\mathbf{V}^T$. Take the matrix completion problem as an example, it can be reformulated as follows, which is a nonconvex problem,

$$\min_{\mathbf{U} \in \mathbb{R}^{m \times r}, \mathbf{V} \in \mathbb{R}^{n \times r}} \frac{1}{2} \sum_{(i, j) \in \Omega} \left\| \mathbf{U}_i \mathbf{V}_j^T - \mathbf{D}_{ij} \right\|_F^2 + \frac{\lambda}{2} \left(\|\mathbf{U}\|_F^2 + \|\mathbf{V}\|_F^2 \right).$$

For more examples of optimization problems in machine learning, one may refer to the survey paper written by Gambella, Ghaddar, and Naoum-Sawaya in 2019 [28].

1.2 First-Order Algorithm

In most machine learning models, a moderate numerical precision of parameters already suffices. Moreover, an iteration needs to be finished in reasonable amount of time. Thus, *first-order* optimization methods are the mainstream algorithms used in the machine learning community. While "first-order" has its rigorous definition in the complexity theory of optimization, which is based on an oracle that only returns $f(\mathbf{x}_k)$ and $\nabla f(\mathbf{x}_k)$ when queried with $\mathbf{x}_k$, here we adopt a much more general sense that higher order derivatives of the objective function are not used (thus allows the closed form solution of a subproblem and the use of proximal mapping (Definition A.19), etc.). However, we do not want to write a book on all first-order algorithms that are commonly used or actively investigated in the machine learning community, which is clearly out of our capability due to the huge amount of literatures. Some excellent reference books, preprints, or surveys include [7, 12–14, 34, 35, 37, 58, 60, 66]. Rather, we focus on the *accelerated* first-order methods only, where "accelerated" means that the convergence rate is improved without making much stronger assumptions and the techniques used are essentially exquisite interpolation and extrapolation.

1.3 Sketch of Representative Works on Accelerated Algorithms

In the above sense of acceleration, the first accelerated optimization algorithm may be Polyak's heavy-ball method [61]. Consider a problem with an L-smooth (Definition A.12) and μ-strongly convex (Definition A.10) objective, and let ε be the error to the optimal solution. The heavy-ball method reduces the complexity $O\left(\frac{L}{\mu} \log \frac{1}{\varepsilon}\right)$ of the usual gradient descent to $O\left(\sqrt{\frac{L}{\mu}} \log \frac{1}{\varepsilon}\right)$. In 1983, Nesterov proposed his accelerated gradient descent (AGD) for L-smooth objective functions, where the complexity is reduced to $O\left(\frac{1}{\sqrt{\varepsilon}}\right)$ as compared with that of usual gradient descent: $O\left(\frac{1}{\varepsilon}\right)$. Nesterov further proposed another accelerated algorithm for L-smooth objective functions in 1988 [53], smoothing techniques for nonsmooth functions with acceleration tricks in 2005 [54], and an accelerated algorithm for composite functions in 2007 [55] (whose formal publication is [57]). Nesterov's seminal work did not catch much attention in the machine learning community, possibly because the objective functions in machine learning models are often nonsmooth, e.g., due to the adoption of sparse and low-rank regularizers which are not differentiable. The accelerated proximal gradient (APG) for composite functions by Beck and Teboulle [8], which was formally published in 2009 and is an extension of [53] and simpler than [55],[1] somehow gained great interest in the machine learning community as it fits well for the sparse and low-rank models which were hot topics at that time. Tseng further provided a unified analysis of existing acceleration techniques [70] and Bubeck proposed a near optimal method for highly smooth convex optimization [16].

Nesterov's AGD is not quite intuitive. There have been some efforts on interpreting his AGD algorithm. Su et al. gave an interpretation from the viewpoint of differential equations [67] and Wibisono et al. further extended it to higher order AGD [71]. Fazlyab et al. proposed a Linear Matrix Inequality (LMI) using the Integral Quadratic Constraints (IQCs) from robust control theory to interpret AGD [42]. Allen-Zhu and Orecchia connected AGD to mirror descent via the linear coupling technique [6]. On the other hand, some researchers work on designing other interpretable accelerated algorithms. Kim and Fessler designed an optimized first-order algorithm whose complexity is only one half of that of Nesterov's accelerated gradient method via the Performance Estimation Problem approach [40]. Bubeck proposed a geometric descent method inspired from the ellipsoid method [15] and Drusvyatskiy et al. showed that the same iterate sequence is generated via computing an optimal average of quadratic lower-models of the function [24].

For linearly constrained convex problems, different from the unconstrained case, both the errors in the objective function value and the constraint should be taken care of. Ideally, both errors should reduce at the same rate. A straightforward way to extend Nesterov's acceleration technique to constrained optimization is to solve its dual problem (Definition A.24) using AGD directly, which leads to the accelerated dual ascent [9] and accelerated augmented Lagrange multiplier method [36], both with the optimal convergence rate in the dual space. Lu [51] and Li [44] further analyzed the complexity in the primal space for the accelerated dual ascent and its variant. One disadvantage of the dual based method is the need to solve a subproblem at each iteration. Linearization is an effective approach to overcome this shortcoming. Specifically, Li et al. proposed an accelerated linearized penalty method that increases the penalty along with the update of variable [45] and Xu proposed an accelerated linearized augmented Lagrangian method [72]. ADMM and the primal-dual method, as the most commonly used methods for constrained optimization, were also accelerated in [59] and [20] for generally convex (Definition A.10) and smooth objectives, respectively. When the strong convexity is assumed, ADMM and the primal-dual method can have faster convergence rates even if no acceleration techniques are used [19, 72].

Nesterov's AGD has also been extended to nonconvex problems. The first analysis of AGD for nonconvex optimization appeared in [31], which minimizes a composite objective with a smooth (Definition A.11) nonconvex part and a

---

[1] In each iteration [8] uses only information from two last iterations and makes one call on proximal mapping, while [55] uses entire history of previous iterations and makes two calls on proximal mapping.

nonsmooth convex (Definition A.7) part. Inspired by [31], Li and Lin proposed AGD variants for minimizing the composition of a smooth nonconvex part and a nonsmooth nonconvex part [43]. Both works in [31, 43] studied the convergence to the first-order critical point (Definition A.34). Carmon et al. further gave an $O\left(\frac{1}{\varepsilon^{7/4}} \log \frac{1}{\varepsilon}\right)$ complexity analysis [17]. For many famous machine learning problems, e.g., matrix sensing and matrix completion, there is no spurious local minimum [11, 30] and the only task is to escape strict saddle points (Definition A.29). The first accelerated method to find the second-order critical point appeared in [18], which alternates between two subroutines: negative curvature descent and Almost Convex AGD, and can be seen as a combination of accelerated gradient descent and the Lanczos method. Jin et al. further proposed a single-loop accelerated method [38]. Agarwal et al. proposed a careful implementation of the Nesterov–Polyak method, using accelerated methods for fast approximate matrix inversion [1]. The complexities established in [1, 18, 38] are all $O\left(\frac{1}{\varepsilon^{7/4}} \log \frac{1}{\varepsilon}\right)$.

As for stochastic algorithms, compared with the deterministic algorithms, the main challenge is that the noise of gradient will not reach zero through updates and this makes the famous stochastic gradient descent (SGD) converge only with a sublinear rate even for strongly convex and smooth problems. Variance reduction (VR) is an efficient technique to reduce the negative effect of noise [22, 39, 52, 63]. With the VR and the momentum technique, Allen-Zhu proposed the first truly accelerated stochastic algorithm, named Katyusha [2]. Katyusha is an algorithm working in the primal space. Another way to accelerate the stochastic algorithms is to solve the problem in the dual space so that we can use the techniques like stochastic coordinate descent (SCD) [27, 48, 56] and stochastic primal-dual method [41, 74]. On the other hand, in 2015 Lin et al. proposed a generic framework, called Catalyst [49], that minimizes a convex objective function via an accelerated proximal point method and gains acceleration, whose idea previously appeared in [65]. Stochastic nonconvex optimization is also an important topic and some excellent works include [3–5, 29, 62, 69, 73]. Particularly, Fang et al. proposed a Stochastic Path-Integrated Differential Estimator (SPIDER) technique and attained the near optimal convergence rate under certain conditions [26].

The acceleration techniques are also applicable to parallel optimization. Parallel algorithms can be implemented in two fashions: asynchronous updates and synchronous updates. For asynchronous update, none of the machines need to wait for the others to finish computing. Representative works include asynchronous accelerated gradient descent (AAGD) [25] and asynchronous accelerated coordinate descent (AACD) [32]. Based on different topologies, synchronous algorithms include centralized and decentralized distributed methods. Typical works for the former organization include the distributed ADMM [13], distributed dual coordinate ascent [75] and their extensions. One bottleneck of centralized topology lies in high communication cost at the central node [47]. Although decentralized algorithms have been widely studied by the control community, the lower bound has not been established until 2017 [64] and a distributed dual ascent with a matching upper bound is given in [64]. Motivated by the lower bound, Li et al. further analyzed the distributed accelerated gradient descent with both optimal communication and computation complexities up to a log factor [46].

1.4 About the Book

In the previous section, we have briefly introduced the representative works on accelerated first-order algorithms. However, due to limited time we do not give details of all of them in the subsequent chapters. Rather, we only introduce results and proofs of part of them, based on our personal flavor and familiarity. The algorithms are organized by their nature: deterministic algorithms for unconstrained convex problems (Chap. 2), constrained convex problems (Chap. 3), and (unconstrained) nonconvex problems (Chap. 4), as well as stochastic algorithms for centralized optimization (Chap. 5) and distributed optimization (Chap. 6). To make our book self-contained, for each introduced algorithm we give the details of its proof. This book serves as a reference to part of the recent advances in optimization. It is appropriate for graduate students and researchers who are interested in machine learning and optimization. Nonetheless, the proofs for achieving critical points (Sect. 4.2), escaping saddle points (Sect. 4.3), and decentralized topology (Sect. 6.2.2) are highly non-trivial. So uninterested readers may skip them.

References

1. N. Agarwal, Z. Allen-Zhu, B. Bullins, E. Hazan, T. Ma, Finding approximate local minima for nonconvex optimization in linear time, in *Proceedings of the 49th Annual ACM SIGACT Symposium on Theory of Computing*, Montreal, (2017), pp. 1195–1200
2. Z. Allen-Zhu, Katyusha: the first truly accelerated stochastic gradient method, in *Proceedings of the 49th Annual ACM SIGACT Symposium on Theory of Computing*, Montreal, (2017), pp. 1200–1206
3. Z. Allen-Zhu, Natasha2: faster non-convex optimization than SGD, in *Advances in Neural Information Processing Systems*, Montreal, vol. 31 (2018), pp. 2675–2686
4. Z. Allen-Zhu, E. Hazan, Variance reduction for faster non-convex optimization, in *Proceedings of the 33th International Conference on Machine Learning*, New York, (2016), pp. 699–707
5. Z. Allen-Zhu, Y. Li, Neon2: finding local minima via first-order oracles, in *Advances in Neural Information Processing Systems*, Montreal, vol. 31 (2018), pp. 3716–3726
6. Z. Allen-Zhu, L. Orecchia, Linear coupling: an ultimate unification of gradient and mirror descent, in *Proceedings of the 8th Innovations in Theoretical Computer Science*, Berkeley, (2017)
7. A. Beck, *First-Order Methods in Optimization*, vol. 25 (SIAM, Philadelphia, 2017)
8. A. Beck, M. Teboulle, A fast iterative shrinkage-thresholding algorithm for linear inverse problems. SIAM J. Imag. Sci. **2**(1), 183–202 (2009)
9. A. Beck, M. Teboulle, A fast dual proximal gradient algorithm for convex minimization and applications. Oper. Res. Lett. **42**(1), 1–6 (2014)
10. J. Berkson, Application of the logistic function to bio-assay. J. Am. Stat. Assoc. **39**(227), 357–365 (1944)

11. S. Bhojanapalli, B. Neyshabur, N. Srebro, Global optimality of local search for low rank matrix recovery, in *Advances in Neural Information Processing Systems*, Barcelona, vol. 29 (2016), pp. 3873–3881
12. L. Bottou, F.E. Curtis, J. Nocedal, Optimization methods for large-scale machine learning. SIAM Rev. **60**(2), 223–311 (2018)
13. S. Boyd, N. Parikh, E. Chu, B. Peleato, J. Eckstein, Distributed optimization and statistical learning via the alternating direction method of multipliers. Found. Trends Mach. Learn. **3**(1), 1–122 (2011)
14. S. Bubeck, Convex optimization: algorithms and complexity. Found. Trends Mach. Learn. **8**(3–4), 231–357 (2015)
15. S. Bubeck, Y.T. Lee, M. Singh, A geometric alternative to Nesterov's accelerated gradient descent (2015). Preprint. arXiv:1506.08187
16. S. Bubeck, Q. Jiang, Y.T. Lee, Y. Li, A. Sidford, Near-optimal method for highly smooth convex optimization, in *Proceedings of the 32th Conference on Learning Theory*, Phoenix, (2019), pp. 492–507
17. Y. Carmon, J.C. Duchi, O. Hinder, A. Sidford, Convex until proven guilty: dimension-free acceleration of gradient descent on non-convex functions, in *Proceedings of the 34th International Conference on Machine Learning*, Sydney, (2017), pp. 654–663
18. Y. Carmon, J.C. Duchi, O. Hinder, A. Sidford, Accelerated methods for nonconvex optimization. SIAM J. Optim. **28**(2), 1751–1772 (2018)
19. A. Chambolle, T. Pock, A first-order primal-dual algorithm for convex problems with applications to imaging. J. Math. Imag. Vis. **40**(1), 120–145 (2011)
20. Y. Chen, G. Lan, Y. Ouyang, Optimal primal-dual methods for a class of saddle point problems. SIAM J. Optim. **24**(4), 1779–1814 (2014)
21. C. Cortes, V. Vapnik, Support-vector networks. Mach. Learn. **20**(3), 273–297 (1995)
22. A. Defazio, F. Bach, S. Lacoste-Julien, SAGA: a fast incremental gradient method with support for non-strongly convex composite objectives, in *Advances in Neural Information Processing Systems*, Montreal, vol. 27 (2014), pp. 1646–1654
23. P.M. Domingos, A few useful things to know about machine learning. Commun. ACM **55**(10), 78–87 (2012)
24. D. Drusvyatskiy, M. Fazel, S. Roy, An optimal first order method based on optimal quadratic averaging. SIAM J. Optim. **28**(1), 251–271 (2018)
25. C. Fang, Y. Huang, Z. Lin, Accelerating asynchronous algorithms for convex optimization by momentum compensation (2018). Preprint. arXiv:1802.09747
26. C. Fang, C.J. Li, Z. Lin, T. Zhang, SPIDER: near-optimal non-convex optimization via stochastic path-integrated differential estimator, in *Advances in Neural Information Processing Systems*, Montreal, vol. 31 (2018), pp. 689–699
27. O. Fercoq, P. Richtárik, Accelerated, parallel, and proximal coordinate descent. SIAM J. Optim. **25**(4), 1997–2023 (2015)
28. C. Gambella, B. Ghaddar, J. Naoum-Sawaya, Optimization models for machine learning: a survey (2019). Preprint. arXiv:1901.05331
29. R. Ge, F. Huang, C. Jin, Y. Yuan, Escaping from saddle points – online stochastic gradient for tensor decomposition, in *Proceedings of the 28th Conference on Learning Theory*, Paris, (2015), pp. 797–842
30. R. Ge, J.D. Lee, T. Ma, Matrix completion has no spurious local minimum, in *Advances in Neural Information Processing Systems*, Barcelona, vol. 29 (2016), pp. 2973–2981
31. S. Ghadimi, G. Lan, Accelerated gradient methods for nonconvex nonlinear and stochastic programming. Math. Program. **156**(1–2), 59–99 (2016)
32. R. Hannah, F. Feng, W. Yin, A2BCD: an asynchronous accelerated block coordinate descent algorithm with optimal complexity, in *Proceedings of the 7th International Conference on Learning Representations*, New Orleans, (2019)
33. S. Haykin, *Neural Networks: A Comprehensive Foundation*, 2nd edn. (Pearson Prentice Hall, Upper Saddle River, 1999)

34. E. Hazan, Introduction to online convex optimization. Found. Trends Optim. **2**(3–4), 157–325 (2016)
35. E. Hazan, Optimization for machine learning. Technical report, Princeton University (2019)
36. B. He, X. Yuan, On the acceleration of augmented Lagrangian method for linearly constrained optimization. Optim. (2010). Preprint. http://www.optimization-online.org/DB_FILE/2010/10/2760.pdf
37. P. Jain, P. Kar, Non-convex optimization for machine learning. Found. Trends Mach. Learn. **10**(3–4), 142–336 (2017)
38. C. Jin, P. Netrapalli, M.I. Jordan, Accelerated gradient descent escapes saddle points faster than gradient descent, in *Proceedings of the 31th Conference On Learning Theory*, Stockholm, (2018), pp. 1042–1085
39. R. Johnson, T. Zhang, Accelerating stochastic gradient descent using predictive variance reduction, in *Advances in Neural Information Processing Systems*, Lake Tahoe, vol. 26 (2013), pp. 315–323
40. D. Kim, J.A. Fessler, Optimized first-order methods for smooth convex minimization. Math. Program. **159**(1–2), 81–107 (2016)
41. G. Lan, Y. Zhou, An optimal randomized incremental gradient method. Math. Program. **171**(1–2), 167–215 (2018)
42. L. Lessard, B. Recht, A. Packard, Analysis and design of optimization algorithms via integral quadratic constraints. SIAM J. Optim. **26**(1), 57–95 (2016)
43. H. Li, Z. Lin, Accelerated proximal gradient methods for nonconvex programming, in *Advances in Neural Information Processing Systems*, Montreal, vol. 28 (2015), pp. 379–387
44. H. Li, Z. Lin, On the complexity analysis of the primal solutions for the accelerated randomized dual coordinate ascent. J. Mach. Learn. Res. (2020). http://jmlr.org/papers/v21/18-425.html
45. H. Li, C. Fang, Z. Lin, Convergence rates analysis of the quadratic penalty method and its applications to decentralized distributed optimization (2017). Preprint. arXiv:1711.10802
46. H. Li, C. Fang, W. Yin, Z. Lin, A sharp convergence rate analysis for distributed accelerated gradient methods (2018). Preprint. arXiv:1810.01053
47. X. Lian, C. Zhang, H. Zhang, C.-J. Hsieh, W. Zhang, J. Liu, Can decentralized algorithms outperform centralized algorithms? A case study for decentralized parallel stochastic gradient descent, in *Advances in Neural Information Processing Systems*, Long Beach, vol. 30 (2017), pp. 5330–5340
48. Q. Lin, Z. Lu, L. Xiao, An accelerated proximal coordinate gradient method, in *Advances in Neural Information Processing Systems*, Montreal, vol. 27 (2014), pp. 3059–3067
49. H. Lin, J. Mairal, Z. Harchaoui, A universal catalyst for first-order optimization, in *Advances in Neural Information Processing Systems*, Montreal, vol. 28 (2015), pp. 3384–3392
50. G. Liu, Z. Lin, Y. Yu, Robust subspace segmentation by low-rank representation, in *Proceedings of the 27th International Conference on Machine Learning*, Haifa, vol. 1 (2010), pp. 663–670
51. J. Lu, M. Johansson, Convergence analysis of approximate primal solutions in dual first-order methods. SIAM J. Optim. **26**(4), 2430–2467 (2016)
52. J. Mairal, Optimization with first-order surrogate functions, in *Proceedings of the 30th International Conference on Machine Learning*, Atlanta, (2013), pp. 783–791
53. Y. Nesterov, On an approach to the construction of optimal methods of minimization of smooth convex functions. Ekonomika I Mateaticheskie Metody **24**(3), 509–517 (1988)
54. Y. Nesterov, Smooth minimization of non-smooth functions. Math. Program. **103**(1), 127–152 (2005)
55. Y. Nesterov, Gradient methods for minimizing composite objective function. Technical Report Discussion Paper #2007/76, CORE (2007)
56. Y. Nesterov, Efficiency of coordinate descent methods on huge-scale optimization problems. SIAM J. Optim. **22**(2), 341–362 (2012)
57. Y. Nesterov, Gradient methods for minimizing composite functions. Math. Program. **140**(1), 125–161 (2013)
58. Y. Nesterov, *Lectures on Convex Optimization* (Springer, New York, 2018)

59. Y. Ouyang, Y. Chen, G. Lan, E. Pasiliao Jr., An accelerated linearized alternating direction method of multipliers. SIAM J. Imag. Sci. **8**(1), 644–681 (2015)
60. N. Parikh, S. Boyd, Proximal algorithms. Found. Trends Optim. **1**(3), 127–239 (2014)
61. B.T. Polyak, Some methods of speeding up the convergence of iteration methods. USSR Comput. Math. Math. Phys. **4**(5), 1–17 (1964)
62. S.J. Reddi, A. Hefny, S. Sra, B. Poczos, A. Smola, Stochastic variance reduction for nonconvex optimization, in *Proceedings of the 33th International Conference on Machine Learning*, New York, (2016), pp. 314–323
63. M. Schmidt, N. Le Roux, F. Bach, Minimizing finite sums with the stochastic average gradient. Math. Program. **162**(1–2), 83–112 (2017)
64. K. Seaman, F. Bach, S. Bubeck, Y.T. Lee, L. Massoulié, Optimal algorithms for smooth and strongly convex distributed optimization in networks, in *Proceedings of the 34th International Conference on Machine Learning*, Sydney, (2017), pp. 3027–3036
65. S. Shalev-Shwartz, T. Zhang, Accelerated proximal stochastic dual coordinate ascent for regularized loss minimization, in *Proceedings of the 31th International Conference on Machine Learning*, Beijing, (2014), pp. 64–72
66. S. Sra, S. Nowozin, S.J. Wright (eds.), *Optimization for Machine Learning* (MIT Press, Cambridge, MA, 2012)
67. W. Su, S. Boyd, E. Candès, A differential equation for modeling Nesterov's accelerated gradient method: theory and insights, in *Advances in Neural Information Processing Systems*, Montreal, vol. 27 (2014), pp. 2510–2518
68. R. Tibshirani, Regression shrinkage and selection via the lasso. J. R. Stat. Soc. Ser. B Methodol. **58**(1), 267–288 (1996)
69. N. Tripuraneni, M. Stern, C. Jin, J. Regier, M.I. Jordan, Stochastic cubic regularization for fast nonconvex optimization, in *Advances in Neural Information Processing Systems*, Montreal, vol. 31 (2018), pp. 2899–2908
70. P. Tseng, On accelerated proximal gradient methods for convex-concave optimization. Technical report, University of Washington, Seattle (2008)
71. A. Wibisono, A.C. Wilson, M.I. Jordan, A variational perspective on accelerated methods in optimization. Proc. Natl. Acad. Sci. **113**(47), 7351–7358 (2016)
72. Y. Xu, Accelerated first-order primal-dual proximal methods for linearly constrained composite convex programming. SIAM J. Optim. **27**(3), 1459–1484 (2017)
73. Y. Xu, J. Rong, T. Yang, First-order stochastic algorithms for escaping from saddle points in almost linear time, in *Advances in Neural Information Processing Systems*, Montreal, vol. 31 (2018), pp. 5530–5540
74. Y. Zhang, L. Xiao, Stochastic primal-dual coordinate method for regularized empirical risk minimization. J. Mach. Learn. Res. **18**(1), 2939–2980 (2017)
75. S. Zheng, J. Wang, F. Xia, W. Xu, T. Zhang, A general distributed dual coordinate optimization framework for regularized loss minimization. J. Mach. Learn. Res. **18**(115), 1–52 (2017)

Chapter 2
Accelerated Algorithms for Unconstrained Convex Optimization

First-order methods have received extensive attention recently because they are effective in solving large-scale optimization problems. The accelerated gradient method may be one of the most widely used first-order methods due to its solid theoretical foundation, effective practical performance, and simple implementation. In this chapter, we summarize the basic accelerated gradient methods for the unconstrained convex optimization. We are interested in algorithms for solving the following problem:

$$\min_{\mathbf{x}} f(\mathbf{x}), \tag{2.1}$$

where the objective function $f(\mathbf{x})$ is convex.

This chapter includes the descriptions of several accelerated gradient methods for smooth and composite optimization, accelerated algorithms with inexact gradient and proximal computing, restart technique for restricted strongly convex optimization, smoothing technique for nonsmooth optimization, and higher order accelerated gradient algorithms and their explanation from the variational perspective.

2.1 Accelerated Gradient Method for Smooth Optimization

In a series of celebrated works [19, 20, 22], Y. Nesterov proposed several accelerated gradient methods to solve the smooth problem (2.1). We introduce the momentum based accelerated gradient descent (AGD) in this section and leave the other algorithms in Sect. 2.2. Accelerated gradient descent is an extension of gradient descent, where the latter has the following iterations:

$$\mathbf{x}_{k+1} = \mathbf{x}_k - \frac{1}{L}\nabla f(\mathbf{x}_k). \tag{2.2}$$

© Springer Nature Singapore Pte Ltd. 2020
Z. Lin et al., *Accelerated Optimization for Machine Learning*,
https://doi.org/10.1007/978-981-15-2910-8_2

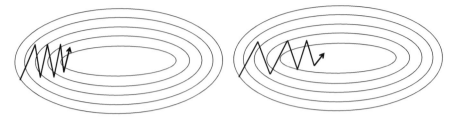

Fig. 2.1 Comparison of gradient descent without momentum (left) and with momentum (right). Reproduced from https://www.willamette.edu/~gorr/classes/cs449/momrate.html

Physically, accelerated gradient descent adds an inertia to the current point to generate an extrapolated point $\mathbf{y}_k$ and then performs a gradient descent step at $\mathbf{y}_k$. The algorithm is described in Algorithm 2.1, where β_k will be specified in Theorem 2.1. Figure 2.1 demonstrates the intuition behind momentum. Intuitively, gradient descent may oscillate across the slopes of the ravine around the local minimum, while only making hesitant progress along the bottom towards the local minimum. Momentum helps accelerate gradient descent in the relevant direction and dampens oscillations.

Algorithm 2.1 Accelerated gradient descent (AGD)

Initialize $\mathbf{x}_0 = \mathbf{x}_{-1}$.
for $k = 0, 1, 2, 3, \cdots$ **do**
    $\mathbf{y}_k = \mathbf{x}_k + \beta_k(\mathbf{x}_k - \mathbf{x}_{k-1})$,
    $\mathbf{x}_{k+1} = \mathbf{y}_k - \frac{1}{L}\nabla f(\mathbf{y}_k)$.
end for

Theoretically, the gradient descent (2.2) has the $O\left(\frac{1}{k}\right)$ convergence rate for generally convex problems and $O\left(\left(1 - \frac{\mu}{L}\right)^k\right)$ convergence rate for μ-strongly convex ones, respectively [21]. As a comparison, the accelerated gradient descent can improve the convergence rates to $O\left(\frac{1}{k^2}\right)$ and $O\left(\left(1 - \sqrt{\frac{\mu}{L}}\right)^k\right)$ for generally convex and strongly convex problems, respectively.

We use the technique of estimate sequence to prove the convergence rate, which was originally proposed in [19] and then some interests in this concept resurrected after the publication of [22]. A more recent introduction of estimate sequence can be found in [4]. We first define the estimate sequence.

Definition 2.1 A pair of sequences $\{\phi_k(\mathbf{x})\}_{k=0}^{\infty}$ and $\{\lambda_k\}_{k=0}^{\infty}$, where $\lambda_k \geq 0$, is called an estimate sequence of function $f(\mathbf{x})$ if $\lambda_k \to 0$ and for any $\mathbf{x}$, we have

$$\phi_k(\mathbf{x}) \leq (1 - \lambda_k)f(\mathbf{x}) + \lambda_k\phi_0(\mathbf{x}). \tag{2.3}$$

The following lemma indicates how estimate sequence can be used for analyzing an optimization algorithm and how fast it would converge.

Lemma 2.1 *If $\phi_k^* \equiv \min_{\mathbf{x}} \phi_k(\mathbf{x}) \geq f(\mathbf{x}_k)$ and (2.3) holds, then we can have*

$$f(\mathbf{x}_k) - f(\mathbf{x}^*) \leq \lambda_k(\phi_0(\mathbf{x}^*) - f(\mathbf{x}^*)).$$

Proof From $\phi_k^* \geq f(\mathbf{x}_k)$ and (2.3), we can have

$$f(\mathbf{x}_k) \leq \phi_k^* \leq \min_{\mathbf{x}} [(1 - \lambda_k)f(\mathbf{x}) + \lambda_k \phi_0(\mathbf{x})] \leq (1 - \lambda_k)f(\mathbf{x}^*) + \lambda_k \phi_0(\mathbf{x}^*),$$

which leads to the conclusion.   □

For μ-strongly convex function f (if f is generally convex, we let $\mu = 0$), we use the following way to construct the estimate sequence. Define two sequences:

$$\lambda_{k+1} = (1 - \theta_k)\lambda_k, \tag{2.4}$$

$$\phi_{k+1}(\mathbf{x}) = (1 - \theta_k)\phi_k(\mathbf{x}) + \theta_k \left(f(\mathbf{y}_k) + \langle \nabla f(\mathbf{y}_k), \mathbf{x} - \mathbf{y}_k \rangle + \frac{\mu}{2}\|\mathbf{x} - \mathbf{y}_k\|^2 \right), \tag{2.5}$$

with $\lambda_0 = 1$ and $\phi_0(\mathbf{x}) = f(\mathbf{x}_0) + \frac{\gamma_0}{2}\|\mathbf{x} - \mathbf{x}_0\|^2$. Then we can prove (2.3) by induction. Indeed, $\phi_0(\mathbf{x}) \leq (1 - \lambda_0)f(\mathbf{x}) + \lambda_0\phi_0(\mathbf{x})$ since $\lambda_0 = 1$. Assume that (2.3) holds for some k. Then from the μ-strong convexity of $f(\mathbf{x})$ and (2.4), we obtain

$$\begin{aligned} \phi_{k+1}(\mathbf{x}) &\leq (1 - \theta_k)\phi_k(\mathbf{x}) + \theta_k f(\mathbf{x}) \\ &\leq (1 - \theta_k)\left[(1 - \lambda_k)f(\mathbf{x}) + \lambda_k\phi_0(\mathbf{x})\right] + \theta_k f(\mathbf{x}) \\ &= (1 - \lambda_{k+1})f(\mathbf{x}) + \lambda_{k+1}\phi_0(\mathbf{x}). \end{aligned} \tag{2.6}$$

Thus (2.3) also holds for $k + 1$.

The following lemma describes the minimum value of $\phi_k(\mathbf{x})$ and its minimizer, which will be used to validate $\phi_k^* \geq f(\mathbf{x}_k)$ that appeared in Lemma 2.1.

Lemma 2.2 *Process (2.5) forms*

$$\phi_k(\mathbf{x}) = \phi_k^* + \frac{\gamma_k}{2}\|\mathbf{x} - \mathbf{z}_k\|^2, \tag{2.7}$$

where

$$\gamma_{k+1} = (1 - \theta_k)\gamma_k + \theta_k\mu, \quad \gamma_0 = \begin{cases} L, & \text{if } \mu = 0, \\ \mu, & \text{if } \mu > 0, \end{cases} \tag{2.8}$$

$$\mathbf{z}_{k+1} = \frac{1}{\gamma_{k+1}}\left[(1 - \theta_k)\gamma_k\mathbf{z}_k + \theta_k\mu\mathbf{y}_k - \theta_k\nabla f(\mathbf{y}_k)\right], \quad \mathbf{z}_0 = \mathbf{x}_0, \tag{2.9}$$

$$\phi_{k+1}^* = (1 - \theta_k)\phi_k^* + \theta_k f(\mathbf{y}_k) - \frac{\theta_k^2}{2\gamma_{k+1}}\|\nabla f(\mathbf{y}_k)\|^2$$

$$+ \frac{\theta_k(1 - \theta_k)\gamma_k}{\gamma_{k+1}}\left(\frac{\mu}{2}\|\mathbf{y}_k - \mathbf{z}_k\|^2 + \langle\nabla f(\mathbf{y}_k), \mathbf{z}_k - \mathbf{y}_k\rangle\right), \quad \phi_0^* = f(\mathbf{x}_0).$$

$$(2.10)$$

Proof From the definition in (2.5), we know that $\phi_k(\mathbf{x})$ is in the quadratic form of (2.7). We only need to get the recursions of γ_k, $\mathbf{z}_k$, and ϕ_k^*. Indeed,

$$\phi_{k+1}(\mathbf{x}) = (1 - \theta_k)\left(\phi_k^* + \frac{\gamma_k}{2}\|\mathbf{x} - \mathbf{z}_k\|^2\right)$$

$$+ \theta_k\left(f(\mathbf{y}_k) + \langle\nabla f(\mathbf{y}_k), \mathbf{x} - \mathbf{y}_k\rangle + \frac{\mu}{2}\|\mathbf{x} - \mathbf{y}_k\|^2\right).$$

Letting $\nabla\phi_{k+1}(\mathbf{z}_{k+1}) = 0$, we obtain (2.8) and (2.9). From (2.5), we also have

$$\phi_{k+1}^* + \frac{\gamma_{k+1}}{2}\|\mathbf{y}_k - \mathbf{z}_{k+1}\|^2 = \phi_{k+1}(\mathbf{y}_k)$$

$$= (1 - \theta_k)\left(\phi_k^* + \frac{\gamma_k}{2}\|\mathbf{y}_k - \mathbf{z}_k\|^2\right) + \theta_k f(\mathbf{y}_k). \quad (2.11)$$

From (2.9), we have

$$\frac{\gamma_{k+1}}{2}\|\mathbf{z}_{k+1} - \mathbf{y}_k\|^2$$

$$= \frac{1}{2\gamma_{k+1}}\left[(1 - \theta_k)^2\gamma_k^2\|\mathbf{z}_k - \mathbf{y}_k\|^2 - 2\theta_k(1 - \theta_k)\gamma_k\langle\nabla f(\mathbf{y}_k), \mathbf{z}_k - \mathbf{y}_k\rangle\right.$$

$$\left. + \theta_k^2\|\nabla f(\mathbf{y}_k)\|^2\right].$$

Substituting it into (2.11), we obtain (2.10). □

Now, we are ready to prove $\phi_k^* \geq f(\mathbf{x}_k)$ and thus get the final conclusions via Lemma 2.1.

Theorem 2.1 *Suppose that $f(\mathbf{x})$ is convex and L-smooth. Let $\theta_{-1} = 1$, $\theta_{k+1} = \frac{\sqrt{\theta_k^4 + 4\theta_k^2} - \theta_k^2}{2}$, and $\beta_k = \frac{\theta_k(1 - \theta_{k-1})}{\theta_{k-1}}$. Then for Algorithm 2.1, we have*

$$f(\mathbf{x}_{K+1}) - f(\mathbf{x}^*) \leq \frac{4}{(K+2)^2}\left(f(\mathbf{x}_0) - f(\mathbf{x}^*) + \frac{L}{2}\|\mathbf{x}_0 - \mathbf{x}^*\|^2\right).$$

Suppose that $f(\mathbf{x})$ is μ-strongly convex and L-smooth. Let $\theta_k = \sqrt{\frac{\mu}{L}}$ and $\beta_k = \frac{\sqrt{L}-\sqrt{\mu}}{\sqrt{L}+\sqrt{\mu}}$. Then for Algorithm 2.1, we have

$$f(\mathbf{x}_{K+1}) - f(\mathbf{x}^*) \leq \left(1 - \sqrt{\frac{\mu}{L}}\right)^{K+1} \left(f(\mathbf{x}_0) - f(\mathbf{x}^*) + \frac{\mu}{2}\|\mathbf{x}_0 - \mathbf{x}^*\|^2\right).$$

Proof We prove $\phi_k^* \geq f(\mathbf{x}_k)$ by induction. Assume that it holds for some k. From (2.10), we have

$$\phi_{k+1}^* \geq (1-\theta_k)f(\mathbf{x}_k) + \theta_k f(\mathbf{y}_k) - \frac{\theta_k^2}{2\gamma_{k+1}}\|\nabla f(\mathbf{y}_k)\|^2$$
$$+ \frac{\theta_k(1-\theta_k)\gamma_k}{\gamma_{k+1}}\langle\nabla f(\mathbf{y}_k), \mathbf{z}_k - \mathbf{y}_k\rangle$$
$$\overset{a}{\geq} f(\mathbf{y}_k) - \frac{\theta_k^2}{2\gamma_{k+1}}\|\nabla f(\mathbf{y}_k)\|^2$$
$$+ (1-\theta_k)\left\langle\nabla f(\mathbf{y}_k), \frac{\theta_k\gamma_k}{\gamma_{k+1}}(\mathbf{z}_k - \mathbf{y}_k) + \mathbf{x}_k - \mathbf{y}_k\right\rangle,$$

where we use the convexity of $f(\mathbf{x})$ in $\overset{a}{\geq}$. From the L-smoothness of $f(\mathbf{x})$, we obtain $f(\mathbf{x}_{k+1}) \leq f(\mathbf{y}_k) - \frac{1}{2L}\|\nabla f(\mathbf{y}_k)\|^2$ by (A.6). Then $\phi_{k+1}^* \geq f(\mathbf{x}_{k+1})$ if $\frac{\theta_k^2}{\gamma_{k+1}} = \frac{1}{L}$ and

$$\frac{\theta_k\gamma_k}{\gamma_{k+1}}(\mathbf{z}_k - \mathbf{y}_k) + \mathbf{x}_k - \mathbf{y}_k = \mathbf{0}. \tag{2.12}$$

Case 1: $\mu = 0$. Then from (2.8), we have $\frac{\theta_k^2}{\gamma_{k+1}} = \frac{1}{L} \Rightarrow \theta_k^2 = \frac{(1-\theta_k)\gamma_k}{L} = (1-\theta_k)\theta_{k-1}^2$, which leads to $\theta_k = \frac{\sqrt{\theta_{k-1}^4 + 4\theta_{k-1}^2} - \theta_{k-1}^2}{2}$, $\theta_k \leq \frac{2}{k+2}$, and $\lambda_{k+1} = \prod_{i=0}^{k}(1-\theta_i) = \prod_{i=0}^{k}\frac{\theta_i^2}{\theta_{i-1}^2} = \theta_k^2 \leq \frac{4}{(k+2)^2}$ via Lemma 2.3. From (2.8) and (2.9), we have

$$\mathbf{z}_{k+1} = \mathbf{z}_k - \frac{\theta_k}{\gamma_{k+1}}\nabla f(\mathbf{y}_k) = \mathbf{z}_k - \frac{1}{L\theta_k}\nabla f(\mathbf{y}_k) = \mathbf{z}_k - \frac{1}{\theta_k}(\mathbf{y}_k - \mathbf{x}_{k+1}). \tag{2.13}$$

From (2.12) and (2.8), we have

$$\mathbf{y}_k = \theta_k\mathbf{z}_k + (1-\theta_k)\mathbf{x}_k. \tag{2.14}$$

From (2.13) and (2.14), we have

$$\mathbf{x}_{k+1} = \theta_k \mathbf{z}_{k+1} + (1 - \theta_k)\mathbf{x}_k.$$

Thus,

$$\mathbf{y}_k = \frac{\theta_k}{\theta_{k-1}}[\mathbf{x}_k - (1 - \theta_{k-1})\mathbf{x}_{k-1}] + (1 - \theta_k)\mathbf{x}_k = \mathbf{x}_k + \frac{\theta_k(1 - \theta_{k-1})}{\theta_{k-1}}(\mathbf{x}_k - \mathbf{x}_{k-1}).$$

From Lemma 2.1, we can get the conclusion.

Case 2: $\mu > 0$. Then $\theta_k = \theta = \sqrt{\frac{\mu}{L}}$ and $\gamma_k = \mu$ satisfy $\frac{\theta_k^2}{\gamma_{k+1}} = \frac{1}{L}$ and (2.8). Equation (2.4) leads to $\lambda_{k+1} = (1 - \theta)^{k+1}$. From (2.9) and $\gamma_k = \mu$, we have

$$\mathbf{z}_{k+1} = (1 - \theta)\mathbf{z}_k + \theta\mathbf{y}_k - \frac{\theta}{\mu}\nabla f(\mathbf{y}_k) = (1 - \theta)\mathbf{z}_k + \theta\mathbf{y}_k + \frac{1}{\theta}(\mathbf{x}_{k+1} - \mathbf{y}_k). \quad (2.15)$$

From (2.12), we have

$$\theta\mathbf{z}_k + \mathbf{x}_k - (\theta + 1)\mathbf{y}_k = \mathbf{0}. \quad (2.16)$$

Thus from (2.15) and (2.16), one can obtain

$$\mathbf{x}_{k+1} \overset{a}{=} \theta\mathbf{z}_{k+1} - \theta(1 - \theta)\mathbf{z}_k - \theta^2\mathbf{y}_k + \mathbf{y}_k$$
$$= \theta\mathbf{z}_{k+1} + \mathbf{y}_k - \theta\mathbf{z}_k + \theta^2(\mathbf{z}_k - \mathbf{y}_k)$$
$$\overset{b}{=} \theta\mathbf{z}_{k+1} + \mathbf{x}_k - \theta\mathbf{y}_k + \theta(\mathbf{y}_k - \mathbf{x}_k)$$
$$= \theta\mathbf{z}_{k+1} + (1 - \theta)\mathbf{x}_k,$$

where $\overset{a}{=}$ uses (2.15) and $\overset{b}{=}$ uses (2.16). So

$$\mathbf{y}_k \overset{a}{=} \frac{1}{\theta + 1}(\theta\mathbf{z}_k + \mathbf{x}_k) = \frac{1}{\theta + 1}[\mathbf{x}_k - (1 - \theta)\mathbf{x}_{k-1} + \mathbf{x}_k]$$
$$= \mathbf{x}_k + \frac{1 - \theta}{1 + \theta}(\mathbf{x}_k - \mathbf{x}_{k-1}) = \mathbf{x}_k + \frac{\sqrt{L} - \sqrt{\mu}}{\sqrt{L} + \sqrt{\mu}}(\mathbf{x}_k - \mathbf{x}_{k-1}),$$

where $\overset{a}{=}$ uses (2.16). $\square$

The following lemma describes some useful properties for the sequence $\{\theta_k\}_{k=0}^{\infty}$, which will be frequently used in this book.

Lemma 2.3 *If sequence $\{\theta_k\}_{k=0}^{\infty}$ satisfies $\frac{1-\theta_k}{\theta_k^2} = \frac{1}{\theta_{k-1}^2}$ and $\theta_0 \leq 1$, then $\frac{1}{k+1/\theta_0} \leq$*
$\theta_k \leq \frac{2}{k+2/\theta_0}$, $\sum_{i=0}^{k} \frac{1}{\theta_i} = \frac{1}{\theta_k^2} - \frac{1}{\theta_{-1}^2}$, *and* $\theta_{k+1} = \frac{\sqrt{\theta_k^4 + 4\theta_k^2} - \theta_k^2}{2}$.

Proof In fact, from $\frac{1-\theta_k}{\theta_k^2} = \frac{1}{\theta_{k-1}^2}$, we can have $\left(\frac{1}{\theta_k} - \frac{1}{2}\right)^2 \geq \frac{1}{\theta_{k-1}^2}$, which leads to $\frac{1}{\theta_k} - \frac{1}{2} \geq \frac{1}{\theta_{k-1}}$. Summing over $k = 1, 2, \cdots, K$, we have $\frac{1}{\theta_K} \geq \frac{1}{\theta_0} + \frac{K}{2}$, which leads to $\theta_K \leq \frac{2}{K+2/\theta_0}$. On the other hand, we know $\theta_k \leq 1$ for all k and thus $\left(\frac{1}{\theta_k} - 1\right)^2 \leq \frac{1}{\theta_{k-1}^2}$, which leads to $\frac{1}{\theta_k} - 1 \leq \frac{1}{\theta_{k-1}}$. Similarly, we have $\frac{1}{\theta_K} \leq \frac{1}{\theta_0} + K$, which leads to $\theta_K \geq \frac{1}{K+1/\theta_0}$. The second conclusion can be obtained by $\frac{1}{\theta_k} = \frac{1}{\theta_k^2} - \frac{1}{\theta_{k-1}^2}$ and the last conclusion can be obtained from $\frac{1-\theta_{k+1}}{\theta_{k+1}^2} = \frac{1}{\theta_k^2}$. $\qquad\square$

2.2 Extension to Composite Optimization

Composite convex optimization consists of the optimization of a convex function with Lipschitz continuous gradients (Definition A.12) and a nonsmooth function, which can be written as

$$\min_{\mathbf{x}} F(\mathbf{x}) \equiv f(\mathbf{x}) + h(\mathbf{x}), \tag{2.17}$$

where $f(\mathbf{x})$ is smooth and we often assume that the proximal mapping of $h(\mathbf{x})$ has a closed form solution or can be computed efficiently. Accelerated gradient descent was extended to the composite optimization in [5, 24] and a unified analysis of acceleration techniques was given in [29]. We follow [29] to describe Nesterov's three methods for solving problem (2.17).

2.2.1 Nesterov's First Scheme

The first method we describe is an extension of Algorithm 2.1, which is described in Algorithm 2.2. It can be easily checked that the momentum parameter $\frac{(L\theta_k - \mu)(1 - \theta_{k-1})}{(L-\mu)\theta_{k-1}}$ is equivalent to the settings in Theorem 2.1. Please see Remark 2.1.

Algorithm 2.2 Accelerated proximal gradient (APG) method 1

Initialize $\mathbf{x}_0 = \mathbf{x}_{-1}$.

for $k = 0, 1, 2, 3, \cdots$ **do**

$\quad \mathbf{y}_k = \mathbf{x}_k + \frac{(L\theta_k - \mu)(1 - \theta_{k-1})}{(L - \mu)\theta_{k-1}}(\mathbf{x}_k - \mathbf{x}_{k-1}),$

$\quad \mathbf{x}_{k+1} = \mathrm{argmin}_{\mathbf{x}} \left(h(\mathbf{x}) + \frac{L}{2} \left\| \mathbf{x} - \mathbf{y}_k + \frac{1}{L}\nabla f(\mathbf{y}_k) \right\|^2 \right).$

end for

We can also use the estimate sequence technique to prove the convergence rate of Algorithm 2.2. However, we introduce the techniques in [29] to enrich the toolbox of this monograph. We first describe the following lemma, which can serve as a starting point for analyzing various first-order methods.

Lemma 2.4 *Suppose that $h(\mathbf{x})$ is convex and $f(\mathbf{x})$ is μ-strongly convex and L-smooth. Then for Algorithm 2.2, we have*

$$F(\mathbf{x}_{k+1}) \leq F(\mathbf{x}) - \frac{\mu}{2}\|\mathbf{x} - \mathbf{y}_k\|^2 - \frac{L}{2}\|\mathbf{x}_{k+1} - \mathbf{y}_k\|^2 + L\langle \mathbf{x}_{k+1} - \mathbf{y}_k, \mathbf{x} - \mathbf{y}_k \rangle, \forall \mathbf{x}.$$

Proof From the optimality condition of the second step, we obtain

$$\mathbf{0} \in \partial h(\mathbf{x}_{k+1}) + L(\mathbf{x}_{k+1} - \mathbf{y}_k) + \nabla f(\mathbf{y}_k).$$

Then from the convexity of $h(\mathbf{x})$, we have

$$h(\mathbf{x}) - h(\mathbf{x}_{k+1}) \geq \langle -L(\mathbf{x}_{k+1} - \mathbf{y}_k) - \nabla f(\mathbf{y}_k), \mathbf{x} - \mathbf{x}_{k+1} \rangle. \qquad (2.18)$$

From the L-smoothness and the μ-strong convexity of $f(\mathbf{x})$ and (2.18), we get

$$\begin{aligned}
F(\mathbf{x}_{k+1}) &\leq f(\mathbf{y}_k) + \langle \nabla f(\mathbf{y}_k), \mathbf{x}_{k+1} - \mathbf{y}_k \rangle + \frac{L}{2}\|\mathbf{x}_{k+1} - \mathbf{y}_k\|^2 + h(\mathbf{x}_{k+1}) \\
&= f(\mathbf{y}_k) + \langle \nabla f(\mathbf{y}_k), \mathbf{x} - \mathbf{y}_k \rangle + \langle \nabla f(\mathbf{y}_k), \mathbf{x}_{k+1} - \mathbf{x} \rangle \\
&\quad + \frac{L}{2}\|\mathbf{x}_{k+1} - \mathbf{y}_k\|^2 + h(\mathbf{x}_{k+1}) \\
&\leq f(\mathbf{x}) - \frac{\mu}{2}\|\mathbf{x} - \mathbf{y}_k\|^2 + \frac{L}{2}\|\mathbf{x}_{k+1} - \mathbf{y}_k\|^2 + h(\mathbf{x}) \\
&\quad + L\langle \mathbf{x}_{k+1} - \mathbf{y}_k, \mathbf{x} - \mathbf{x}_{k+1} \rangle \\
&= F(\mathbf{x}) - \frac{\mu}{2}\|\mathbf{x} - \mathbf{y}_k\|^2 - \frac{L}{2}\|\mathbf{x}_{k+1} - \mathbf{y}_k\|^2 + L\langle \mathbf{x}_{k+1} - \mathbf{y}_k, \mathbf{x} - \mathbf{y}_k \rangle.
\end{aligned}$$

The proof is complete.   □

We define the Lyapunov function

$$\ell_{k+1} = \frac{F(\mathbf{x}_{k+1}) - F(\mathbf{x}^*)}{\theta_k^2} + \frac{L}{2} \left\| \mathbf{z}_{k+1} - \mathbf{x}^* \right\|^2 \tag{2.19}$$

for the case of $\mu = 0$ and

$$\ell_{k+1} = \frac{1}{\left(1 - \sqrt{\mu/L}\right)^{k+1}} \left(F(\mathbf{x}_{k+1}) - F(\mathbf{x}^*) + \frac{\mu}{2} \left\| \mathbf{z}_{k+1} - \mathbf{x}^* \right\|^2 \right) \tag{2.20}$$

for $\mu > 0$, where

$$\mathbf{z}_{k+1} \equiv \frac{1}{\theta_k} \mathbf{x}_{k+1} - \frac{1 - \theta_k}{\theta_k} \mathbf{x}_k, \qquad \mathbf{z}_0 = \mathbf{x}_0. \tag{2.21}$$

From the definitions of $\mathbf{w}^{k+1}$ and $\mathbf{y}^k$, we can have the following easy-to-verify identities.

Lemma 2.5 *For Algorithm 2.1, we have*

$$\mathbf{x}^* + \frac{(1 - \theta_k)L}{L\theta_k - \mu} \mathbf{x}_k - \frac{L - \mu}{L\theta_k - \mu} \mathbf{y}_k = \mathbf{x}^* - \mathbf{z}_k, \tag{2.22}$$

$$\theta_k \mathbf{x}^* + (1 - \theta_k)\mathbf{x}_k - \mathbf{x}_{k+1} = \theta_k \left(\mathbf{x}^* - \mathbf{z}_{k+1} \right).$$

We will show $\ell_{k+1} \leq \ell_k$ for all $k = 0, 1, \cdots$ and establish the convergence rates in the following theorem.

Theorem 2.2 *Suppose that $f(\mathbf{x})$ and $h(\mathbf{x})$ are convex and $f(\mathbf{x})$ is L-smooth. Let $\theta_0 = 1$ and $\theta_{k+1} = \frac{\sqrt{\theta_k^4 + 4\theta_k^2} - \theta_k^2}{2}$. Then for Algorithm 2.2, we have*

$$F(\mathbf{x}_{K+1}) - F(\mathbf{x}^*) \leq \frac{2L}{(K+2)^2} \left\| \mathbf{x}_0 - \mathbf{x}^* \right\|^2.$$

Suppose that $h(\mathbf{x})$ is convex and $f(\mathbf{x})$ is μ-strongly convex and L-smooth. Let $\theta_k = \sqrt{\frac{\mu}{L}}$ for all k. Then for Algorithm 2.2, we have

$$F(\mathbf{x}_{K+1}) - F(\mathbf{x}^*) \leq \left(1 - \sqrt{\frac{\mu}{L}}\right)^{K+1} \left(F(\mathbf{x}_0) - F(\mathbf{x}^*) + \frac{\mu}{2} \left\| \mathbf{x}_0 - \mathbf{x}^* \right\|^2 \right).$$

Proof We apply Lemma 2.4, first with $\mathbf{x} = \mathbf{x}_k$ and then with $\mathbf{x} = \mathbf{x}^*$, to obtain two inequalities

$$F(\mathbf{x}_{k+1}) \le F(\mathbf{x}_k) - \frac{L}{2}\|\mathbf{x}_{k+1} - \mathbf{y}_k\|^2 + L\langle \mathbf{x}_{k+1} - \mathbf{y}_k, \mathbf{x}_k - \mathbf{y}_k \rangle,$$

$$F(\mathbf{x}_{k+1}) \le F(\mathbf{x}^*) - \frac{\mu}{2}\|\mathbf{x}^* - \mathbf{y}_k\|^2 - \frac{L}{2}\|\mathbf{x}_{k+1} - \mathbf{y}_k\|^2 + L\langle \mathbf{x}_{k+1} - \mathbf{y}_k, \mathbf{x}^* - \mathbf{y}_k \rangle.$$

Multiplying the first inequality by $(1 - \theta_k)$ and the second by θ_k and adding them together, we have

$$F(\mathbf{x}_{k+1}) - F(\mathbf{x}^*)$$

$$\le (1 - \theta_k)(F(\mathbf{x}_k) - F(\mathbf{x}^*)) - \frac{L}{2}\|\mathbf{x}_{k+1} - \mathbf{y}_k\|^2 - \frac{\theta_k \mu}{2}\|\mathbf{x}^* - \mathbf{y}_k\|^2$$

$$+ L\langle \mathbf{x}_{k+1} - \mathbf{y}_k, (1 - \theta_k)\mathbf{x}_k + \theta_k \mathbf{x}^* - \mathbf{y}_k \rangle$$

$$\overset{a}{=} (1 - \theta_k)(F(\mathbf{x}_k) - F(\mathbf{x}^*)) - \frac{L}{2}\|\mathbf{x}_{k+1} - \mathbf{y}_k\|^2 - \frac{\theta_k \mu}{2}\|\mathbf{x}^* - \mathbf{y}_k\|^2$$

$$+ \frac{L}{2}\left(\|\mathbf{x}_{k+1} - \mathbf{y}_k\|^2 + \|(1 - \theta_k)\mathbf{x}_k + \theta_k \mathbf{x}^* - \mathbf{y}_k\|^2 \right.$$

$$\left. - \|(1 - \theta_k)\mathbf{x}_k + \theta_k \mathbf{x}^* - \mathbf{x}_{k+1}\|^2 \right)$$

$$= (1 - \theta_k)(F(\mathbf{x}_k) - F(\mathbf{x}^*)) - \frac{\theta_k \mu}{2}\|\mathbf{x}^* - \mathbf{y}_k\|^2$$

$$+ \frac{L\theta_k^2}{2}\left(\left\| \mathbf{x}^* - \frac{1}{\theta_k}\mathbf{y}_k + \frac{1 - \theta_k}{\theta_k}\mathbf{x}_k \right\|^2 - \|\mathbf{x}^* - \mathbf{z}_{k+1}\|^2 \right),$$

where $\overset{a}{=}$ uses (A.1). By reorganizing the terms in $\mathbf{x}^* - \frac{1}{\theta_k}\mathbf{y}^k + \frac{1-\theta_k}{\theta_k}\mathbf{x}^k$ carefully, we can have

$$\frac{L\theta_k^2}{2}\left\| \mathbf{x}^* - \frac{1}{\theta_k}\mathbf{y}_k + \frac{1 - \theta_k}{\theta_k}\mathbf{x}_k \right\|^2$$

$$= \frac{L\theta_k^2}{2}\left\| \frac{\mu}{L\theta_k}(\mathbf{x}^* - \mathbf{y}_k) + \frac{L\theta_k - \mu}{L\theta_k}\left(\mathbf{x}^* + \frac{L(1 - \theta_k)}{L\theta_k - \mu}\mathbf{x}_k - \frac{L - \mu}{L\theta_k - \mu}\mathbf{y}_k \right) \right\|^2$$

$$\overset{a}{\le} \frac{\mu\theta_k}{2}\|\mathbf{x}^* - \mathbf{y}_k\|^2 + \frac{\theta_k(L\theta_k - \mu)}{2}\left\| \mathbf{x}^* + \frac{L(1 - \theta_k)}{L\theta_k - \mu}\mathbf{x}_k - \frac{L - \mu}{L\theta_k - \mu}\mathbf{y}_k \right\|^2$$

$$\overset{b}{=} \frac{\mu\theta_k}{2}\|\mathbf{x}^* - \mathbf{y}_k\|^2 + \frac{\theta_k(L\theta_k - \mu)}{2}\|\mathbf{z}_k - \mathbf{x}^*\|^2, \tag{2.23}$$

where we let $0 \leq \frac{\mu}{L\theta_k} < 1$, use the convexity of $\|\cdot\|^2$ in $\overset{a}{\leq}$, and use (2.22) in $\overset{b}{=}$. Thus we can have

$$F(\mathbf{x}_{k+1}) - F(\mathbf{x}^*) + \frac{L\theta_k^2}{2} \left\| \mathbf{z}_{k+1} - \mathbf{x}^* \right\|^2$$

$$\leq (1 - \theta_k)(F(\mathbf{x}_k) - F(\mathbf{x}^*)) + \frac{\theta_k(L\theta_k - \mu)}{2} \left\| \mathbf{z}_k - \mathbf{x}^* \right\|^2. \qquad (2.24)$$

Case 1: $\mu = 0$. Dividing both sides of (2.24) by θ_k^2 and using $\frac{1-\theta_k}{\theta_k^2} = \frac{1}{\theta_{k-1}^2}$, we obtain $\ell_{k+1} \leq \ell_k$, which leads to the first conclusion, where we use $\frac{1}{\theta_{-1}^2} = 0$.

Case 2: $\mu > 0$. Letting $\theta(L\theta - \mu) = L\theta^2(1 - \theta)$, we have $\theta = \sqrt{\frac{\mu}{L}}$. Dividing both sides of (2.24) by $(1 - \theta)^{k+1}$, we obtain $\ell_{k+1} \leq \ell_k$, which leads to the second conclusion. $\qquad \square$

Remark 2.1 When $\mu = 0$, $\frac{(L\theta_k - \mu)(1-\theta_{k-1})}{(L-\mu)\theta_{k-1}} = \frac{\theta_k(1-\theta_{k-1})}{\theta_{k-1}}$. When $\mu \neq 0$ and $\theta_k = \sqrt{\frac{\mu}{L}}, \forall k$, $\frac{(L\theta_k - \mu)(1-\theta_{k-1})}{(L-\mu)\theta_{k-1}} = \frac{\sqrt{L}-\sqrt{\mu}}{\sqrt{L}+\sqrt{\mu}}$.

Remark 2.2 In Theorem 2.1, we start from $\theta_{-1} = 1$, while in Theorem 2.2 we start from $\theta_0 = 1$.

2.2.2 Nesterov's Second Scheme

Besides Algorithm 2.2, another accelerated algorithm is also widely used in practice and we describe it in Algorithm 2.3. Algorithm 2.3 is equivalent to Algorithm 2.1 when $h(\mathbf{x}) = 0$. In fact, in this case the two algorithms produce the same sequences of $\{\mathbf{x}_k\}$, $\{\mathbf{y}_k\}$, and $\{\mathbf{z}_k\}$. However, this is not true for nonsmooth problems.

Algorithm 2.3 Accelerated proximal gradient (APG) method 2

Initialize $\mathbf{z}_0 = \mathbf{x}_0$.
for $k = 0, 1, 2, 3, \cdots$ **do**
$\quad \mathbf{y}_k = \frac{L\theta_k - \mu}{L-\mu} \mathbf{z}_k + \frac{L-L\theta_k}{L-\mu} \mathbf{x}_k,$
$\quad \mathbf{z}_{k+1} = \text{argmin}_{\mathbf{z}} \left(h(\mathbf{z}) + \langle \nabla f(\mathbf{y}_k), \mathbf{z} \rangle + \frac{\theta_k L}{2} \left\| \mathbf{z} - \frac{1}{\theta_k} \mathbf{y}_k + \frac{1-\theta_k}{\theta_k} \mathbf{x}_k \right\|^2 \right),$
$\quad \mathbf{x}_{k+1} = (1 - \theta_k)\mathbf{x}_k + \theta_k \mathbf{z}_{k+1}.$
end for

We give the convergence rates of Algorithm 2.3 in Theorem 2.3. The proof is similar to that of Algorithm 2.2 and we only show the difference. We use the Lyapunov functions defined in (2.19) and (2.20), and we can easily verify that $\mathbf{x}_k$, $\mathbf{y}_k$, and $\mathbf{z}_k$ satisfy the relations in Lemma 2.5.

Theorem 2.3 *Suppose that $f(\mathbf{x})$ and $h(\mathbf{x})$ are convex and $f(\mathbf{x})$ is L-smooth. Let* $\theta_0 = 1$ *and* $\theta_{k+1} = \frac{\sqrt{\theta_k^4 + 4\theta_k^2} - \theta_k^2}{2}$. *Then for Algorithm 2.3, we have*

$$F(\mathbf{x}_{K+1}) - F(\mathbf{x}^*) \leq \frac{2L}{(K+2)^2} \|\mathbf{x}_0 - \mathbf{x}^*\|^2.$$

Suppose that $h(\mathbf{x})$ is convex and $f(\mathbf{x})$ is μ-strongly convex and L-smooth. Let $\theta_k = \sqrt{\frac{\mu}{L}}$ *for all k. Then for Algorithm 2.3, we have*

$$F(\mathbf{x}_{K+1}) - F(\mathbf{x}^*) \leq \left(1 - \sqrt{\frac{\mu}{L}}\right)^{K+1} \left(F(\mathbf{x}_0) - F(\mathbf{x}^*) + \frac{\mu}{2}\|\mathbf{x}_0 - \mathbf{x}^*\|^2\right).$$

Proof From the optimality condition of the second step, we have

$$\mathbf{0} \in \partial h(\mathbf{z}_{k+1}) + \nabla f(\mathbf{y}_k) + \theta_k L \left(\mathbf{z}_{k+1} - \frac{1}{\theta_k}\mathbf{y}_k + \frac{1 - \theta_k}{\theta_k}\mathbf{x}_k\right).$$

Then from the convexity of $h(\mathbf{x})$, we have

$$h(\mathbf{x}) - h(\mathbf{z}_{k+1}) \geq \left\langle -\theta_k L \left(\mathbf{z}_{k+1} - \frac{1}{\theta_k}\mathbf{y}_k + \frac{1 - \theta_k}{\theta_k}\mathbf{x}_k\right) - \nabla f(\mathbf{y}_k), \mathbf{x} - \mathbf{z}_{k+1}\right\rangle.$$
$$(2.25)$$

From the smoothness and the strong convexity of $f(\mathbf{x})$, we have

$$F(\mathbf{x}_{k+1})$$
$$\leq f(\mathbf{y}_k) + \langle \nabla f(\mathbf{y}_k), \mathbf{x}_{k+1} - \mathbf{y}_k\rangle + \frac{L}{2}\|\mathbf{x}_{k+1} - \mathbf{y}_k\|^2 + h(\mathbf{x}_{k+1})$$
$$= f(\mathbf{y}_k) + \langle \nabla f(\mathbf{y}_k), (1 - \theta_k)\mathbf{x}_k + \theta_k \mathbf{z}_{k+1} - \mathbf{y}_k\rangle$$
$$+ \frac{L\theta_k^2}{2}\left\|\frac{1 - \theta_k}{\theta_k}\mathbf{x}_k + \mathbf{z}_{k+1} - \frac{1}{\theta_k}\mathbf{y}_k\right\|^2 + h((1 - \theta_k)\mathbf{x}_k + \theta_k \mathbf{z}_{k+1})$$
$$\leq (1 - \theta_k)(f(\mathbf{y}_k) + \langle \nabla f(\mathbf{y}_k), \mathbf{x}_k - \mathbf{y}_k\rangle + h(\mathbf{x}_k))$$
$$+ \theta_k (f(\mathbf{y}_k) + \langle \nabla f(\mathbf{y}_k), \mathbf{z}_{k+1} - \mathbf{y}_k\rangle + h(\mathbf{z}_{k+1}))$$
$$+ \frac{L\theta_k^2}{2}\left\|\frac{1 - \theta_k}{\theta_k}\mathbf{x}_k + \mathbf{z}_{k+1} - \frac{1}{\theta_k}\mathbf{y}_k\right\|^2$$
$$\leq (1 - \theta_k)F(\mathbf{x}_k) + \theta_k \left(f(\mathbf{y}_k) + \langle \nabla f(\mathbf{y}_k), \mathbf{x}^* - \mathbf{y}_k\rangle + \langle \nabla f(\mathbf{y}_k), \mathbf{z}_{k+1} - \mathbf{x}^*\rangle\right.$$
$$+ h(\mathbf{z}_{k+1})) + \frac{L\theta_k^2}{2}\left\|\frac{1 - \theta_k}{\theta_k}\mathbf{x}_k + \mathbf{z}_{k+1} - \frac{1}{\theta_k}\mathbf{y}_k\right\|^2$$

$$\overset{a}{\le} (1 - \theta_k) F(\mathbf{x}_k) + \theta_k \left(f(\mathbf{x}^*) - \frac{\mu}{2} \|\mathbf{y}_k - \mathbf{x}^*\|^2 + h(\mathbf{x}^*) \right.$$

$$+ \theta_k L \left\langle \mathbf{z}_{k+1} - \frac{1}{\theta_k} \mathbf{y}_k + \frac{1 - \theta_k}{\theta_k} \mathbf{x}_k, \mathbf{x}^* - \mathbf{z}_{k+1} \right\rangle \bigg)$$

$$+ \frac{L \theta_k^2}{2} \left\| \frac{1 - \theta_k}{\theta_k} \mathbf{x}_k + \mathbf{z}_{k+1} - \frac{1}{\theta_k} \mathbf{y}_k \right\|^2$$

$$\overset{b}{\le} (1 - \theta_k) F(\mathbf{x}_k) + \theta_k F(\mathbf{x}^*) - \frac{\mu \theta_k}{2} \|\mathbf{y}_k - \mathbf{x}^*\|^2$$

$$+ \frac{\theta_k^2 L}{2} \left(\left\| \frac{1}{\theta_k} \mathbf{y}_k - \frac{1 - \theta_k}{\theta_k} \mathbf{x}_k - \mathbf{x}^* \right\|^2 - \|\mathbf{z}_{k+1} - \mathbf{x}^*\|^2 \right)$$

$$\overset{c}{\le} (1 - \theta_k) F(\mathbf{x}_k) + \theta_k F(\mathbf{x}^*) + \frac{\theta_k (L\theta_k - \mu)}{2} \|\mathbf{z}_k - \mathbf{x}^*\|^2 - \frac{\theta_k^2 L}{2} \|\mathbf{z}_{k+1} - \mathbf{x}^*\|^2,$$

where $\overset{a}{\le}$ uses (2.25), $\overset{b}{\le}$ uses (A.2), and $\overset{c}{\le}$ uses (2.23). Following the same inductions in the proof of Theorem 2.2, we can have the conclusions. $\qquad\square$

2.2.2.1 A Primal-Dual Perspective

Several different explanations have been proposed for Algorithm 2.3. Allen-Zhu and Orecchia [2] considered it as a linear coupling of gradient descent and mirror descent and [15] gave an explanation from the perspective of the primal-dual method. We introduce the idea in [15] here. We only consider the case of $\mu = 0$ and in this case the iterations of Algorithm 2.3 become

$$\mathbf{y}_k = \theta_k \mathbf{z}_k + (1 - \theta_k) \mathbf{x}_k, \tag{2.26}$$

$$\mathbf{z}_{k+1} = \underset{\mathbf{x}}{\operatorname{argmin}} \left(h(\mathbf{x}) + \langle \nabla f(\mathbf{y}_k), \mathbf{x} \rangle + \frac{\theta_k L}{2} \|\mathbf{x} - \mathbf{z}_k\|^2 \right),$$

$$\mathbf{x}_{k+1} = (1 - \theta_k) \mathbf{x}_k + \theta_k \mathbf{z}_{k+1}. \tag{2.27}$$

Let the Fenchel conjugate (Definition A.21) of $f(\mathbf{z})$ be $f^*(\mathbf{u})$. Since $f^{**} = f$, we can rewrite problem (2.17) in the following min-max form:

$$\min_{\mathbf{x}} \max_{\mathbf{u}} \left(h(\mathbf{x}) + \langle \mathbf{x}, \mathbf{u} \rangle - f^*(\mathbf{u}) \right). \tag{2.28}$$

Since f is L-smooth, f^* is $\frac{1}{L}$-strongly convex (point 4 of Proposition A.12). We can define the Bregman distance (Definition A.20) induced by f^*:

$$D_{f^*}(\mathbf{u}, \mathbf{v}) = f^*(\mathbf{u}) - \left(f^*(\mathbf{v}) + \left\langle \hat{\nabla} f^*(\mathbf{v}), \mathbf{u} - \mathbf{v} \right\rangle \right),$$

where $\hat{\nabla} f^*(\mathbf{v}) \in \partial f^*(\mathbf{v})$. We can use the primal-dual method [8, 9] to find the saddle point (Definition A.29) of problem (2.28), which updates $\mathbf{x}$ and $\mathbf{u}$ in the primal space and the dual space, respectively:

$$\hat{\mathbf{z}}_k = \alpha_k(\mathbf{z}_k - \mathbf{z}_{k-1}) + \mathbf{z}_k, \tag{2.29}$$

$$\mathbf{u}_{k+1} = \operatorname*{argmax}_{\mathbf{u}} \left(\langle \hat{\mathbf{z}}_k, \mathbf{u} \rangle - f^*(\mathbf{u}) - \tau_k D_{f^*}(\mathbf{u}, \mathbf{u}_k) \right),$$

$$\mathbf{z}_{k+1} = \operatorname*{argmin}_{\mathbf{z}} \left(h(\mathbf{z}) + \langle \mathbf{z}, \mathbf{u}_{k+1} \rangle + \frac{\eta_k}{2} \|\mathbf{z} - \mathbf{z}_k\|^2 \right). \tag{2.30}$$

In the following Lemma, we establish the equivalence between algorithms (2.26)–(2.27) and (2.29)–(2.30).

Lemma 2.6 *Let* $\mathbf{u}_0 = \nabla f(\mathbf{z}_0)$, $\mathbf{z}_{-1} = \mathbf{z}_0$, $\tau_k = \frac{1-\theta_k}{\theta_k}$, $\alpha_k = \frac{\theta_{k-1}(1-\theta_k)}{\theta_k}$, $\eta_k = L\theta_k$, *and* $\theta_0 = 1$. *Then Algorithms (2.26)–(2.27) and (2.29)–(2.30) are equivalent.*

Proof Denoting $\mathbf{y}_{-1} = \mathbf{z}_0$, by point 5 of Proposition A.12 we can see from $\mathbf{u}_0 = \nabla f(\mathbf{z}_0)$ that $\mathbf{y}_{-1} \in \partial f^*(\mathbf{u}_0)$. Letting $\mathbf{y}_{k-1} = \hat{\nabla} f^*(\mathbf{u}_k) \in \partial f^*(\mathbf{u}_k)$ and defining $\mathbf{y}_k = \frac{1}{1+\tau_k} \left(\hat{\mathbf{z}}_k + \tau_k \mathbf{y}_{k-1} \right)$, we have

$$\mathbf{u}_{k+1} = \operatorname*{argmin}_{\mathbf{u}} \left(-\left\langle \hat{\mathbf{z}}_k + \tau_k \hat{\nabla} f^*(\mathbf{u}_k), \mathbf{u} \right\rangle + (1 + \tau_k) f^*(\mathbf{u}) \right)$$

$$= \operatorname*{argmax}_{\mathbf{u}} \left(\langle \mathbf{y}_k, \mathbf{u} \rangle - f^*(\mathbf{u}) \right).$$

Hence $\mathbf{y}_k \in \partial f^*(\mathbf{u}_{k+1})$. So by mathematical induction, we have $\mathbf{y}_k \in \partial f^*(\mathbf{u}_{k+1})$, $\forall k$. Thus again by point 5 of Proposition A.12, we have $\mathbf{u}_{k+1} = \nabla f(\mathbf{y}_k)$, $\forall k$, and (2.29)–(2.30) are equivalent to

$$\hat{\mathbf{z}}_k = \alpha_k(\mathbf{z}_k - \mathbf{z}_{k-1}) + \mathbf{z}_k,$$

$$\mathbf{y}_k = \frac{1}{1+\tau_k} \left(\hat{\mathbf{z}}_k + \tau_k \mathbf{y}_{k-1} \right),$$

$$\mathbf{z}_{k+1} = \operatorname*{argmin}_{\mathbf{z}} \left(h(\mathbf{z}) + \langle \mathbf{z}, \nabla f(\mathbf{y}_k) \rangle + \frac{\eta_k}{2} \|\mathbf{z} - \mathbf{z}_k\|^2 \right).$$

Letting $\tau_k = \frac{1-\theta_k}{\theta_k}$, $\alpha_k = \frac{\theta_{k-1}(1-\theta_k)}{\theta_k}$, and $\eta_k = L\theta_k$, we have

$$\mathbf{y}_k = \frac{1}{1+\tau_k} \left[\alpha_k(\mathbf{z}_k - \mathbf{z}_{k-1}) + \mathbf{z}_k + \tau_k \mathbf{y}_{k-1} \right]$$

$$= \theta_{k-1}(1 - \theta_k)(\mathbf{z}_k - \mathbf{z}_{k-1}) + \theta_k \mathbf{z}_k + (1 - \theta_k)\mathbf{y}_{k-1}$$

$$= \theta_k \mathbf{z}_k + (1 - \theta_k)(\theta_{k-1}\mathbf{z}_k + \mathbf{y}_{k-1} - \theta_{k-1}\mathbf{z}_{k-1}),$$

which satisfies the relations in (2.26) and (2.27).   □

2.2.3 Nesterov's Third Scheme

The third method we describe is remarkably different from the previous two in that it uses a weighted sum of previous gradients. We only consider the generally convex case in this section and describe the method in Algorithm 2.4. We can see that the first step and the last step of Algorithm 2.4 are the same as (2.26) and (2.27). The difference lies in the way that $\mathbf{z}_{k+1}$ is generated. Accordingly, their proofs of convergence rates are also different. Note that the $\mathbf{z}_{k+1}$ in Algorithm 2.4 is different from the $\mathbf{z}_{k+1}$ in Algorithm (2.26)–(2.27). They are the same when $h(\mathbf{x}) = 0$.

Algorithm 2.4 Accelerated proximal gradient (APG) method 3

Initialize $\mathbf{x}_0 = \mathbf{z}_0$, $\phi_0(\mathbf{x}) = 0$, and $\theta_0 = 1$.
for $k = 0, 1, 2, 3, \cdots$ **do**
$\quad \mathbf{y}_k = \theta_k \mathbf{z}_k + (1 - \theta_k)\mathbf{x}_k$,
$\quad \phi_{k+1}(\mathbf{x}) = \sum_{i=0}^{k} \frac{f(\mathbf{y}_i) + \langle \nabla f(\mathbf{y}_i), \mathbf{x} - \mathbf{y}_i \rangle + h(\mathbf{x})}{\theta_i}$,
$\quad \mathbf{z}_{k+1} = \operatorname{argmin}_{\mathbf{z}} \left(\phi_{k+1}(\mathbf{z}) + \frac{L}{2} \|\mathbf{z} - \mathbf{x}_0\|^2 \right)$,
$\quad \mathbf{x}_{k+1} = (1 - \theta_k)\mathbf{x}_k + \theta_k \mathbf{z}_{k+1}$.
end for

We define the Lyapunov function

$$\ell_{k+1} = \frac{F(\mathbf{x}_{k+1})}{\theta_k^2} - \phi_{k+1}(\mathbf{z}_{k+1}) - \frac{L}{2} \|\mathbf{z}_{k+1} - \mathbf{x}_0\|^2.$$

We can see that it is different from the one in (2.19) used for the previous two methods. Then by exploiting the non-increment of ℓ_k, we can establish the convergence rate in the following theorem.

Theorem 2.4 *Suppose that $f(\mathbf{x})$ and $h(\mathbf{x})$ are convex and $f(\mathbf{x})$ is L-smooth. Let $\theta_0 = 1$ and $\theta_{k+1} = \frac{\sqrt{\theta_k^4 + 4\theta_k^2} - \theta_k^2}{2}$. Then for Algorithm 2.4, we have*

$$F(\mathbf{x}_{K+1}) - F(\mathbf{x}^*) \leq \frac{2L}{(K+2)^2} \|\mathbf{x}^* - \mathbf{x}_0\|^2.$$

Proof From the optimality condition of the third step, we have

$$\mathbf{0} \in \partial \phi_k(\mathbf{z}_k) + L(\mathbf{z}_k - \mathbf{x}_0).$$

Then from the convexity of $\phi_k(\mathbf{x})$, we have

$$\phi_k(\mathbf{z}) - \phi_k(\mathbf{z}_k) \geq -L \langle \mathbf{z}_k - \mathbf{x}_0, \mathbf{z} - \mathbf{z}_k \rangle$$

$$\overset{a}{=} \frac{L}{2} \|\mathbf{z}_k - \mathbf{x}_0\|^2 - \frac{L}{2} \|\mathbf{z} - \mathbf{x}_0\|^2 + \frac{L}{2} \|\mathbf{z}_k - \mathbf{z}\|^2,$$

where $\stackrel{a}{=}$ uses (A.2). Letting $\mathbf{z} = \mathbf{z}_{k+1}$, we obtain

$$\frac{L}{2}\|\mathbf{z}_k - \mathbf{z}_{k+1}\|^2$$

$$\leq \left(\phi_k(\mathbf{z}_{k+1}) + \frac{L}{2}\|\mathbf{z}_{k+1} - \mathbf{x}_0\|^2\right) - \left(\phi_k(\mathbf{z}_k) + \frac{L}{2}\|\mathbf{z}_k - \mathbf{x}_0\|^2\right). \quad (2.31)$$

Following the same induction in the proof of Theorem 2.3, we have

$$F(\mathbf{x}_{k+1})$$

$$\leq (1 - \theta_k)F(\mathbf{x}_k) + \theta_k\left(f(\mathbf{y}_k) + \langle \nabla f(\mathbf{y}_k), \mathbf{z}_{k+1} - \mathbf{y}_k \rangle + h(\mathbf{z}_{k+1})\right)$$

$$+ \frac{L\theta_k^2}{2}\|\mathbf{z}_{k+1} - \mathbf{z}_k\|^2$$

$$\stackrel{a}{=} (1 - \theta_k)F(\mathbf{x}_k) + \theta_k^2\left(\phi_{k+1}(\mathbf{z}_{k+1}) - \phi_k(\mathbf{z}_{k+1})\right) + \frac{L\theta_k^2}{2}\|\mathbf{z}_k - \mathbf{z}_{k+1}\|^2$$

$$\stackrel{b}{\leq} (1 - \theta_k)F(\mathbf{x}_k) + \theta_k^2\phi_{k+1}(\mathbf{z}_{k+1}) + \frac{L\theta_k^2}{2}\|\mathbf{z}_{k+1} - \mathbf{x}_0\|^2 - \theta_k^2\phi_k(\mathbf{z}_k)$$

$$- \frac{L\theta_k^2}{2}\|\mathbf{z}_k - \mathbf{x}_0\|^2,$$

where $\stackrel{a}{=}$ uses the definition of $\phi_{k+1}(\mathbf{x})$ in Step 2 of Algorithm 2.4 and $\stackrel{b}{\leq}$ uses (2.31). The above also holds for $k = 0$ due to $\phi_0(\mathbf{x}) = 0$. Dividing both sides by θ_k^2 and using $\frac{1-\theta_k}{\theta_k^2} = \frac{1}{\theta_{k-1}^2}$, where $\frac{1}{\theta_{-1}^2} = 0$, we obtain $\ell_{k+1} \leq \ell_k$, i.e.,

$$\frac{F(\mathbf{x}_{K+1})}{\theta_K^2} \leq \phi_{K+1}(\mathbf{z}_{K+1}) + \frac{L}{2}\|\mathbf{z}_{K+1} - \mathbf{x}_0\|^2 - \phi_0(\mathbf{z}_0) - \frac{L}{2}\|\mathbf{z}_0 - \mathbf{x}_0\|^2$$

$$\stackrel{a}{\leq} \phi_{K+1}(\mathbf{x}^*) + \frac{L}{2}\|\mathbf{x}^* - \mathbf{x}_0\|^2$$

$$\stackrel{b}{\leq} \sum_{i=0}^{K}\frac{F(\mathbf{x}^*)}{\theta_i} + \frac{L}{2}\|\mathbf{x}^* - \mathbf{x}_0\|^2$$

$$\stackrel{c}{=} \frac{F(\mathbf{x}^*)}{\theta_K^2} + \frac{L}{2}\|\mathbf{x}^* - \mathbf{x}_0\|^2,$$

where we use Step 3 of Algorithm 2.4 in $\stackrel{a}{\leq}$, the convexity of $f(\mathbf{x})$ in $\stackrel{b}{\leq}$, and Lemma 2.3 and $\frac{1}{\theta_{-1}^2} = \frac{1-\theta_0}{\theta_0^2} = 0$ in $\stackrel{c}{=}$. □

Remark 2.3 When $h(\mathbf{x}) = 0$, Algorithms 2.3 and 2.4 are equivalent. In this case, from Step 2 of Algorithm 2.3, we have

$$\nabla f(\mathbf{y}_k) + \theta_k L(\mathbf{z}_{k+1} - \mathbf{z}_k) = \mathbf{0}. \tag{2.32}$$

From Steps 2 and 3 of Algorithm 2.4, we have

$$\sum_{i=0}^{k} \frac{1}{\theta_i} \nabla f(\mathbf{y}_i) + L(\mathbf{z}_{k+1} - \mathbf{x}_0) = \mathbf{0}. \tag{2.33}$$

Dividing both sides of (2.32) by θ_k and summing over $k = 0, 1, \cdots$, we have (2.33).

2.3 Inexact Proximal and Gradient Computing

In Sect. 2.2, we prove the convergence rate under the assumption that the proximal mapping of h is easily computable, e.g., having a closed form solution. This is the case for several notable choices of h, e.g., the ℓ_1-regularization [3]. However, in many scenarios the proximal mapping may not have an analytic solution, or it may be very expensive to compute this solution exactly. This includes important problems such as total-variation regularization [12], graph-guided-fused LASSO [10], and overlapping group ℓ_1-regularization with general groups [14]. Moreover, in some scenarios the gradient may be corrupted by noise.

Motivated by these problems, several works, e.g., [11, 27], study the case that both the gradient and proximal mapping are computed inexactly. We only analyze a variant of Algorithm 2.2 and describe it in Algorithm 2.5. In Algorithm 2.5, $\widetilde{\nabla} f(\mathbf{y}_k)$ means the gradient with noise, i.e.,

$$\widetilde{\nabla} f(\mathbf{y}_k) = \nabla f(\mathbf{y}_k) + \mathbf{e}_k.$$

We consider the inexact proximal mapping with error ϵ_k, which is described as

$$h(\mathbf{x}_{k+1}) + \frac{L}{2}\|\mathbf{x}_{k+1} - \mathbf{w}_k\|^2 \leq \min_{\mathbf{x}} \left(h(\mathbf{x}) + \frac{L}{2}\|\mathbf{x} - \mathbf{w}_k\|^2 \right) + \epsilon_k. \tag{2.34}$$

Algorithm 2.5 Inexact accelerated proximal gradient method

Initialize $\mathbf{x}_0 = \mathbf{x}_{-1}$.
for $k = 0, 1, 2, 3, \cdots$ **do**
$\quad \mathbf{y}_k = \mathbf{x}_k + \frac{(L\theta_k - \mu_1 - \mu_2(1-\theta_k))(1-\theta_{k-1})}{(L-\mu_1)\theta_{k-1}}(\mathbf{x}_k - \mathbf{x}_{k-1})$,
$\quad \mathbf{w}_k = \mathbf{y}_k - \frac{1}{L}\widetilde{\nabla} f(\mathbf{y}_k)$,
$\quad \mathbf{x}_{k+1} \approx \operatorname{argmin}_{\mathbf{x}} \left(h(\mathbf{x}) + \frac{L}{2}\|\mathbf{x} - \mathbf{w}_k\|^2 \right)$.
end for

We first give a crucial property when the proximal mapping is computed inexactly. When $\epsilon_k = 0$, from the optimality condition of the proximal mapping and the strong convexity of h, we can immediately have

$$h(\mathbf{x}) - h(\mathbf{x}_{k+1}) \geq L \langle \mathbf{x}_{k+1} - \mathbf{w}_k, \mathbf{x}_{k+1} - \mathbf{x} \rangle + \frac{\mu}{2} \|\mathbf{x} - \mathbf{x}_{k+1}\|^2. \qquad (2.35)$$

However, when $\epsilon_k \neq 0$ we need to modify (2.35) accordingly and it is described in the following lemma. Specially, in Lemma 2.7 h can be generally convex, i.e., $\mu = 0$.

Lemma 2.7 *Assume that $h(\mathbf{x})$ is μ-strongly convex. Let $\mathbf{x}_{k+1}$ be an inexact proximal mapping of $h(\mathbf{x})$ such that (2.34) holds. Then there exists $\boldsymbol{\sigma}_k$ satisfying* $\|\boldsymbol{\sigma}_k\| \leq \sqrt{\frac{2(L+\mu)\epsilon_k}{L^2}}$ *such that*

$$h(\mathbf{x}) - h(\mathbf{x}_{k+1}) \geq -\epsilon_k + L \langle \mathbf{x}_{k+1} - \mathbf{w}_k + \boldsymbol{\sigma}_k, \mathbf{x}_{k+1} - \mathbf{x} \rangle + \frac{\mu}{2} \|\mathbf{x} - \mathbf{x}_{k+1}\|^2. \quad (2.36)$$

Proof Let

$$\mathbf{x}_{k+1}^* = \underset{\mathbf{x}}{\text{argmin}} \left(h(\mathbf{x}) + \frac{L}{2} \|\mathbf{x} - \mathbf{w}_k\|^2 \right).$$

From the strong convexity of $h(\mathbf{x})$ and the definition of $\mathbf{x}_{k+1}$, we have

$$\mathbf{0} \in \partial h(\mathbf{x}_{k+1}^*) + L(\mathbf{x}_{k+1}^* - \mathbf{w}_k),$$

$$h(\mathbf{x}) - h(\mathbf{x}_{k+1}^*) \geq -L \langle \mathbf{x}_{k+1}^* - \mathbf{w}_k, \mathbf{x} - \mathbf{x}_{k+1}^* \rangle + \frac{\mu}{2} \|\mathbf{x} - \mathbf{x}_{k+1}^*\|^2,$$

$$h(\mathbf{x}_{k+1}^*) + \frac{L}{2} \|\mathbf{x}_{k+1}^* - \mathbf{w}_k\|^2 + \epsilon_k \geq h(\mathbf{x}_{k+1}) + \frac{L}{2} \|\mathbf{x}_{k+1} - \mathbf{w}_k\|^2.$$

So we can have

$$h(\mathbf{x}) - h(\mathbf{x}_{k+1})$$

$$\geq -\epsilon_k - L \langle \mathbf{x}_{k+1}^* - \mathbf{w}_k, \mathbf{x} - \mathbf{x}_{k+1}^* \rangle + \frac{\mu}{2} \|\mathbf{x} - \mathbf{x}_{k+1}^*\|^2 + \frac{L}{2} \|\mathbf{x}_{k+1} - \mathbf{w}_k\|^2$$

$$\quad - \frac{L}{2} \|\mathbf{x}_{k+1}^* - \mathbf{w}_k\|^2$$

$$\overset{a}{=} -\epsilon_k + \frac{L}{2} \left(\|\mathbf{x}_{k+1}^* - \mathbf{w}_k\|^2 + \|\mathbf{x}_{k+1}^* - \mathbf{x}\|^2 - \|\mathbf{w}_k - \mathbf{x}\|^2 \right)$$

$$\quad + \frac{L}{2} \|\mathbf{x}_{k+1} - \mathbf{w}_k\|^2 - \frac{L}{2} \|\mathbf{x}_{k+1}^* - \mathbf{w}_k\|^2 + \frac{\mu}{2} \|\mathbf{x} - \mathbf{x}_{k+1}^*\|^2$$

$$= -\epsilon_k + \frac{L}{2}\left(\|\mathbf{x}_{k+1} - \mathbf{w}_k\|^2 + \|\mathbf{x}^*_{k+1} - \mathbf{x}\|^2 - \|\mathbf{w}_k - \mathbf{x}\|^2\right) + \frac{\mu}{2}\|\mathbf{x} - \mathbf{x}^*_{k+1}\|^2$$

$$= -\epsilon_k + \frac{L}{2}\left(\|\mathbf{x}_{k+1} - \mathbf{w}_k\|^2 + \|\mathbf{x}_{k+1} - \mathbf{x}\|^2 - \|\mathbf{w}_k - \mathbf{x}\|^2\right)$$

$$+ \frac{L}{2}\|\mathbf{x}^*_{k+1} - \mathbf{x}\|^2 - \frac{L}{2}\|\mathbf{x}_{k+1} - \mathbf{x}\|^2 + \frac{\mu}{2}\|\mathbf{x} - \mathbf{x}^*_{k+1}\|^2$$

$$\overset{b}{=} -\epsilon_k + L\langle \mathbf{x}_{k+1} - \mathbf{w}_k, \mathbf{x}_{k+1} - \mathbf{x}\rangle + \frac{L+\mu}{2}\|\mathbf{x}^*_{k+1} - \mathbf{x}\|^2$$

$$- \frac{L+\mu}{2}\|\mathbf{x}_{k+1} - \mathbf{x}\|^2 + \frac{\mu}{2}\|\mathbf{x} - \mathbf{x}_{k+1}\|^2$$

$$= -\epsilon_k + L\langle \mathbf{x}_{k+1} - \mathbf{w}_k, \mathbf{x}_{k+1} - \mathbf{x}\rangle - (L+\mu)\langle \mathbf{x}_{k+1} - \mathbf{x}^*_{k+1}, \mathbf{x}_{k+1} - \mathbf{x}\rangle$$

$$+ \frac{L+\mu}{2}\|\mathbf{x}_{k+1} - \mathbf{x}^*_{k+1}\|^2 + \frac{\mu}{2}\|\mathbf{x} - \mathbf{x}_{k+1}\|^2$$

$$= -\epsilon_k + L\langle \mathbf{x}_{k+1} - \mathbf{w}_k + \boldsymbol{\sigma}_k, \mathbf{x}_{k+1} - \mathbf{x}\rangle + \frac{L+\mu}{2}\|\mathbf{x}_{k+1} - \mathbf{x}^*_{k+1}\|^2$$

$$+ \frac{\mu}{2}\|\mathbf{x} - \mathbf{x}_{k+1}\|^2,$$

where $\overset{a}{=}$ uses (A.2), $\overset{b}{=}$ uses (A.1), and we define $\boldsymbol{\sigma}_k = \frac{L+\mu}{L}(\mathbf{x}^*_{k+1} - \mathbf{x}_{k+1})$. Letting $\mathbf{x} = \mathbf{x}_{k+1}$, we have $\epsilon_k \geq \frac{L+\mu}{2}\|\mathbf{x}^*_{k+1} - \mathbf{x}_{k+1}\|^2 = \frac{L^2}{2(L+\mu)}\|\boldsymbol{\sigma}_k\|^2$. $\qquad\square$

The following lemma gives a useful tool to analyze the convergence of algorithms with inexact computing.

Lemma 2.8 *Assume that sequence $\{S_k\}$ is increasing, $\{u_k\}$ and $\{\alpha_i\}$ are nonnegative, and $u_0^2 \leq S_0$. If*

$$u_k^2 \leq S_k + \sum_{i=1}^{k} \alpha_i u_i, \tag{2.37}$$

then

$$S_k + \sum_{i=1}^{k} \alpha_i u_i \leq \left(\sqrt{S_k} + \sum_{i=1}^{k} \alpha_i\right)^2. \tag{2.38}$$

Proof Let b_k^2 be the right-hand side of (2.37). We have for all $k \geq 1$, $u_k \leq b_k$, and

$$b_k^2 = S_k + \sum_{i=1}^{k} \alpha_i u_i \leq S_k + \sum_{i=1}^{k} \alpha_i b_i \leq S_k + \left(\sum_{i=1}^{k} \alpha_i\right) b_k.$$

So

$$b_k \leq \frac{1}{2} \sum_{i=1}^{k} \alpha_i + \sqrt{\left(\frac{1}{2} \sum_{i=1}^{k} \alpha_i\right)^2 + S_k}.$$

Using the inequality $\sqrt{x+y} \leq \sqrt{x} + \sqrt{y}$, we have

$$b_k \leq \frac{1}{2} \sum_{i=1}^{k} \alpha_i + \sqrt{\left(\frac{1}{2} \sum_{i=1}^{k} \alpha_i\right)^2 + \sqrt{S_k}} = \sqrt{S_k} + \sum_{i=1}^{k} \alpha_i.$$

Hence

$$S_k + \sum_{i=1}^{k} \alpha_i u_i = b_k^2 \leq \left(\sqrt{S_k} + \sum_{i=1}^{k} \alpha_i\right)^2,$$

finishing the proof.   □

Chaining (2.37) and (2.38) we immediately conclude the following.

Corollary 2.1 *With the assumptions in Lemma 2.8, we have*

$$u_k \leq \sum_{i=1}^{k} \alpha_i + \sqrt{S_k},$$

and thus

$$u_k^2 \leq 2 \left(\sum_{i=1}^{k} \alpha_i\right)^2 + 2S_k. \tag{2.39}$$

Similar to Lemma 2.5, we can give the following easy-to-verify identities. The difference from Lemma 2.5 is that Lemma 2.5 only considers the case that only f is strongly convex, while in this section we consider the case that both f and h are strongly convex with moduli μ_1 and μ_2, respectively.

Lemma 2.9 *Also define z_k as in (2.21). For Algorithm 2.5, we have*

$$\mathbf{x}^* + \frac{L - L\theta_k + \mu_2(1 - \theta_k)}{L\theta_k - \mu_1 - \mu_2(1 - \theta_k)} \mathbf{x}_k - \frac{L - \mu_1}{L\theta_k - \mu_1 - \mu_2(1 - \theta_k)} \mathbf{y}_k = \mathbf{x}^* - \mathbf{z}_k, \tag{2.40}$$

$$\theta_k \mathbf{x}^* + (1 - \theta_k)\mathbf{x}_k - \mathbf{x}_{k+1} = \theta_k \left(\mathbf{x}^* - \mathbf{z}_{k+1}\right).$$

Following the proof in Sect. 2.2.1, we first give the following lemma, which is the counterpart of Lemma 2.4.

Lemma 2.10 *Suppose that $f(\mathbf{x})$ is μ_1-strongly convex and L-smooth and $h(\mathbf{x})$ is μ_2-strongly convex. Then for Algorithm 2.5, we have*

$$F(\mathbf{x}_{k+1}) - F(\mathbf{x})$$

$$\leq L\langle \mathbf{x}_{k+1} - \mathbf{y}_k, \mathbf{x} - \mathbf{y}_k \rangle - \frac{\mu_1}{2}\|\mathbf{x} - \mathbf{y}_k\|^2 - \frac{\mu_2}{2}\|\mathbf{x} - \mathbf{x}_{k+1}\|^2$$

$$- \frac{L}{2}\|\mathbf{x}_{k+1} - \mathbf{y}_k\|^2 + \langle L\boldsymbol{\sigma}_k + \mathbf{e}_k, \mathbf{x} - \mathbf{x}_{k+1} \rangle + \epsilon_k. \qquad (2.41)$$

Proof From the L-smoothness and the μ-strong convexity of $f(\mathbf{x})$, we have

$$f(\mathbf{x}_{k+1}) \leq f(\mathbf{y}_k) + \langle \nabla f(\mathbf{y}_k), \mathbf{x}_{k+1} - \mathbf{y}_k \rangle + \frac{L}{2}\|\mathbf{x}_{k+1} - \mathbf{y}_k\|^2$$

$$= f(\mathbf{y}_k) + \langle \nabla f(\mathbf{y}_k), \mathbf{x} - \mathbf{y}_k \rangle + \langle \nabla f(\mathbf{y}_k), \mathbf{x}_{k+1} - \mathbf{x} \rangle + \frac{L}{2}\|\mathbf{x}_{k+1} - \mathbf{y}_k\|^2$$

$$\leq f(\mathbf{x}) - \frac{\mu_1}{2}\|\mathbf{x} - \mathbf{y}_k\|^2 + \frac{L}{2}\|\mathbf{x}_{k+1} - \mathbf{y}_k\|^2 + \langle \widetilde{\nabla} f(\mathbf{y}_k), \mathbf{x}_{k+1} - \mathbf{x} \rangle$$

$$+ \langle \mathbf{e}_k, \mathbf{x} - \mathbf{x}_{k+1} \rangle.$$

Adding (2.36), we can see that there exists $\boldsymbol{\sigma}_k$ satisfying $\|\boldsymbol{\sigma}_k\| \leq \sqrt{\frac{2(L+\mu)\epsilon_k}{L^2}}$ such that

$$F(\mathbf{x}_{k+1}) - F(\mathbf{x})$$

$$\leq L\langle \mathbf{x}_{k+1} - \mathbf{y}_k, \mathbf{x} - \mathbf{x}_{k+1} \rangle - \frac{\mu_1}{2}\|\mathbf{x} - \mathbf{y}_k\|^2 - \frac{\mu_2}{2}\|\mathbf{x} - \mathbf{x}_{k+1}\|^2$$

$$+ \frac{L}{2}\|\mathbf{x}_{k+1} - \mathbf{y}_k\|^2 + \langle L\boldsymbol{\sigma}_k + \mathbf{e}_k, \mathbf{x} - \mathbf{x}_{k+1} \rangle + \epsilon_k$$

$$\leq L\langle \mathbf{x}_{k+1} - \mathbf{y}_k, \mathbf{x} - \mathbf{y}_k \rangle - \frac{\mu_1}{2}\|\mathbf{x} - \mathbf{y}_k\|^2 - \frac{\mu_2}{2}\|\mathbf{x} - \mathbf{x}_{k+1}\|^2 - \frac{L}{2}\|\mathbf{x}_{k+1} - \mathbf{y}_k\|^2$$

$$+ \langle L\boldsymbol{\sigma}_k + \mathbf{e}_k, \mathbf{x} - \mathbf{x}_{k+1} \rangle + \epsilon_k.$$

The proof is complete. $\qquad\qquad\qquad\qquad\qquad\qquad\qquad\qquad\qquad\qquad\qquad\square$

In the following lemma, we give a progress in one iteration of Algorithm 2.5.

Lemma 2.11 *Suppose that $f(\mathbf{x})$ is μ_1-strongly convex and L-smooth and $h(\mathbf{x})$ is μ_2-strongly convex. Let $\theta_k \in (0, 1]$ and $\frac{\mu_1}{L\theta_k} + \frac{\mu_2(1-\theta_k)}{L\theta_k} \leq 1$. Then for Algorithm 2.5, we have*

$$F(\mathbf{x}_{k+1}) - F(\mathbf{x}^*) - (1 - \theta_k)(F(\mathbf{x}_k) - F(\mathbf{x}^*))$$

$$\leq \left(\frac{L\theta_k^2}{2} - \frac{\mu_1\theta_k}{2} - \frac{\mu_2\theta_k(1-\theta_k)}{2} \right)\|\mathbf{z}_k - \mathbf{x}^*\|^2 - \frac{L\theta_k^2}{2}\|\mathbf{x}^* - \mathbf{z}_{k+1}\|^2$$

$$+\frac{\mu_2\theta_k(1-\theta_k)}{2}\|\mathbf{x}^*-\mathbf{x}_k\|^2-\frac{\theta_k\mu_2}{2}\|\mathbf{x}^*-\mathbf{x}_{k+1}\|^2$$

$$+\theta_k\left(\sqrt{2(L+\mu_2)\epsilon_k}+\|\mathbf{e}_k\|\right)\|\mathbf{x}^*-\mathbf{z}_{k+1}\|+\epsilon_k. \tag{2.42}$$

Proof Following the same proof of Theorem 2.2, i.e., applying (2.41) first with $\mathbf{x}=\mathbf{x}_k$ and then with $\mathbf{x}=\mathbf{x}^*$ to obtain two inequalities, multiplying the first inequality by $(1-\theta_k)$, multiplying the second by θ_k, and adding them together, we can have

$$F(\mathbf{x}_{k+1})-F(\mathbf{x}^*)-(1-\theta_k)(F(\mathbf{x}_k)-F(\mathbf{x}^*))$$

$$\leq L\langle\mathbf{x}_{k+1}-\mathbf{y}_k,(1-\theta_k)\mathbf{x}_k+\theta_k\mathbf{x}^*-\mathbf{y}_k\rangle-\frac{\theta_k\mu_1}{2}\|\mathbf{x}^*-\mathbf{y}_k\|^2$$

$$-\frac{\theta_k\mu_2}{2}\|\mathbf{x}^*-\mathbf{x}_{k+1}\|^2-\frac{L}{2}\|\mathbf{x}_{k+1}-\mathbf{y}_k\|^2$$

$$+\langle L\boldsymbol{\sigma}_k+\mathbf{e}_k,(1-\theta_k)\mathbf{x}_k+\theta_k\mathbf{x}^*-\mathbf{x}_{k+1}\rangle+\epsilon_k$$

$$\overset{a}{=}\frac{L\theta_k^2}{2}\left(\left\|\mathbf{x}^*-\frac{1}{\theta_k}\mathbf{y}_k+\frac{1-\theta_k}{\theta_k}\mathbf{x}_k\right\|^2-\|\mathbf{x}^*-\mathbf{z}_{k+1}\|^2\right)-\frac{\theta_k\mu_1}{2}\|\mathbf{x}^*-\mathbf{y}_k\|^2$$

$$-\frac{\theta_k\mu_2}{2}\|\mathbf{x}^*-\mathbf{x}_{k+1}\|^2+\theta_k\langle L\boldsymbol{\sigma}_k+\mathbf{e}_k,\mathbf{x}^*-\mathbf{z}_{k+1}\rangle+\epsilon_k,$$

where $\overset{a}{=}$ uses (2.21). By reorganizing the terms in $\mathbf{x}^*-\frac{1}{\theta_k}\mathbf{y}_k+\frac{1-\theta_k}{\theta_k}\mathbf{x}_k$ carefully, we can have

$$\frac{L\theta_k^2}{2}\left\|\mathbf{x}^*-\frac{1}{\theta_k}\mathbf{y}_k+\frac{1-\theta_k}{\theta_k}\mathbf{x}_k\right\|^2$$

$$=\frac{L\theta_k^2}{2}\left\|\frac{\mu_1}{L\theta_k}(\mathbf{x}^*-\mathbf{y}_k)+\frac{\mu_2(1-\theta_k)}{L\theta_k}(\mathbf{x}^*-\mathbf{x}_k)+\left(1-\frac{\mu_1}{L\theta_k}-\frac{\mu_2(1-\theta_k)}{L\theta_k}\right)\right.$$

$$\left.\left(\mathbf{x}^*+\frac{\frac{1}{\theta_k}-1+\frac{\mu_2(1-\theta_k)}{L\theta_k}}{1-\frac{\mu_1}{L\theta_k}-\frac{\mu_2(1-\theta_k)}{L\theta_k}}\mathbf{x}_k-\frac{\frac{L-\mu_1}{L\theta_k}}{1-\frac{\mu_1}{L\theta_k}-\frac{\mu_2(1-\theta_k)}{L\theta_k}}\mathbf{y}_k\right)\right\|^2$$

$$\leq\frac{\mu_1\theta_k}{2}\|\mathbf{x}^*-\mathbf{y}_k\|^2+\frac{\mu_2\theta_k(1-\theta_k)}{2}\|\mathbf{x}^*-\mathbf{x}_k\|^2$$

$$+\left(\frac{L\theta_k^2}{2}-\frac{\mu_1\theta_k}{2}-\frac{\mu_2\theta_k(1-\theta_k)}{2}\right)\left\|\mathbf{x}^*+\frac{L-L\theta_k+\mu_2(1-\theta_k)}{L\theta_k-\mu_1-\mu_2(1-\theta_k)}\mathbf{x}_k\right.$$

$$\left.-\frac{L-\mu_1}{L\theta_k-\mu_1-\mu_2(1-\theta_k)}\mathbf{y}_k\right\|^2$$

$$\overset{a}{=} \frac{\mu_1\theta_k}{2}\|\mathbf{x}^* - \mathbf{y}_k\|^2 + \frac{\mu_2\theta_k(1-\theta_k)}{2}\|\mathbf{x}^* - \mathbf{x}_k\|^2$$

$$+ \left(\frac{L\theta_k^2}{2} - \frac{\mu_1\theta_k}{2} - \frac{\mu_2\theta_k(1-\theta_k)}{2}\right)\|\mathbf{x}^* - \mathbf{z}_k\|^2,$$

where we let $\frac{\mu_1}{L\theta_k} + \frac{\mu_2(1-\theta_k)}{L\theta_k} \leq 1$ and $\overset{a}{=}$ uses (2.40). Plugging it into the above inequality and using $\|\boldsymbol{\sigma}_k\| \leq \sqrt{\frac{2(L+\mu_2)\epsilon_k}{L^2}}$, we can have the conclusion. $\square$

Now, we are ready to prove the convergence of Algorithm 2.5 in the following theorem. It describes that when we control the error to be very small, which is dependent on the iteration k, the accelerated convergence can still be maintained. Different from Theorem 2.2, in Theorem 2.5 we consider three scenarios: both f and h are generally convex, only one of f and h is strongly convex, and both f and h are strongly convex.

Theorem 2.5 *Suppose that $f(\mathbf{x})$ and $h(\mathbf{x})$ are convex and $f(\mathbf{x})$ is L-smooth. Let $\theta_0 = 1$, $\theta_{k+1} = \frac{\sqrt{\theta_k^4 + 4\theta_k^2} - \theta_k^2}{2}$, $\|\mathbf{e}_k\| \leq \frac{1}{(k+1)^{2+\delta}}\sqrt{\frac{L}{2}}$, and (2.34) is satisfied with $\epsilon_k \leq \frac{1}{(k+1)^{4+2\delta}}$, where δ can be any small positive constant. Then for Algorithm 2.5, we have*

$$F(\mathbf{x}_{K+1}) - F(\mathbf{x}^*) \leq \frac{4}{(K+2)^2}\left(L\|\mathbf{x}_0 - \mathbf{x}^*\|^2 + \frac{2}{1+2\delta} + \frac{18}{\delta^2}\right).$$

Suppose that $f(\mathbf{x})$ is μ_1-strongly convex and L-smooth and $h(\mathbf{x})$ is μ_2-strongly convex (we allow $\mu_1 = 0$ or $\mu_2 = 0$ but require $\mu_1 + \mu_2 > 0$). Let $\theta_k = \theta \equiv$
$$\frac{1}{\frac{\mu_2}{2(\mu_1+\mu_2)} + \sqrt{\left[\frac{\mu_2}{2(\mu_1+\mu_2)}\right]^2 + \frac{L}{\mu_1+\mu_2}}}$$
for all k, $\|\mathbf{e}_k\| \leq [1 - (1-\delta)\theta]^{\frac{k+1}{2}}$, and (2.34) is satisfied with $\epsilon_k \leq [1 - (1-\delta)\theta]^{k+1}$. Then for Algorithm 2.5, we have

$$F(\mathbf{x}_{K+1}) - F(\mathbf{x}^*) \leq C[1 - (1-\delta)\theta]^{K+1},$$

where $C = 2(F(\mathbf{x}_0) - F(\mathbf{x}^*)) + (L\theta^2 + \theta\mu_2)\|\mathbf{x}_0 - \mathbf{x}^*\|^2 + \left(\frac{2}{\delta\theta} + \frac{8}{\delta^2\theta^2}\right)\left(2\sqrt{\frac{L+\mu_2}{L}} + \sqrt{\frac{2}{L}}\right)^2$.

Proof Case 1: $\mu_1 = \mu_2 = 0$. We use the Lyapunov function defined in (2.19). Dividing both sides of (2.42) by θ_k^2 and using $\frac{1-\theta_k}{\theta_k^2} = \frac{1}{\theta_{k-1}^2}$, we have

$$\ell_{k+1} \leq \ell_k + \frac{2\sqrt{\epsilon_k} + \sqrt{2/L}\|\mathbf{e}_k\|}{\theta_k}\sqrt{\ell_{k+1}} + \frac{\epsilon_k}{\theta_k^2}.$$

Summing over $k = 0, 1, 2, \cdots, K - 1$, we have

$$\ell_K \leq \ell_0 + \sum_{k=1}^{K} \frac{\epsilon_{k-1}}{\theta_{k-1}^2} + \sum_{k=1}^{K} \frac{2\sqrt{\epsilon_{k-1}} + \sqrt{2/L}\|\mathbf{e}_{k-1}\|}{\theta_{k-1}}\sqrt{\ell_k}.$$

From (2.39) of Corollary 2.1, we have

$$\ell_K \leq 2\left(\ell_0 + \sum_{k=1}^{K} \frac{\epsilon_{k-1}}{\theta_{k-1}^2}\right) + 2\left(\sum_{k=1}^{K} \frac{2\sqrt{\epsilon_{k-1}} + \sqrt{2/L}\|\mathbf{e}_{k-1}\|}{\theta_{k-1}}\right)^2$$

$$\leq 2\ell_0 + 2\sum_{k=1}^{K} k^2\epsilon_{k-1} + 2\left[\sum_{k=1}^{K}\left(2k\sqrt{\epsilon_{k-1}} + \sqrt{2/L}k\|\mathbf{e}_{k-1}\|\right)\right]^2$$

$$\leq L\|\mathbf{x}_0 - \mathbf{x}^*\|^2 + \frac{2}{1 + 2\delta} + \frac{18}{\delta^2},$$

where we use $\theta_k \geq \frac{1}{k+1}$ from Lemma 2.3, $\epsilon_k \leq \frac{1}{(k+1)^{4+2\delta}}$, and $\|\mathbf{e}_k\| \leq \frac{1}{(k+1)^{2+\delta}}\sqrt{\frac{L}{2}}$. The conclusion can be obtained by the definition of ℓ_K.

Case 2: $\mu_1 + \mu_2 > 0$. Let $\theta_k = \theta, \forall k$, such that θ satisfies $L\theta^2 - \mu_1\theta - \mu_2\theta(1 - \theta) = (1 - \theta)L\theta^2$, which leads to

$$\frac{\mu_1}{L\theta} + \frac{\mu_2(1 - \theta)}{L\theta} = \theta$$

and

$$\theta = \frac{1}{\frac{\mu_2}{2(\mu_1+\mu_2)} + \sqrt{\left[\frac{\mu_2}{2(\mu_1+\mu_2)}\right]^2 + \frac{L}{\mu_1+\mu_2}}} \leq \sqrt{\frac{\mu_1 + \mu_2}{L}}.$$

We can easily check that $\theta < 1$ and the conditions in Lemma 2.11 hold. Define the Lyapunov function

$$\ell_{k+1} = \frac{1}{(1 - \theta)^{k+1}}\left(F(\mathbf{x}_{k+1}) - F(\mathbf{x}^*) + \frac{L\theta^2}{2}\|\mathbf{z}_{k+1} - \mathbf{x}^*\|^2\right.$$

$$\left. + \frac{\mu_2\theta}{2}\|\mathbf{x}_{k+1} - \mathbf{x}^*\|^2\right).$$

Dividing both sides of (2.42) by $(1 - \theta)^{k+1}$, we have

$$\ell_{k+1} \leq \ell_k + \frac{2\sqrt{\frac{(L+\mu_2)\epsilon_k}{L}} + \sqrt{\frac{2}{L}}\|\mathbf{e}_k\|}{(1 - \theta)^{\frac{k+1}{2}}}\sqrt{\ell_{k+1}} + \frac{\epsilon_k}{(1 - \theta)^{k+1}}.$$

Summing over $k = 0, 1, 2, \cdots, K - 1$, we have

$$\ell_K \le \ell_0 + \sum_{k=1}^{K} \frac{\epsilon_{k-1}}{(1-\theta)^k} + \sum_{k=1}^{K} \frac{2\sqrt{\frac{(L+\mu_2)\epsilon_{k-1}}{L}} + \sqrt{\frac{2}{L}}\|\mathbf{e}_{k-1}\|}{(1-\theta)^{\frac{k}{2}}} \sqrt{\ell_k}.$$

From (2.39) of Corollary 2.1, we have

$$\ell_K \le 2\ell_0 + 2\sum_{k=1}^{K} \frac{\epsilon_{k-1}}{(1-\theta)^k} + 2\left(\sum_{k=1}^{K} \frac{2\sqrt{\frac{(L+\mu_2)\epsilon_{k-1}}{L}} + \sqrt{\frac{2}{L}}\|\mathbf{e}_{k-1}\|}{(1-\theta)^{\frac{k}{2}}} \right)^2$$

$$\le 2(F(\mathbf{x}_0) - F(\mathbf{x}^*)) + (L\theta^2 + \theta\mu_2)\|\mathbf{x}_0 - \mathbf{x}^*\|^2 + 2\sum_{k=1}^{K} \left[\frac{1 - (1-\delta)\theta}{1-\theta} \right]^k$$

$$+2\left\{ 2\sqrt{\frac{L+\mu_2}{L}} \sum_{k=1}^{K} \left[\frac{1 - (1-\delta)\theta}{1-\theta} \right]^{\frac{k}{2}} + \sqrt{\frac{2}{L}} \sum_{k=1}^{K} \left[\frac{1 - (1-\delta)\theta}{1-\theta} \right]^{\frac{k}{2}} \right\}^2$$

$$\overset{a}{\le} 2(F(\mathbf{x}_0) - F(\mathbf{x}^*)) + (L\theta^2 + \theta\mu_2)\|\mathbf{x}_0 - \mathbf{x}^*\|^2$$

$$+ \left(\frac{2}{\delta\theta} + \frac{8}{\delta^2\theta^2} \right)\left(2\sqrt{\frac{L+\mu_2}{L}} + \sqrt{\frac{2}{L}} \right)^2 \left[\frac{1 - (1-\delta)\theta}{1-\theta} \right]^K$$

$$\le C\left[\frac{1 - (1-\delta)\theta}{1-\theta} \right]^K,$$

where in $\overset{a}{\le}$ we use $\epsilon_k \le [1 - (1-\delta)\theta]^{k+1}$, $\|\mathbf{e}_k\| \le [1 - (1-\delta)\theta]^{\frac{k+1}{2}}$,

$$\sum_{k=1}^{K} \left[\frac{1 - (1-\delta)\theta}{1-\theta} \right]^k = \frac{1 - (1-\delta)\theta}{1-\theta} \frac{\left[\frac{1-(1-\delta)\theta}{1-\theta} \right]^K - 1}{\frac{1-(1-\delta)\theta}{1-\theta} - 1}$$

$$\le \frac{1 - (1-\delta)\theta}{\delta\theta} \left[\frac{1 - (1-\delta)\theta}{1-\theta} \right]^K$$

$$\le \frac{1}{\delta\theta} \left[\frac{1 - (1-\delta)\theta}{1-\theta} \right]^K,$$

and

$$\sum_{k=1}^{K} \left[\frac{1 - (1-\delta)\theta}{1-\theta} \right]^{k/2} = \frac{q^K - 1}{q - 1} \quad \left(\text{where } q = \sqrt{\frac{1 - (1-\delta)\theta}{1-\theta}} \right)$$

$$\le \frac{q}{q-1} q^K$$

$$= \frac{\sqrt{1 - (1 - \delta)\theta}}{\sqrt{1 - (1 - \delta)\theta} - \sqrt{1 - \theta}} q^K$$

$$= \frac{\sqrt{1 - (1 - \delta)\theta} \left(\sqrt{1 - (1 - \delta)\theta} + \sqrt{1 - \theta} \right)}{\delta\theta} q^K$$

$$\leq \frac{2}{\delta\theta} q^K.$$

The conclusion can be obtained by the definition of ℓ_K.   □

2.3.1 Inexact Accelerated Gradient Descent

In this section, we specify $h = 0$. Namely, we consider the inexact variant of Algorithm 2.1. In this case, $\mu_2 = 0$ and Theorem 2.5 reduces to the following theorem.

Algorithm 2.6 Inexact accelerated gradient descent

Initialize $\mathbf{x}_0 = \mathbf{x}_{-1}$.
for $k = 0, 1, 2, 3, \cdots$ **do**
  $\mathbf{y}_k = \mathbf{x}_k + \frac{(L\theta_k - \mu_1)(1 - \theta_{k-1})}{(L - \mu_1)\theta_{k-1}} (\mathbf{x}_k - \mathbf{x}_{k-1})$,
  $\mathbf{x}_{k+1} = \mathbf{y}_k - \frac{1}{L} \tilde{\nabla} f(\mathbf{y}_k)$.
end for

Theorem 2.6 *Suppose that $f(\mathbf{x})$ is convex and L-smooth. Let $\theta_0 = 1$, $\theta_{k+1} = \frac{\sqrt{\theta_k^4 + 4\theta_k^2} - \theta_k^2}{2}$, and $\|\mathbf{e}_k\| \leq \frac{1}{(k+1)^{2+\delta}} \sqrt{\frac{L}{2}}$, where δ can be any small positive constant. Then for Algorithm 2.6, we have*

$$F(\mathbf{x}_{K+1}) - F(\mathbf{x}^*) \leq \frac{4}{(K + 2)^2} \left(L\|\mathbf{x}_0 - \mathbf{x}^*\|^2 + \frac{2}{1 + 2\delta} + \frac{18}{\delta^2} \right).$$

Suppose that $f(\mathbf{x})$ is μ_1-strongly convex and L-smooth. Let $\theta_k = \theta \equiv \sqrt{\frac{\mu_1}{L}}$ for all k, $\|\mathbf{e}_k\| \leq [1 - (1 - \delta)\theta]^{\frac{k+1}{2}}$. Then for Algorithm 2.6, we have

$$F(\mathbf{x}_{K+1}) - F(\mathbf{x}^*) \leq C[1 - (1 - \delta)\theta]^{K+1},$$

where $C = 2(F(\mathbf{x}_0) - F(\mathbf{x}^*)) + \mu\|\mathbf{x}_0 - \mathbf{x}^*\|^2 + \left(\frac{2}{\delta\theta} + \frac{8}{\delta^2\theta^2} \right) \left(2 + \sqrt{\frac{2}{L}} \right)^2$.

2.3.2 Inexact Accelerated Proximal Point Method

Now we consider the case that $f = 0$. In this case, Algorithm 2.5 reduces to the well-known accelerated proximal point algorithm, which is described in Algorithm 2.7.

Algorithm 2.7 Inexact accelerated proximal point method

Initialize $\mathbf{x}_0 = \mathbf{x}_{-1}$.
for $k = 0, 1, 2, 3, \cdots$ **do**
$\qquad \mathbf{y}_k = \mathbf{x}_k + \frac{[\tau\theta_k - \mu_2(1-\theta_k)](1-\theta_{k-1})}{\tau\theta_{k-1}}(\mathbf{x}_k - \mathbf{x}_{k-1})$,
$\qquad \mathbf{x}_{k+1} \approx \operatorname{argmin}_{\mathbf{x}} \left(h(\mathbf{x}) + \frac{\tau}{2}\|\mathbf{x} - \mathbf{y}_k\|^2\right)$.
end for

Accordingly, $\mu_1 = 0$ in this scenario and Theorem 2.5 reduces to the following Theorem.

Theorem 2.7 *Suppose that $h(\mathbf{x})$ is convex. Let $\theta_0 = 1$, $\theta_{k+1} = \frac{\sqrt{\theta_k^4 + 4\theta_k^2} - \theta_k^2}{2}$, and*

$$h(\mathbf{x}_{k+1}) + \frac{\tau}{2}\|\mathbf{x}_{k+1} - \mathbf{y}_k\|^2 \le \min_{\mathbf{x}}\left(h(\mathbf{x}) + \frac{\tau}{2}\|\mathbf{x} - \mathbf{y}_k\|^2\right) + \epsilon_k \qquad (2.43)$$

is satisfied with $\epsilon_k \le \frac{1}{(k+1)^{4+2\delta}}$, where δ can be any small positive constant. Then for Algorithm 2.7, we have

$$F(\mathbf{x}_{K+1}) - F(\mathbf{x}^*) \le \frac{4}{(K+2)^2}\left(\tau\|\mathbf{x}_0 - \mathbf{x}^*\|^2 + \frac{2}{1+2\delta} + \frac{18}{\delta^2}\right).$$

Suppose that $h(\mathbf{x})$ is μ_2-strongly convex. Let $\theta_k = \theta \equiv \frac{1}{0.5 + \sqrt{0.25 + \frac{\tau}{\mu_2}}}$ for all k and (2.43) is satisfied with $\epsilon_k \le [1 - (1-\delta)\theta]^{k+1}$. Then for Algorithm 2.7, we have

$$F(\mathbf{x}_{K+1}) - F(\mathbf{x}^*) \le C[1 - (1-\delta)\theta]^{K+1},$$

*where $C = 2(F(\mathbf{x}_0) - F(\mathbf{x}^*)) + (\tau\theta^2 + \theta\mu_2)\|\mathbf{x}_0 - \mathbf{x}^*\|^2 + \left(\frac{2}{\delta\theta} + \frac{8}{\delta^2\theta^2}\right)$
$\left(2\sqrt{\frac{\tau+\mu_2}{\tau}} + \sqrt{\frac{2}{\tau}}\right)^2$.*

The accelerated proximal point method can be seen as a general algorithm framework and has been widely used in stochastic optimization, which leads to the popular framework of Catalyst [17] (see Algorithm 5.4) for the acceleration of first-order methods. The details will be described in Sect. 5.1.4.

2.4 Restart

In this section, we consider the restart technique, which was firstly proposed in [26]. Namely, we run Algorithm 2.2 for a few iterations and then restart it with warm start. We describe the method in Algorithm 2.8.

Besides its practical use, a beauty of the restart technique is that we can prove a faster rate even when the objective function is generally convex, e.g., see [18]. Specifically, we introduce the Hölderian error bound condition. Consider problem (2.17).

Definition 2.2 The Hölderian error bound condition is $\|\mathbf{x} - \bar{\mathbf{x}}\| \leq \nu \left(F(\mathbf{x}) - F^*\right)^\vartheta$, where $0 < \nu < \infty$, $\vartheta \in (0, 1]$, and $\bar{\mathbf{x}}$ is the projection of $\mathbf{x}$ onto the optimal solution set $\mathcal{X}^*$.

If $F(\mathbf{x})$ has a unique minimizer, then $\bar{\mathbf{x}} = \mathbf{x}^*$, $\forall \mathbf{x}$.

When $\vartheta = 0.5$, $\nu = \sqrt{\frac{2}{\mu}}$, and $F(\mathbf{x})$ has a unique minimizer, the Hölderian error bound condition reduces to the strong convexity (see (A.10)).

Algorithm 2.8 Accelerated proximal gradient (APG) method with restart

---

Initialize $\mathbf{x}_{K_{-1}} = \mathbf{x}_0$.
for $k = 1, 2, 3, \cdots, T$ **do**
  Minimize $F(\mathbf{x})$ using Algorithm 2.2 for K_t iterations with $\mathbf{x}_{K_{t-1}}$ being the initializer and output $\mathbf{x}_{K_t}$.
end for

From the results in Theorem 2.1, we can give the total runtime of Algorithm 2.8 in the following theorem.

Theorem 2.8 *Suppose that $F(\mathbf{x})$ is convex and L-smooth and satisfies the Hölderian error bound condition. Let $K_t \geq 2\nu\sqrt{2L} \left(\frac{\sqrt{C}}{2^{t-1}}\right)^{2\vartheta - 1}$, where $C = F(\mathbf{x}_0) - F^*$. To achieve an $\mathbf{x}_{K_T}$ such that $F(\mathbf{x}_{K_T}) - F(\mathbf{x}^*) \leq \epsilon$, Algorithm 2.8 needs*

1. $O\left(\frac{\nu\sqrt{L}}{\epsilon^{0.5-\vartheta}}\right)$ *runtime when* $\vartheta < 0.5$,

2. $O\left(\nu\sqrt{L}\log\frac{1}{\epsilon}\right)$ *runtime when* $\vartheta = 0.5$,

3. $O\left(\nu\sqrt{L}\left(\sqrt{C}\right)^{2\vartheta-1}\right)$ *runtime when* $\vartheta > 0.5$.

Proof We prove $F(\mathbf{x}_{K_t}) - F^* \leq \frac{C}{4^t}$ by induction.

Assume that $F(\mathbf{x}_{K_{t-1}}) - F^* \leq \frac{C}{4^{t-1}}$. From Theorem 2.2 and the Hölderian error bound condition, we have

$$
\begin{aligned}
F(\mathbf{x}_{K_t}) - F^* &\leq \frac{2L}{(K_t + 1)^2} \|\mathbf{x}_{K_{t-1}} - \bar{\mathbf{x}}_{K_{t-1}}\|^2 \\
&\leq \frac{2L}{(K_t + 1)^2} \nu^2 \left(f(\mathbf{x}_{K_{t-1}}) - f^* \right)^{2\vartheta} \\
&\leq \frac{2L}{(K_t + 1)^2} \nu^2 \left(\frac{C}{4^{t-1}} \right)^{2\vartheta} \leq \frac{C}{4^t},
\end{aligned}
$$

where we use $K_t + 1 \geq \sqrt{8L\nu^2 \left(\frac{C}{4^{t-1}} \right)^{2\vartheta-1}}$. So we only need $4^T = \frac{C}{\epsilon}$. The total runtime is

$$
\sum_{t=1}^{T} K_t \geq 2\nu\sqrt{2L} \left(\sqrt{C} \right)^{2\vartheta-1} \sum_{t=0}^{T-1} \left(2^{1-2\vartheta} \right)^t
$$

$$
= \begin{cases}
2\nu\sqrt{2L} \left(\sqrt{C} \right)^{2\vartheta-1} \frac{(2^T)^{1-2\vartheta}-1}{2^{1-2\vartheta}-1}, & \text{if } \vartheta < 0.5, \\
2\nu\sqrt{2L} \log_4 \frac{C}{\epsilon}, & \text{if } \vartheta = 0.5, \\
2\nu\sqrt{2L} \left(\sqrt{C} \right)^{2\vartheta-1} \frac{1-(2^T)^{1-2\vartheta}}{1-2^{1-2\vartheta}}, & \text{if } \vartheta > 0.5.
\end{cases}
$$

$$
\geq \begin{cases}
2\nu\sqrt{2L} \left(2\sqrt{C} \right)^{2\vartheta-1} \left[\left(\sqrt{\frac{C}{\epsilon}} \right)^{1-2\vartheta} - 1 \right], & \text{if } \vartheta < 0.5, \\
2\nu\sqrt{2L} \log_4 \frac{C}{\epsilon}, & \text{if } \vartheta = 0.5, \\
2\nu\sqrt{2L} \left(\sqrt{C} \right)^{2\vartheta-1} \left[1 - \left(\sqrt{\frac{\epsilon}{C}} \right)^{2\vartheta-1} \right], & \text{if } \vartheta > 0.5.
\end{cases}
$$

The proof is complete. □

2.5 Smoothing for Nonsmooth Optimization

When f is nonsmooth in problem (2.1) and only the subgradient is available, we can only obtain the $O\left(\frac{1}{\sqrt{K}} \right)$ convergence rate when the subgradient-type methods are used. In 2003, Nesterov proposed a first-order smoothing method in his seminal paper [22], where the accelerated gradient descent is used to solve a smoothed problem and the $O\left(\frac{1}{K} \right)$ total convergence rate can be obtained by carefully designing the smoothing parameter.

We use the tool of Fenchel conjugate to smooth a nonsmooth function. Let

$$f^*(\mathbf{y}) = \max_{\mathbf{x}} \left(\langle \mathbf{x}, \mathbf{y} \rangle - f(\mathbf{x}) \right) \tag{2.44}$$

be the Fenchel conjugate of $f(\mathbf{x})$. Then we have

$$\mathbf{y} \in \partial f(\mathbf{x}^*), \quad \text{where } \mathbf{x}^* = \underset{\mathbf{x}}{\operatorname{argmax}} (\langle \mathbf{x}, \mathbf{y} \rangle - f(\mathbf{x})). \tag{2.45}$$

Define a regularized function of $f^*(\mathbf{y})$ as

$$f_\delta^*(\mathbf{y}) = f^*(\mathbf{y}) + \frac{\delta}{2} \|\mathbf{y}\|^2. \tag{2.46}$$

Let

$$f_\delta(\mathbf{x}) = \max_{\mathbf{y}} \left(\langle \mathbf{x}, \mathbf{y} \rangle - f_\delta^*(\mathbf{y}) \right).$$

The following proposition describes that $f_\delta(\mathbf{x})$ is a smoothed approximation of $f(\mathbf{x})$.

Proposition 2.1 *Suppose that $f(\mathbf{x})$ is convex, then*

1. *$f_\delta(\mathbf{x})$ is $\frac{1}{\delta}$-smooth.*
2. *$f_{\delta_1}(\mathbf{x}) \leq f_{\delta_2}(\mathbf{x})$ if $\delta_1 \geq \delta_2$.*
3. *Suppose that $f(\mathbf{x})$ has bounded subgradient: $\|\partial f(\mathbf{x})\| \leq M$. Then $f(\mathbf{x}) - \frac{\delta M^2}{2} \leq f_\delta(\mathbf{x}) \leq f(\mathbf{x})$.*

Proof The first conclusion is a direct consequence of point 4 of Proposition A.12. For the second conclusion, we know

$$f_{\delta_1}^*(\mathbf{y}) = f^*(\mathbf{y}) + \frac{\delta_1}{2} \|\mathbf{y}\|^2 \geq f^*(\mathbf{y}) + \frac{\delta_2}{2} \|\mathbf{y}\|^2 = f_{\delta_2}^*(\mathbf{y})$$

and

$$f_{\delta_1}(\mathbf{x}) = \max_{\mathbf{y}} \left(\langle \mathbf{x}, \mathbf{y} \rangle - f_{\delta_1}^*(\mathbf{y}) \right) \leq \max_{\mathbf{y}} \left(\langle \mathbf{x}, \mathbf{y} \rangle - f_{\delta_2}^*(\mathbf{y}) \right) = f_{\delta_2}(\mathbf{x}).$$

From (2.45) and $\|\partial f(\mathbf{x})\| \leq M$, we have $\|\mathbf{y}\| \leq M$, i.e., $f_\delta^*(\mathbf{y})$ has a bounded domain. From (2.46), we have $f^*(\mathbf{y}) \leq f_\delta^*(\mathbf{y}) \leq f^*(\mathbf{y}) + \frac{\delta M^2}{2}$. Thus we have

$$f_\delta(\mathbf{x}) \geq \max_{\mathbf{y}} \left[\langle \mathbf{x}, \mathbf{y} \rangle - \left(f^*(\mathbf{y}) + \frac{\delta M^2}{2} \right) \right] = f(\mathbf{x}) - \frac{\delta M^2}{2}$$

and

$$f_\delta(\mathbf{x}) \leq \max_{\mathbf{y}} \left(\langle \mathbf{x}, \mathbf{y} \rangle - f^*(\mathbf{y}) \right) = f(\mathbf{x}). \qquad \square$$

We can use Algorithm 2.2 to minimize $f_\delta(\mathbf{x})$ and have the following convergence rate theorem. Let $\hat{\mathbf{x}}_\delta^* = \text{argmin}_\mathbf{x} f_\delta(\mathbf{x})$.

Theorem 2.9 *Suppose that $f(\mathbf{x})$ is convex and has bounded subgradient: $\|\partial f(\mathbf{x})\| \leq M$. Run Algorithm 2.2 with $L = \frac{M^2}{\epsilon}$ to minimize $f_\delta(\mathbf{x})$, where $\delta = \frac{\epsilon}{M^2}$.*

1. If $f(\mathbf{x})$ is generally convex, then we only need $K = \frac{2M\|\mathbf{x}_0 - \hat{\mathbf{x}}_\delta^*\|}{\epsilon}$ iterations such that $f(\mathbf{x}_{K+1}) - f(\mathbf{x}^*) \leq \epsilon$.
2. If $f(\mathbf{x})$ is μ-strongly convex, then we only need

$$K = \frac{M}{\sqrt{\mu\epsilon}} \log \frac{2(f(\mathbf{x}_0) - f(\mathbf{x}^*)) + \delta M^2 + \mu\|\mathbf{x}_0 - \hat{\mathbf{x}}_\delta^*\|^2}{\epsilon}$$

iterations such that $f(\mathbf{x}_{K+1}) - f(\mathbf{x}^*) \leq \epsilon$.

Proof From Proposition 2.1, we know

$$f(\mathbf{x}_{K+1}) - f(\mathbf{x}^*) \leq f_\delta(\mathbf{x}_{K+1}) + \frac{\delta M^2}{2} - f_\delta(\mathbf{x}^*) \leq f_\delta(\mathbf{x}_{K+1}) - f_\delta(\hat{\mathbf{x}}_\delta^*) + \frac{\delta M^2}{2}.$$

When $f(\mathbf{x})$ is generally convex, $f_\delta(\mathbf{x})$ is also generally convex. From Theorem 2.2 with $h(\mathbf{x}) = 0$, we have

$$f_\delta(\mathbf{x}_{K+1}) - f_\delta(\hat{\mathbf{x}}_\delta^*) \leq \frac{2}{\delta(K+2)^2} \|\mathbf{x}_0 - \hat{\mathbf{x}}_\delta^*\|^2.$$

Thus we have

$$f(\mathbf{x}_{K+1}) - f(\mathbf{x}^*) \leq \frac{2}{\delta(K+2)^2} \|\mathbf{x}_0 - \hat{\mathbf{x}}_\delta^*\|^2 + \frac{\delta M^2}{2}.$$

Letting $\delta = \frac{\epsilon}{M^2}$, we only need $K = \frac{2M\|\mathbf{x}_0 - \hat{\mathbf{x}}_\delta^*\|}{\epsilon}$ iterations such that $f(\mathbf{x}_{K+1}) - f(\mathbf{x}^*) \leq \epsilon$.

When $f(\mathbf{x})$ is μ-strongly convex, $f_\delta(\mathbf{x})$ is also μ-strongly convex. From Theorem 2.2, we have

$$f_\delta(\mathbf{x}_{K+1}) - f_\delta(\hat{\mathbf{x}}_\delta^*) \leq \left(1 - \sqrt{\mu\delta}\right)^{K+1} \left(f_\delta(\mathbf{x}_0) - f_\delta(\hat{\mathbf{x}}_\delta^*) + \frac{\mu}{2}\|\mathbf{x}_0 - \hat{\mathbf{x}}_\delta^*\|^2\right)$$

$$\leq \left(1 - \sqrt{\mu\delta}\right)^{K+1} \left(f(\mathbf{x}_0) - f(\hat{\mathbf{x}}_\delta^*) + \frac{\delta M^2}{2} + \frac{\mu}{2}\|\mathbf{x}_0 - \hat{\mathbf{x}}_\delta^*\|^2\right)$$

$$\leq \exp\left(-(K+1)\sqrt{\mu\delta}\right)$$

$$\times \left(f(\mathbf{x}_0) - f(\mathbf{x}^*) + \frac{\delta M^2}{2} + \frac{\mu}{2}\|\mathbf{x}_0 - \hat{\mathbf{x}}_\delta^*\|^2\right).$$

Letting $\delta = \frac{\epsilon}{M^2}$, we only need $K = \frac{M}{\sqrt{\mu\epsilon}} \log \frac{2(f(\mathbf{x}_0)-f(\mathbf{x}^*))+\delta M^2+\mu\|\mathbf{x}_0-\hat{\mathbf{x}}_\delta^*\|^2}{\epsilon}$ iterations such that $f(\mathbf{x}_{K+1}) - f(\mathbf{x}^*) \leq \epsilon$. $\qquad\square$

From Theorem 2.9 we know that when f is strongly convex, the accelerated gradient descent with the smoothing technique needs $O\left(\frac{\log 1/\epsilon}{\sqrt{\epsilon}}\right)$ time to achieve an $\mathbf{x}$ such that $f(\mathbf{x}) - f(\mathbf{x}^*) \leq \epsilon$. [1] proposed an efficient reduction which solves a sequence of smoothed problems with decreasing smoothing parameter δ such that the convergence rate can be improved to $O\left(\frac{1}{\sqrt{\epsilon}}\right)$. We describe the method in Algorithm 2.9 and the convergence rate in Theorem 2.10.

Algorithm 2.9 Accelerated proximal gradient (APG) method with smoothing

---

Initialize $\mathbf{x}_0$ and δ_0.
for $k = 1, 2, 3, \cdots, K$ **do**
 Minimize $f_{\delta_{k-1}}(\mathbf{x})$ using Algorithm 2.2 for $\frac{\log 8}{\sqrt{\mu\delta_{k-1}}}$ iterations with $\mathbf{x}_{k-1}$ being the initializer
 and output $\mathbf{x}_k$,
 $\delta_k = \delta_{k-1}/2$.
end for

Theorem 2.10 *Suppose that $f(\mathbf{x})$ is μ-strongly convex and has bounded subgradient: $\|\partial f(\mathbf{x})\| \leq M$. Let $\delta_0 = \frac{f(\mathbf{x}_0)-f(\mathbf{x}^*)}{M^2}$, then Algorithm 2.9 needs $O\left(\frac{M}{\sqrt{\mu\epsilon}}\right)$ time such that $f(\mathbf{x}_K) - f(\mathbf{x}^*) \leq \epsilon$.*

Proof Let $\hat{\mathbf{x}}_{\delta_k}^* = \operatorname{argmin}_{\mathbf{x}} f_{\delta_k}(\mathbf{x})$ and $t = \frac{\log 8}{\sqrt{\mu\delta_{k-1}}}$. From Theorem 2.2, we have

$$
\begin{aligned}
f_{\delta_{k-1}}(\mathbf{x}_t) &- f_{\delta_{k-1}}(\hat{\mathbf{x}}_{\delta_{k-1}}^*) \\
&\leq \left(1 - \sqrt{\mu\delta_{k-1}}\right)^t \left(f_{\delta_{k-1}}(\mathbf{x}_{k-1}) - f_{\delta_{k-1}}(\hat{\mathbf{x}}_{\delta_{k-1}}^*) + \frac{\mu}{2}\|\mathbf{x}_{k-1} - \hat{\mathbf{x}}_{\delta_{k-1}}^*\|^2\right) \\
&\overset{a}{\leq} 2\left(1 - \sqrt{\mu\delta_{k-1}}\right)^t \left(f_{\delta_{k-1}}(\mathbf{x}_{k-1}) - f_{\delta_{k-1}}(\hat{\mathbf{x}}_{\delta_{k-1}}^*)\right) \\
&\leq 2\exp\left(-t\sqrt{\mu\delta_{k-1}}\right) \left(f_{\delta_{k-1}}(\mathbf{x}_{k-1}) - f_{\delta_{k-1}}(\hat{\mathbf{x}}_{\delta_{k-1}}^*)\right) \\
&= \frac{f_{\delta_{k-1}}(\mathbf{x}_{k-1}) - f_{\delta_{k-1}}(\hat{\mathbf{x}}_{\delta_{k-1}}^*)}{4},
\end{aligned}
$$

where we use the fact that $f_{\delta_{k-1}}(\mathbf{x})$ is μ-strongly convex in $\overset{a}{\leq}$. Let $D_{\delta_k} = f_{\delta_k}(\mathbf{x}_k) - f_{\delta_k}(\hat{\mathbf{x}}_{\delta_k}^*)$. Then we have

$$
\begin{aligned}
D_{\delta_k} &= f_{\delta_k}(\mathbf{x}_k) - f_{\delta_k}(\hat{\mathbf{x}}_{\delta_k}^*) \\
&\leq f(\mathbf{x}_k) - f_{\delta_{k-1}}(\hat{\mathbf{x}}_{\delta_k}^*)
\end{aligned}
$$

$$\leq f_{\delta_{k-1}}(\mathbf{x}_k) + \frac{\delta_{k-1} M^2}{2} - f_{\delta_{k-1}}(\hat{\mathbf{x}}^*_{\delta_{k-1}})$$

$$\leq \frac{D_{\delta_{k-1}}}{4} + \frac{\delta_{k-1} M^2}{2},$$

and

$$D_{\delta_0} \leq f(\mathbf{x}_0) - f(\hat{\mathbf{x}}^*_{\delta_0}) + \frac{\delta_0 M^2}{2} \leq f(\mathbf{x}_0) - f(\mathbf{x}^*) + \frac{\delta_0 M^2}{2},$$

where we use Proposition 2.1, $f_{\delta_{k-1}}(\hat{\mathbf{x}}^*_{\delta_k}) \geq f_{\delta_{k-1}}(\hat{\mathbf{x}}^*_{\delta_{k-1}})$, and $f(\hat{\mathbf{x}}^*_{\delta_0}) \geq f(\mathbf{x}^*)$. From $\delta_{k-1} = 2\delta_k$, we have

$$D_{\delta_K} \leq \frac{D_{\delta_0}}{4^K} + \frac{M^2}{2}\left(\delta_{K-1} + \frac{\delta_{K-2}}{4} + \frac{\delta_{K-3}}{4^2} + \cdots + \frac{\delta_0}{4^{K-1}}\right)$$

$$\leq \frac{f(\mathbf{x}_0) - f(\mathbf{x}^*)}{4^K} + \frac{M^2}{2}\left(\delta_{K-1} + \frac{\delta_{K-2}}{4} + \frac{\delta_{K-3}}{4^2} + \cdots + \frac{\delta_0}{4^{K-1}} + \frac{\delta_0}{4^K}\right)$$

$$= \frac{f(\mathbf{x}_0) - f(\mathbf{x}^*)}{4^K} + M^2 \delta_K \left(1 + \frac{1}{2} + \frac{1}{4} + \cdots + \frac{1}{2^{K-1}} + \frac{1}{2^{K+1}}\right)$$

$$\leq \frac{f(\mathbf{x}_0) - f(\mathbf{x}^*)}{4^K} + 2M^2 \delta_K$$

and

$$f(\mathbf{x}_K) - f(\mathbf{x}^*) \leq f_{\delta_K}(\mathbf{x}_K) - f_{\delta_K}(\hat{\mathbf{x}}^*_{\delta_K}) + \frac{\delta_K M^2}{2}$$

$$= D_{\delta_K} + \frac{\delta_K M^2}{2}$$

$$\leq \frac{f(\mathbf{x}_0) - f(\mathbf{x}^*)}{4^K} + 2M^2 \delta_K + \frac{\delta_K M^2}{2}$$

$$= \frac{f(\mathbf{x}_0) - f(\mathbf{x}^*)}{4^K} + \frac{5M^2 \delta_0}{2^K}$$

$$\leq 6\frac{f(\mathbf{x}_0) - f(\mathbf{x}^*)}{2^K}.$$

Thus we only need $K = \log_2 \frac{6(f(\mathbf{x}_0) - f(\mathbf{x}^*))}{\epsilon}$ such that $f(\mathbf{x}_K) - f(\mathbf{x}^*) \leq \epsilon$. So the total time is

$$\sum_{k=1}^{K} \frac{\sqrt{2^{k-1}}\log 8}{\sqrt{\mu\delta_0}} = \frac{\log 8}{\sqrt{\mu\delta_0}}\frac{\sqrt{\frac{6(f(\mathbf{x}_0) - f(\mathbf{x}^*))}{\epsilon}} - 1}{\sqrt{2} - 1} \leq \frac{3M}{\sqrt{\mu\epsilon}}. \qquad \square$$

2.6 Higher Order Accelerated Method

When the objective function $f(\mathbf{x})$ is twice continuously differentiable and has Lipschitz continuous Hessians (Definition A.14), the accelerated method has a faster convergence rate by applying the cubic regularization, which was originally proposed in [25] and further extended in [23]. Baes [4] and Bubeck et al. [7] extended the cubic regularization to higher regularization and studied the higher order accelerated method, which includes the cubic regularization as a special case. We introduce the study in [4] in this section. The method in [4] is described in Algorithm 2.10, where $f^{(m)}(\mathbf{x})[\cdot]$ is the multilinear operator that corresponds to the m-th order derivative of f and maps a tuple of $m - 1$ vectors to a vector. For consistency, we write $\nabla f(\mathbf{x})$ as $f^{(1)}(\mathbf{x})$. Then the Taylor expansion of $f(\mathbf{x})$ at $\mathbf{x}_k$ can be written as

$$f(\mathbf{x}) = f(\mathbf{x}_k) + \sum_{i=1}^{m} \frac{1}{m!} \left\langle f^{(m)}(\mathbf{x}_k)[\mathbf{x} - \mathbf{x}_k, \cdots, \mathbf{x} - \mathbf{x}_k], \mathbf{x} - \mathbf{x}_k \right\rangle + \cdots.$$

Algorithm 2.10 High order accelerated gradient descent (AGD)

Initialize $\mathbf{x}_0 = \mathbf{z}_0$.

for $k = 1, 2, 3, \cdots$ **do**

$\quad \lambda_{k+1} = (1 - \theta_k)\lambda_k,$

$\quad \mathbf{y}_k = \theta_k \mathbf{z}_k + (1 - \theta_k)\mathbf{x}_k,$

$\quad \mathbf{x}_{k+1} = \operatorname{argmin}_{\mathbf{x}} \left(\sum_{i=1}^{m} \frac{1}{m!} \left\langle f^{(m)}(\mathbf{x}_k)[\mathbf{x} - \mathbf{x}_k, \cdots, \mathbf{x} - \mathbf{x}_k], \mathbf{x} - \mathbf{x}_k \right\rangle + \frac{N}{(m+1)!} \|\mathbf{x} - \mathbf{y}_k\|^{m+1} \right),$

$\quad$ Define $\phi_{k+1}(\mathbf{x})$ in (2.48),

$\quad \mathbf{z}_{k+1} = \operatorname{argmin}_{\mathbf{x}} \phi_{k+1}(\mathbf{x}).$

end for

We make the following high order assumption:

Assumption 2.1 *Assume that the m-th derivative of $f(\mathbf{x})$ is Lipschitz continuous:*

$$\|f^{(m)}(\mathbf{y}) - f^{(m)}(\mathbf{x})\| \le M\|\mathbf{y} - \mathbf{x}\|.$$

By integrating several times the above inequality, we can easily deduce that for all $\mathbf{x}$ and $\mathbf{y}$, we have

$$\left\| f^{(1)}(\mathbf{y}) - f^{(1)}(\mathbf{x}) - \sum_{i=2}^{m} \frac{1}{(m-1)!} f^{(m)}(\mathbf{x})[\mathbf{y} - \mathbf{x}, \cdots, \mathbf{y} - \mathbf{x}] \right\| \le \frac{M}{m!} \|\mathbf{y} - \mathbf{x}\|^m.$$

We use the estimate sequence to prove the convergence. Define two sequences:

$$\lambda_{k+1} = (1 - \theta_k)\lambda_k, \tag{2.47}$$

$$\phi_{k+1}(\mathbf{x}) = (1 - \theta_k)\phi_k(\mathbf{x}) + \theta_k \left[f(\mathbf{x}_{k+1}) + \langle \nabla f(\mathbf{x}_{k+1}), \mathbf{x} - \mathbf{x}_{k+1} \rangle \right], \tag{2.48}$$

with $\lambda_0 = 1$ and $\phi_0(\mathbf{x}) = f(\mathbf{x}_0) + \frac{M}{(m+1)!}\|\mathbf{x} - \mathbf{x}_0\|^{m+1}$. Then similar to the proof of (2.6), we can easily see that (2.3) holds. Thus it is an estimate sequence. From Lemma 2.1, we have that if $\phi_k^* \geq f(\mathbf{x}_k)$, then $f(\mathbf{x}_k) - f(\mathbf{x}^*) \leq \lambda_k(\phi_0(\mathbf{x}^*) - f(\mathbf{x}^*))$. Define

$$\mathbf{z}_k = \underset{\mathbf{x}}{\operatorname{argmin}} \, \phi_k(\mathbf{x}). \tag{2.49}$$

We only need to ensure $f(\mathbf{x}_k) \leq \phi_k(\mathbf{z}_k)$.

The following lemma describes the Bregman distance induced by the higher order form of $\|\mathbf{x} - \mathbf{x}_0\|^p$.

Lemma 2.12 *Let $p \geq 2$ and $\rho(\mathbf{x}) = \|\mathbf{x} - \mathbf{x}_0\|^p$. Then*

$$\rho(\mathbf{y}) - \rho(\mathbf{x}) - \langle \nabla \rho(\mathbf{x}), \mathbf{y} - \mathbf{x} \rangle \geq c_p \|\mathbf{y} - \mathbf{x}\|^p,$$

where $c_p = \dfrac{p-1}{\left[1 + (2p-3)^{1/(p-2)}\right]^{p-2}}$ for $p > 2$ and $c_p = 1$ for $p = 2$.

From the fact of (2.49), we can have the following lemma, which establishes the relation between $\phi_k(\mathbf{x})$ and $\phi_k(\mathbf{z}_k)$ for any $\mathbf{x}$.

Lemma 2.13 *For the definitions of (2.48), we have*

$$\phi_k(\mathbf{x}) \geq \phi_k(\mathbf{z}_k) + \lambda_k \chi(\mathbf{x}, \mathbf{z}_k), \forall \mathbf{x}, \tag{2.50}$$

where $\chi(\mathbf{x}, \mathbf{y}) = \frac{M c_{m+1}}{(m+1)!}\|\mathbf{y} - \mathbf{x}\|^{m+1}$.

Proof From the definition of $\phi_0(\mathbf{x})$ and Lemma 2.12, we have

$$\phi_0(\mathbf{x}) \geq \phi_0(\mathbf{z}_k) + \langle \nabla \phi_0(\mathbf{z}_k), \mathbf{x} - \mathbf{z}_k \rangle + \chi(\mathbf{x}, \mathbf{z}_k).$$

From the definition in (2.48), we know

$$\phi_k(\mathbf{x}) = \lambda_k \phi_0(\mathbf{x}) + l_k(\mathbf{x}),$$

where $l_k(\mathbf{x})$ is an affine function. Thus we have

$$\phi_k(\mathbf{x}) \geq \lambda_k \phi_0(\mathbf{z}_k) + \langle \lambda_k \nabla \phi_0(\mathbf{z}_k), \mathbf{x} - \mathbf{z}_k \rangle + \lambda_k \chi(\mathbf{x}, \mathbf{z}_k) + l_k(\mathbf{x}) - l_k(\mathbf{z}_k) + l_k(\mathbf{z}_k)$$

$$= \phi_k(\mathbf{z}_k) + \langle \nabla \phi_k(\mathbf{z}_k), \mathbf{x} - \mathbf{z}_k \rangle + \lambda_k \chi(\mathbf{x}, \mathbf{z}_k)$$

$$\overset{a}{=} \phi_k(\mathbf{z}_k) + \lambda_k \chi(\mathbf{x}, \mathbf{z}_k),$$

where $\overset{a}{=}$ uses (2.49). $\square$

Now we can provide an intermediate inequality that we will use to prove $f(\mathbf{x}_k) \leq \phi_k(\mathbf{z}_k)$ for all k.

Lemma 2.14 *Denote by $\{\mathbf{x}_k\}_{k\geq 0}$ a sequence satisfying $f(\mathbf{x}_k) \leq \phi_k(\mathbf{z}_k)$, then*

$$\phi_{k+1}(\mathbf{z}_{k+1}) \geq f(\mathbf{x}_{k+1}) + \langle \nabla f(\mathbf{x}_{k+1}), (1-\theta_k)\mathbf{x}_k + \theta_k\mathbf{z}_k - \mathbf{x}_{k+1}\rangle$$
$$+ \min_{\mathbf{x}} \left(\theta_k \langle \nabla f(\mathbf{x}_{k+1}), \mathbf{x} - \mathbf{z}_k\rangle + \lambda_{k+1}\chi(\mathbf{x}, \mathbf{z}_k)\right). \quad (2.51)$$

Proof From (2.48), (2.50), $f(\mathbf{x}_k) \leq \phi_k(\mathbf{z}_k)$, and the convexity of $f(\mathbf{x})$, we have

$$\phi_{k+1}(\mathbf{x}) \geq (1-\theta_k)[\phi_k(\mathbf{z}_k) + \lambda_k\chi(\mathbf{x}, \mathbf{z}_k)] + \theta_k[f(\mathbf{x}_{k+1}) + \langle \nabla f(\mathbf{x}_{k+1}), \mathbf{x} - \mathbf{x}_{k+1}\rangle]$$
$$\geq (1-\theta_k)[f(\mathbf{x}_k) + \lambda_k\chi(\mathbf{x}, \mathbf{z}_k)] + \theta_k[f(\mathbf{x}_{k+1}) + \langle \nabla f(\mathbf{x}_{k+1}), \mathbf{x} - \mathbf{x}_{k+1}\rangle]$$
$$\geq (1-\theta_k)[f(\mathbf{x}_{k+1}) + \langle \nabla f(\mathbf{x}_{k+1}), \mathbf{x}_k - \mathbf{x}_{k+1}\rangle + \lambda_k\chi(\mathbf{x}, \mathbf{z}_k)]$$
$$+\theta_k[f(\mathbf{x}_{k+1}) + \langle \nabla f(\mathbf{x}_{k+1}), \mathbf{x} - \mathbf{x}_{k+1}\rangle]$$
$$= f(\mathbf{x}_{k+1}) + \langle \nabla f(\mathbf{x}_{k+1}), (1-\theta_k)\mathbf{x}_k + \theta_k\mathbf{z}_k - \mathbf{x}_{k+1}\rangle$$
$$+\theta_k\langle \nabla f(\mathbf{x}_{k+1}), \mathbf{x} - \mathbf{z}_k\rangle + \lambda_{k+1}\chi(\mathbf{x}, \mathbf{z}_k).$$

(2.51) follows immediately. □

Our goal is to make the sum of the last two terms on the right-hand side of (2.51) positive. We start our analysis with an easy lemma. The proof is simple and we omit it.

Lemma 2.15 *From the definition of $\mathbf{y}_k$, we have*

$$\min_{\mathbf{x}} \left(\theta_k \langle \nabla f(\mathbf{x}_{k+1}), \mathbf{x} - \mathbf{z}_k\rangle + \lambda_{k+1}\chi(\mathbf{x}, \mathbf{z}_k)\right)$$
$$\geq \min_{\mathbf{x}} \left(\langle \nabla f(\mathbf{x}_{k+1}), \mathbf{x} - \mathbf{y}_k\rangle + \frac{\lambda_{k+1}}{\theta_k^{m+1}}\chi(\mathbf{x}, \mathbf{y}_k)\right). \quad (2.52)$$

The next lemma plays a crucial role in the validation of the desired inequality.

Lemma 2.16 *For any $\mathbf{x}$, we have*

$$\langle \nabla f(\mathbf{x}_{k+1}), \mathbf{x} - \mathbf{x}_{k+1}\rangle$$
$$\geq -\frac{M+N}{m!}\|\mathbf{y}_k - \mathbf{x}_{k+1}\|^m \|\mathbf{x} - \mathbf{y}_k\| + \frac{N-M}{m!}\|\mathbf{y}_k - \mathbf{x}_{k+1}\|^{m+1}.$$

Proof From the optimality condition for $\mathbf{x}_{k+1}$, we have

$$\left\langle f^{(1)}(\mathbf{y}_k) + \cdots + \frac{1}{(m-1)!}f^{(m)}(\mathbf{y}_k)[\mathbf{x}_{k+1} - \mathbf{y}_k, \cdots, \mathbf{x}_{k+1} - \mathbf{y}_k], \mathbf{x} - \mathbf{x}_{k+1}\right\rangle$$
$$+ \frac{N\|\mathbf{y}_k - \mathbf{x}_{k+1}\|^{m-1}}{m!}\langle \mathbf{x}_{k+1} - \mathbf{y}_k, \mathbf{x} - \mathbf{x}_{k+1}\rangle \geq 0, \forall \mathbf{x}.$$

On the other hand, from the high order smoothness of $f(\mathbf{x})$, we have

$$
\Big\langle f^{(1)}(\mathbf{y}_k) + \cdots + \frac{1}{(m-1)!} f^{(m)}(\mathbf{y}_k)[\mathbf{x}_{k+1} - \mathbf{y}_k, \cdots, \mathbf{x}_{k+1} - \mathbf{y}_k]
$$

$$
- f^{(1)}(\mathbf{x}_{k+1}), \mathbf{x} - \mathbf{x}_{k+1} \Big\rangle
$$

$$
\leq \frac{M}{m!} \|\mathbf{y}_k - \mathbf{x}_{k+1}\|^m \|\mathbf{x} - \mathbf{x}_{k+1}\|
$$

$$
\leq \frac{M}{m!} \|\mathbf{y}_k - \mathbf{x}_{k+1}\|^m \left(\|\mathbf{y}_k - \mathbf{x}_{k+1}\| + \|\mathbf{x} - \mathbf{y}_k\| \right).
$$

Thus we have

$$
0 \leq \frac{M}{m!} \|\mathbf{y}_k - \mathbf{x}_{k+1}\|^m \left(\|\mathbf{y}_k - \mathbf{x}_{k+1}\| + \|\mathbf{x} - \mathbf{y}_k\| \right)
$$

$$
+ \Big\langle f^{(1)}(\mathbf{x}_{k+1}), \mathbf{x} - \mathbf{x}_{k+1} \Big\rangle + \frac{N \|\mathbf{y}_k - \mathbf{x}_{k+1}\|^{m-1}}{m!} \langle \mathbf{x}_{k+1} - \mathbf{y}_k, \mathbf{x} - \mathbf{x}_{k+1} \rangle.
$$

Since $\langle \mathbf{x}_{k+1} - \mathbf{y}_k, \mathbf{x} - \mathbf{x}_{k+1} \rangle = \langle \mathbf{x}_{k+1} - \mathbf{y}_k, (\mathbf{x} - \mathbf{y}_k) - (\mathbf{x}_{k+1} - \mathbf{y}_k) \rangle \leq \|\mathbf{x}_{k+1} - \mathbf{y}_k\| \|\mathbf{x} - \mathbf{y}_k\| - \|\mathbf{x}_{k+1} - \mathbf{y}_k\|^2$, we can have the conclusion. $\qquad\square$

Based on the previous lemmas, we are now ready to prove the convergence rate.

Theorem 2.11 Let $N = (2m + 1)M$ and $\theta_k \in (0, 1]$ is obtained by solving the equation $(2m + 2)\theta_k^{m+1} = c_{m+1}(1 - \theta_k)\lambda_k$, then

$$
f(\mathbf{x}_K) - f(\mathbf{x}^*) \leq \frac{2m+2}{c_{m+1}} \left(\frac{m+1}{K} \right)^{m+1} \left(f(\mathbf{x}_0) - f(\mathbf{x}^*) + \frac{M}{(m+1)!} \|\mathbf{x}_0 - \mathbf{x}^*\|^{m+1} \right).
$$

Proof From (2.51), (2.52), and the definition of $\mathbf{y}_k$, we only need

$$
\langle \nabla f(\mathbf{x}_{k+1}), \mathbf{y}_k - \mathbf{x}_{k+1} \rangle + \min_{\mathbf{x}} \left(\langle \nabla f(\mathbf{x}_{k+1}), \mathbf{x} - \mathbf{y}_k \rangle + \frac{\lambda_{k+1}}{\theta_k^{m+1}} \chi(\mathbf{x}, \mathbf{y}_k) \right) \geq 0
$$

to ensure $f(\mathbf{x}_{k+1}) \leq \phi_{k+1}(\mathbf{z}_{k+1})$. Since

$$
\langle \nabla f(\mathbf{x}_{k+1}), \mathbf{y}_k - \mathbf{x}_{k+1} \rangle + \min_{\mathbf{x}} \left(\langle \nabla f(\mathbf{x}_{k+1}), \mathbf{x} - \mathbf{y}_k \rangle + \frac{\lambda_{k+1}}{\theta_k^{m+1}} \chi(\mathbf{x}, \mathbf{y}_k) \right)
$$

$$
= \min_{\mathbf{x}} \left(\langle \nabla f(\mathbf{x}_{k+1}), \mathbf{x} - \mathbf{x}_{k+1} \rangle + \frac{\lambda_{k+1}}{\theta_k^{m+1}} \chi(\mathbf{x}, \mathbf{y}_k) \right)
$$

$$\geq \min_{\mathbf{x}} \left(-\frac{M+N}{m!} \|\mathbf{y}_k - \mathbf{x}_{k+1}\|^m \|\mathbf{x} - \mathbf{y}_k\| \right.$$

$$\left. + \frac{N-M}{m!} \|\mathbf{y}_k - \mathbf{x}_{k+1}\|^{m+1} + \frac{\lambda_{k+1}}{\theta_k^{m+1}} \chi(\mathbf{x}, \mathbf{y}_k) \right)$$

$$= \frac{N-M}{m!} \|\mathbf{y}_k - \mathbf{x}_{k+1}\|^{m+1}$$

$$+ \min_{t \geq 0} \left(-\frac{M+N}{m!} \|\mathbf{y}_k - \mathbf{x}_{k+1}\|^m t + \frac{\lambda_{k+1}}{\theta_k^{m+1}} \frac{M c_{m+1}}{(m+1)!} t^{m+1} \right)$$

$$= \frac{\|\mathbf{y}_k - \mathbf{x}_{k+1}\|^{m+1}}{m!} \left[N - M - \frac{m}{m+1} \left(\frac{(M+N)^{m+1} \theta_k^{m+1}}{M c_{m+1} \lambda_{k+1}} \right)^{1/m} \right].$$

This quantity is nonnegative as long as

$$c_{m+1} \frac{M(N-M)^m}{(M+N)^{m+1}} \left(\frac{m+1}{m} \right)^m \geq \frac{\theta_k^{m+1}}{\lambda_{k+1}}.$$

Maximizing the right-hand side with respect to N, we get a value of $\frac{c_{m+1}}{2m+2}$ obtained for $N = (2m+1)M$. From the definition of θ_k, we have $\frac{\theta_k^{m+1}}{\lambda_{k+1}} = \frac{c_{m+1}}{2m+2}$. From Lemma 2.17, we have $\lambda_K \leq \frac{2m+2}{c_{m+1}} \left(\frac{m+1}{K} \right)^{m+1}$. □

The following lemma is a higher order counterpart of Lemma 2.3 with $p = 1$.

Lemma 2.17 *If there exists $\delta > 0$ such that $\frac{\theta_k^p}{\lambda_{k+1}} \geq \delta$, where λ_{k+1} is defined in (2.47), then*

$$\lambda_K \leq \left(\frac{p}{p + K \sqrt[p]{\delta}} \right)^p \leq \frac{1}{\delta} \left(\frac{p}{K} \right)^p.$$

Proof Since $\lambda_k \leq 1$ is decreasing, we have

$$\lambda_k - \lambda_{k+1} = \left(\sqrt[p]{\lambda_k^{p-1}} + \sqrt[p]{\lambda_k^{p-2} \lambda_{k+1}} + \cdots + \sqrt[p]{\lambda_{k+1}^{p-1}} \right) \left(\sqrt[p]{\lambda_k} - \sqrt[p]{\lambda_{k+1}} \right)$$

$$\leq p \sqrt[p]{\lambda_k^{p-1}} \left(\sqrt[p]{\lambda_k} - \sqrt[p]{\lambda_{k+1}} \right).$$

This inequality implies

$$\frac{1}{\sqrt[p]{\lambda_{k+1}}} - \frac{1}{\sqrt[p]{\lambda_k}} = \frac{\sqrt[p]{\lambda_k} - \sqrt[p]{\lambda_{k+1}}}{\sqrt[p]{\lambda_k \lambda_{k+1}}} \geq \frac{\lambda_k - \lambda_{k+1}}{p \lambda_k \sqrt[p]{\lambda_{k+1}}} = \frac{\theta_k \lambda_k}{p \lambda_k \sqrt[p]{\lambda_{k+1}}} \geq \frac{\sqrt[p]{\delta}}{p}.$$

Summing over $k = 0, \cdots, K - 1$, we have

$$\frac{1}{\sqrt[p]{\lambda_K}} - \frac{1}{\sqrt[p]{\lambda_0}} = \frac{1}{\sqrt[p]{\lambda_K}} - 1 \geq K \frac{\sqrt[p]{\delta}}{p},$$

which is equivalent to the desired result. □

2.7 Explanation: A Variational Perspective

Despite the compelling evidence of the convergence rate of the accelerated algorithms, it remains something of a conceptual mystery. In recent years, a number of explanations and interpretations of acceleration have been proposed [2, 6, 13, 16, 28]. In this section, we introduce the work in [30], which gives a variational perspective.

Define the Bregman Lagrangian

$$L(X, V, t) = \frac{p}{t} t^p \left[D_h \left(X + \frac{t}{p} V, X \right) - C t^p f(X) \right],$$

which is a function of position X, velocity V, and time t. $D_h(\cdot, \cdot)$ is the Bregman distance (Definition A.20) induced by some convex function $h(x)$.

Given a general Lagrangian $L(X, V, t)$, we define a function on the curve X_t via integration of the Lagrangian $J(X) = \int L(X_t, \dot{X}_t, t) dt$. From the calculus of variations, a necessary condition for a curve to minimize this functional is that it solves the Euler–Lagrange equation:

$$\frac{d}{dt} \left[\frac{\partial L}{\partial V}(X_t, \dot{X}_t, t) \right] = \frac{\partial L}{\partial X}(X_t, \dot{X}_t, t).$$

Specifically, for the Bregman Lagrangian the partial derivatives are

$$\frac{\partial L}{\partial X}(X, V, t) = \frac{p}{t} t^p \left[\nabla h \left(X + \frac{t}{p} V \right) - \nabla h(X) - \frac{t}{p} \nabla^2 h(X) V - C t^p \nabla f(X) \right],$$

$$\frac{\partial L}{\partial V}(X, V, t) = t^p \left[\nabla h \left(X + \frac{t}{p} V \right) - \nabla h(X) \right].$$

Thus the Euler–Lagrange equation for the Bregman Lagrangian is a second-order differential equation given by

$$\ddot{X}_t + \frac{p+1}{t} \dot{X}_t + C \frac{p^2}{t^2} t^p \left[\nabla^2 h \left(X_t + \frac{t}{p} \dot{X}_t \right) \right]^{-1} \nabla f(X_t) = \mathbf{0}. \qquad (2.53)$$

We can also write (2.53) in the following way, which only requires that ∇h is differentiable,

$$\frac{d}{dt}\nabla h\left(X_t + \frac{t}{p}\dot{X}_t\right) = -C\frac{p}{t}t^p\nabla f(X_t). \tag{2.54}$$

To establish a convergence rate associated with solutions to the Euler–Lagrange equation, we take a Lyapunov function approach. Defining the energy function

$$\varepsilon_t = D_h\left(\mathbf{x}^*, X_t + \frac{p}{t}\dot{X}_t\right) + Ct^p(f(X_t) - f(\mathbf{x}^*)).$$

Then we have the following theorem to give the convergence rate.

Theorem 2.12 *Solutions to the Euler–Lagrange equation (2.54) satisfy*

$$f(X_t) - f(\mathbf{x}^*) \leq O\left(\frac{1}{Ct^p}\right).$$

Proof The time derivative of the energy function is

$$\dot{\varepsilon}_t = -\left\langle\frac{d}{dt}\nabla h\left(X_t + \frac{t}{p}\dot{X}_t\right), \mathbf{x}^* - X_t - \frac{t}{p}\dot{X}_t\right\rangle + C\frac{p}{t}t^p(f(X_t) - f(\mathbf{x}^*))$$
$$+ Ct^p\left\langle\nabla f(X_t), \dot{X}_t\right\rangle.$$

If X_t satisfies (2.54), then the time derivative simplifies to

$$\dot{\varepsilon}_t = -C\frac{p}{t}t^p\left(f(\mathbf{x}^*) - f(X_t) - \left\langle\nabla f(X_t), \mathbf{x}^* - X_t\right\rangle\right) \leq 0.$$

That $D_h\left(\mathbf{x}^*, X_t + \frac{p}{t}\dot{X}_t\right) \geq 0$ implies that for any t, $Ct^p(f(X_t) - f(\mathbf{x}^*)) \leq \varepsilon_t \leq \varepsilon_{t_0}$. Thus $f(X_t) - f(\mathbf{x}^*) \leq \frac{\varepsilon_{t_0}}{Ct^p}$. $\qquad\square$

2.7.1 Discretization

We now turn to discretize the differential equation (2.54). We write the second-order equation (2.54) as the following system of first-order equations:

$$Z_t = X_t + \frac{t}{p}\dot{X}_t, \tag{2.55}$$

$$\frac{d}{dt}\nabla h(Z_t) = -Cpt^{p-1}\nabla f(X_t). \tag{2.56}$$

Now we discretize X_t and Z_t into sequences $\mathbf{x}_t$ and $\mathbf{z}_t$ and set $\mathbf{x}_t = X_t$, $\mathbf{x}_{t+1} = X_t + \dot{X}_t$, $\mathbf{z}_t = Z_t$, and $\mathbf{z}_{t+1} = Z_t + \dot{Z}_t$. Applying the forward Euler method to (2.55) gives the equation

$$\mathbf{z}_t = \mathbf{x}_t + \frac{t}{p}(\mathbf{x}_{t+1} - \mathbf{x}_t). \tag{2.57}$$

Similarly, applying the backward Euler method to Eq. (2.56) gives $\nabla h(\mathbf{z}_t) - \nabla h(\mathbf{z}_{t-1}) = -Cpt^{p-1}\nabla f(\mathbf{x}_t)$, which can be written as the optimality condition of the following mirror descent:

$$\mathbf{z}_t = \operatorname*{argmin}_{\mathbf{z}} \left\{ Cpt^{p-1} \langle \nabla f(\mathbf{x}_t), \mathbf{z} \rangle + D_h(\mathbf{z}, \mathbf{z}_{t-1}) \right\}. \tag{2.58}$$

However, we cannot prove the convergence for the algorithm in (2.57) and (2.58). Inspired by Algorithm 2.3, which maintains three sequences, we introduce a third sequence $\mathbf{y}_t$ and consider the following iterates:

$$\mathbf{z}_t = \operatorname*{argmin}_{\mathbf{z}} \left\{ Cpt^{\langle p-1 \rangle} \langle \nabla f(\mathbf{y}_t), \mathbf{z} \rangle + D_h(\mathbf{z}, \mathbf{z}_{t-1}) \right\}, \tag{2.59}$$

$$\mathbf{x}_{t+1} = \frac{p}{t+p}\mathbf{z}_t + \frac{t}{t+p}\mathbf{y}_t, \tag{2.60}$$

where $t^{\langle p-1 \rangle} = t(t+1)\cdots(t+p-2)$ is the rising factorial. A sufficient condition for algorithm (2.59)–(2.60) to have an $O(1/t^p)$ convergence rate is that the new sequence $\mathbf{y}_t$ satisfies the inequality:

$$\langle \nabla f(\mathbf{y}_t), \mathbf{x}_t - \mathbf{y}_t \rangle \geq M\|\nabla f(\mathbf{y}_t)\|^{p/(p-1)} \tag{2.61}$$

for some constant $M > 0$.

Theorem 2.13 *Assume that h is 1-uniformly convex of order $p \geq 2$ (i.e., the Bregman distance satisfies $D_h(\mathbf{y}, \mathbf{x}) \geq \frac{1}{p}\|\mathbf{y} - \mathbf{x}\|^p$) and $\mathbf{y}_t$ satisfies (2.61), then the algorithm (2.59)–(2.60) with the constant $C \leq M^{p-1}/p^p$ and initial condition $\mathbf{z}_0 = \mathbf{x}_0$ has the convergence rate*

$$f(\mathbf{y}_t) - f(\mathbf{x}^*) \leq O\left(\frac{1}{t^p}\right).$$

We define the following function, which can be recognized as Nesterov's estimate function,

$$\psi_t(\mathbf{x}) = Cp \sum_{i=0}^{t} i^{\langle p-1 \rangle} [f(\mathbf{y}_i) + \langle \nabla f(\mathbf{y}_i), \mathbf{x} - \mathbf{y}_i \rangle] + D_h(\mathbf{x}, \mathbf{x}_0). \tag{2.62}$$

The optimality condition for (2.59) is

$$\nabla h(\mathbf{z}_t) = \nabla h(\mathbf{z}_{t-1}) - Cpt^{\langle p-1 \rangle} \nabla f(\mathbf{y}_t).$$

By unrolling the recursion, we can write

$$\nabla h(\mathbf{z}_t) = \nabla h(\mathbf{z}_0) - Cp \sum_{i=0}^{t} i^{\langle p-1 \rangle} \nabla f(\mathbf{y}_i),$$

and because $\mathbf{x}_0 = \mathbf{z}_0$, we can write this equation as $\nabla \psi_t(\mathbf{z}_t) = \mathbf{0}$. Thus we have

$$\mathbf{z}_t = \underset{\mathbf{z}}{\operatorname{argmin}}\, \psi_t(\mathbf{z}).$$

For proving Theorem 2.13, we have the following property.

Lemma 2.18 *For all $t \geq 0$, we have*

$$\psi_t(\mathbf{z}_t) \geq Ct^{\langle p \rangle} f(\mathbf{y}_t). \tag{2.63}$$

Proof We prove by induction. The base case $k = 0$ is true because both sides equal 0. Now assume (2.63) holds for some t, we will show it also holds for $t + 1$.

Because h is 1-uniformly convex of order p, the Bregman distance $D_h(\mathbf{x}, \mathbf{x}_0)$ is 1-uniformly convex. Thus the estimate function ψ_t is also 1-uniformly convex of order p. Because $\nabla \psi_t(\mathbf{z}_t) = \mathbf{0}$, $D_{\psi_t}(\mathbf{x}, \mathbf{z}_t) = \psi_t(\mathbf{x}) - \psi_t(\mathbf{z}_t)$. So for all $\mathbf{x}$ we have

$$\psi_t(\mathbf{x}) = \psi_t(\mathbf{z}_t) + D_{\psi_t}(\mathbf{x}, \mathbf{z}_t) \geq \psi_t(\mathbf{z}_t) + \frac{1}{p} \|\mathbf{x} - \mathbf{z}_t\|^p.$$

Applying the inductive hypothesis (2.63) and using the convexity of f gives

$$\psi_t(\mathbf{x}) \geq Ct^{\langle p \rangle} [f(\mathbf{y}_{t+1}) + \langle \nabla f(\mathbf{y}_{t+1}), \mathbf{y}_t - \mathbf{y}_{t+1} \rangle] + \frac{1}{p} \|\mathbf{x} - \mathbf{z}_t\|^p.$$

We now add $Cp(t + 1)^{\langle p-1 \rangle} [f(\mathbf{y}_{t+1}) + \langle \nabla f(\mathbf{y}_{t+1}), \mathbf{x} - \mathbf{y}_{t+1} \rangle]$ to both sides of the equation to obtain

$$\begin{aligned} \psi_{t+1}(\mathbf{x}) &= \psi_t(\mathbf{x}) + Cp(t + 1)^{\langle p-1 \rangle} [f(\mathbf{y}_{t+1}) + \langle \nabla f(\mathbf{y}_{t+1}), \mathbf{x} - \mathbf{y}_{t+1} \rangle] \\ &\geq C(t + 1)^{\langle p \rangle} [f(\mathbf{y}_{t+1}) + \langle \nabla f(\mathbf{y}_{t+1}), \mathbf{x}_{t+1} - \mathbf{y}_{t+1} + \tau_t(\mathbf{x} - \mathbf{z}_t) \rangle] \\ &\quad + \frac{1}{p} \|\mathbf{x} - \mathbf{z}_t\|^p, \end{aligned}$$

from the definition of $\psi_{t+1}(\mathbf{x})$ in (2.62) and the definition of $\mathbf{x}_{t+1}$ in (2.60), where $\tau_t = \frac{p(t+1)^{\langle p-1 \rangle}}{(t+1)^{\langle p \rangle}} = \frac{p}{t+p}$.

Applying (2.61) to the term $\langle \nabla f(\mathbf{y}_{t+1}), \mathbf{x}_{t+1} - \mathbf{y}_{t+1} \rangle$, we have

$$\psi_{t+1}(\mathbf{x}) \geq C(t+1)^{\langle p \rangle} f(\mathbf{y}_{t+1}) + C(t+1)^{\langle p \rangle} M \|\nabla f(\mathbf{y}_{t+1})\|^{p/(p-1)}$$

$$+ Cp(t+1)^{\langle p-1 \rangle} \langle \nabla f(\mathbf{y}_{t+1}), \mathbf{x} - \mathbf{z}_t \rangle + \frac{1}{p} \|\mathbf{x} - \mathbf{z}_t\|^p. \qquad (2.64)$$

Next, we apply the Fenchel-Young inequality (Proposition A.13) to have

$$\langle \mathbf{s}, \mathbf{u} \rangle + \frac{1}{p} \|\mathbf{u}\|^p \geq -\frac{p-1}{p} \|\mathbf{s}\|^{p/(p-1)}$$

with the choices of $\mathbf{u} = \mathbf{x} - \mathbf{z}_t$ and $\mathbf{s} = Cp(t+1)^{\langle p-1 \rangle} \nabla f(\mathbf{y}_{t+1})$. Then from (2.64), we have

$$\psi_{t+1}(\mathbf{x}) \geq C(t+1)^{\langle p \rangle} \left\{ f(\mathbf{y}_{t+1}) \right.$$

$$+ \left[M - \frac{p-1}{p} p^{p/(p-1)} C^{1/(p-1)} \frac{((t+1)^{\langle p-1 \rangle})^{p/(p-1)}}{(t+1)^{\langle p \rangle}} \right]$$

$$\left. \times \|\nabla f(\mathbf{y}_{t+1})\|^{p/(p-1)} \right\}.$$

Note that $[(t+1)^{\langle p-1 \rangle}]^{p/(p-1)} \leq (t+1)^{\langle p \rangle}$. Then from the assumption $C \leq M^{p-1}/p^p$, we see that the second term inside the parentheses is nonnegative. Hence we conclude the desired inequality $\psi_{t+1}(\mathbf{x}) \geq C(t+1)^{\langle p \rangle} f(\mathbf{y}_{t+1})$. Because $\mathbf{x}$ is arbitrary, it also holds for the minimizer $\mathbf{x} = \mathbf{z}_{t+1}$ of ψ_{t+1}, finishing the induction.

□

With Lemma 2.18, we can complete the proof of Theorem 2.13.

Proof Because f is convex, we can bound the estimate function ψ_t by

$$\psi_t(\mathbf{x}) \leq Cp \sum_{i=0}^{t} i^{\langle p-1 \rangle} f(\mathbf{x}) + D_h(\mathbf{x}, \mathbf{x}_0) = Ct^{\langle p \rangle} f(\mathbf{x}) + D_h(\mathbf{x}, \mathbf{x}_0),$$

where we can prove $p \sum_{i=0}^{t} i^{\langle p-1 \rangle} = t^{\langle p \rangle}$ by mathematical induction, e.g., $\sum_{j=p-1}^{t+p-2} C_j^{p-1} = C_{t+p-1}^p$ holds for $t = 1, 2, \cdots$. The above inequality holds for all $\mathbf{x}$ and in particular for the minimizer $\mathbf{x}^*$ of f. Combining the bound with result of Lemma 2.18 and recalling that $\mathbf{z}_t$ is the minimizer of ψ_t, we get

$$Ct^{\langle p \rangle} f(\mathbf{y}_t) \leq \psi_t(\mathbf{z}_t) \leq \psi_t(\mathbf{x}^*) \leq Ct^{\langle p \rangle} f(\mathbf{x}^*) + D_h(\mathbf{x}^*, \mathbf{x}_0).$$

Rearranging and dividing by $Ct^{\langle p \rangle}$ gives the desired convergence rate.

Till now, the remaining thing is to find some $\mathbf{y}_t$ such that (2.61) holds. For simplicity, we only consider the case of $p = 2$ and assume that f is $(p - 1)$-order smooth, i.e., L-smooth when $p = 2$. The higher order smooth case can be analyzed with the similar inductions [30]. Let

$$\mathbf{y}_t = \underset{\mathbf{x}}{\operatorname{argmin}} \left(\langle \nabla f(\mathbf{x}_t), \mathbf{x} \rangle + \frac{1}{4M} \|\mathbf{x} - \mathbf{x}_t\|^2 \right).$$

Then from the optimality condition, we know $\nabla f(\mathbf{x}_t) + \frac{1}{2M}(\mathbf{y}_t - \mathbf{x}_t) = \mathbf{0}$. Since

$$L^2 \|\mathbf{x}_t - \mathbf{y}_t\|^2 \geq \|\nabla f(\mathbf{x}_t) - \nabla f(\mathbf{y}_t)\|^2 = \left\| \frac{1}{2M}(\mathbf{y}_t - \mathbf{x}_t) + \nabla f(\mathbf{y}_t) \right\|^2,$$

letting $2M \leq 1/L$ the above gives

$$\frac{1}{M} \langle \nabla f(\mathbf{y}_t), \mathbf{x}_t - \mathbf{y}_t \rangle \geq \|\nabla f(\mathbf{y}_t)\|^2 + \frac{1}{4M^2} \|\mathbf{y}_t - \mathbf{x}_t\|^2 - L^2 \|\mathbf{x}_t - \mathbf{y}_t\|^2$$

$$\geq \|\nabla f(\mathbf{y}_t)\|^2,$$

which leads to (2.61) with $p = 2$. $\qquad\square$

References

1. Z. Allen-Zhu, E. Hazan, Optimal black-box reductions between optimization objectives, in *Advances in Neural Information Processing Systems*, Barcelona, vol. 29 (2016), pp. 1614–1622
2. Z. Allen-Zhu, L. Orecchia, Linear coupling: an ultimate unification of gradient and mirror descent, in *Proceedings of the 8th Innovations in Theoretical Computer Science (ITCS)*, Berkeley, (2017)
3. F. Bach, R. Jenatton, J. Mairal, G. Obozinski, Convex optimization with sparsity-inducing norms, in *Optimization for Machine Learning* (MIT Press, Cambridge, 2012), pp. 19–53
4. M. Baes, Estimate sequence methods: extensions and approximations. Technical report, Institute for Operations Research, ETH, Zürich (2009)
5. A. Beck, M. Teboulle, A fast iterative shrinkage-thresholding algorithm for linear inverse problems. SIAM J. Imag. Sci. **2**(1), 183–202 (2009)
6. S. Bubeck, Y.T. Lee, M. Singh, A geometric alternative to Nesterov's accelerated gradient descent (2015). Preprint. arXiv:1506.08187
7. S. Bubeck, Q. Jiang, Y.T. Lee, Y. Li, A. Sidford, Near-optimal method for highly smooth convex optimization, in *Proceedings of the 36th Conference on Learning Theory*, Long Beach, (2019), pp. 492–507
8. A. Chambolle, T. Pock, A first-order primal-dual algorithm for convex problems with applications to imaging. J. Math. Imag. Vis. **40**(1), 120–145 (2011)
9. A. Chambolle, T. Pock, On the ergodic convergence rates of a first-order primal-dual algorithm. Math. Program. **159**(1–2), 253–287 (2016)
10. X. Chen, S. Kim, Q. Lin, J.G. Carbonell, E.P. Xing, Graph-structured multi-task regression and an efficient optimization method for general fused lasso (2010). Preprint. arXiv:1005.3579

11. O. Devolder, F. Glineur, Y. Nesterov, First-order methods of smooth convex optimization with inexact oracle. Math. Program. **146**(1–2), 37–75 (2014)
12. J.M. Fadili, G. Peyré, Total variation projection with first order schemes. IEEE Trans. Image Process. **20**(3), 657–669 (2010)
13. N. Flammarion, F. Bach, From averaging to acceleration, there is only a step-size, in *Proceedings of the 28th Conference on Learning Theory*, Paris, (2015), pp. 658–695
14. L. Jacob, G. Obozinski, J.-P. Vert, Group lasso with overlap and graph lasso, in *Proceedings of the 26th International Conference on Machine Learning*, Montreal, (2009), pp. 433–440
15. G. Lan, Y. Zhou, An optimal randomized incremental gradient method. Math. Program. **171**(1–2), 167–215 (2018)
16. L. Lessard, B. Recht, A. Packard, Analysis and design of optimization algorithms via integral quadratic constraints. SIAM J. Optim. **26**(1), 57–95 (2016)
17. H. Lin, J. Mairal, Z. Harchaoui, Catalyst acceleration for first-order convex optimization: from theory to practice. J. Mach. Learn. Res. **18**(212), 1–54 (2018)
18. I. Necoara, Y. Nesterov, F. Glineur, Linear convergence of first order methods for non-strongly convex optimization. Math. Program. **175**(1–2), 69–107 (2019)
19. Y. Nesterov, A method for unconstrained convex minimization problem with the rate of convergence $O(1/k^2)$. Sov. Math. Dokl. **27**(2), 372–376 (1983)
20. Y. Nesterov, On an approach to the construction of optimal methods of minimization of smooth convex functions. Ekonomika I Mateaticheskie Metody **24**(3), 509–517 (1988)
21. Y. Nesterov, *Introductory Lectures on Convex Optimization: A Basic Course* (Springer, New York, 2004)
22. Y. Nesterov, Smooth minimization of non-smooth functions. Math. Program. **103**(1), 127–152 (2005)
23. Y. Nesterov, Accelerating the cubic regularization of Newton's method on convex problems. Math. Program. **181**(1), 112–159 (2008)
24. Y. Nesterov, Gradient methods for minimizing composite functions. Math. Program. **140**(1), 125–161 (2013)
25. Y. Nesterov, B.T. Polyak, Cubic regularization of Newton's method and its global performance. Math. Program. **108**(1), 177–205 (2006)
26. B. O'Donoghue, E. Candès, Adaptive restart for accelerated gradient schemes. Found. Comput. Math. **15**(3), 715–732 (2015)
27. M. Schmidt, N.L. Roux, F.R. Bach, Convergence rates of inexact proximal-gradient methods for convex optimization, in *Advances in Neural Information Processing Systems*, Granada, vol. 24 (2011), pp. 1458–1466
28. W. Su, S. Boyd, E. Candès, A differential equation for modeling Nesterov's accelerated gradient method: theory and insights, in *Advances in Neural Information Processing Systems*, Montreal, vol. 27 (2014), pp. 2510–2518
29. P. Tseng, On accelerated proximal gradient methods for convex-concave optimization. Technical report, University of Washington, Seattle (2008)
30. A. Wibisono, A.C. Wilson, M.I. Jordan, A variational perspective on accelerated methods in optimization. Proc. Natl. Acad. Sci. **113**(47), 7351–7358 (2016)

Chapter 3
Accelerated Algorithms for Constrained Convex Optimization

Besides the unconstrained optimization, the acceleration techniques can also be applied to constrained optimization. In this chapter, we will introduce how to apply the acceleration techniques to constrained optimization. Formally, we consider the following general constrained convex optimization problem:

$$\min_{\mathbf{x} \in \mathbb{R}^n} \ f(\mathbf{x}), \tag{3.1}$$

$$s.t. \ \mathbf{Ax} = \mathbf{b},$$

$$g_i(\mathbf{x}) \leq 0, i = 1, \cdots, p,$$

where both $f(\mathbf{x})$ and $g_i(\mathbf{x})$ are convex and $\mathbf{A} \in \mathbb{R}^{m \times n}$.

This chapter introduces the penalty method, the Lagrange multiplier method, the augmented Lagrange multiplier method, the alternating direction method of multiplier, and the primal-dual method.

3.1 Some Facts for the Case of Linear Equality Constraint

We first consider a simple case of problem (3.1), with linear equality constraint only:

$$\min_{\mathbf{x}} \ f(\mathbf{x}), \tag{3.2}$$

$$s.t. \ \mathbf{Ax} = \mathbf{b}.$$

The introduction in this section is mostly for problem (3.2).

© Springer Nature Singapore Pte Ltd. 2020
Z. Lin et al., *Accelerated Optimization for Machine Learning*,
https://doi.org/10.1007/978-981-15-2910-8_3

From the definition of Fenchel conjugate (Definition A.21), we know that the dual problem (Definition A.24) of (3.2) is

$$\max_{\mathbf{u}} d(\mathbf{u}), \text{ where } d(\mathbf{u}) = -f^*(\mathbf{A}^T \mathbf{u}) - \langle \mathbf{u}, \mathbf{b} \rangle .$$

Let $\sigma_1 \geq \sigma_2 \geq \cdots \geq \sigma_r > 0$ be the nonzero singular values (Definition A.1) of $\mathbf{A}$, we have the following lemma for $d(\mathbf{u})$.

Lemma 3.1

1. *If f is μ-strongly convex, then $d(\mathbf{u})$ is $\frac{\sigma_1^2}{\mu}$-smooth.*
2. *If f is L-smooth, then $-d(\mathbf{u})$ is $\frac{\sigma_r^2}{L}$-strongly convex for all $\mathbf{u} \in Span(\mathbf{A})$.*

Proof

1. Since f is μ-strongly convex, we know that f^* is $1/\mu$-smooth (Point 4 of Proposition A.12). So

$$\|\nabla d(\mathbf{u}) - \nabla d(\mathbf{v})\| = \|\mathbf{A} \nabla f^*(\mathbf{A}^T \mathbf{u}) - \mathbf{A} \nabla f^*(\mathbf{A}^T \mathbf{v})\|$$
$$\leq \|\mathbf{A}\|_2 \|\nabla f^*(\mathbf{A}^T \mathbf{u}) - \nabla f^*(\mathbf{A}^T \mathbf{v})\|$$
$$\leq \frac{\|\mathbf{A}\|_2}{\mu} \|\mathbf{A}^T \mathbf{u} - \mathbf{A}^T \mathbf{v}\|$$
$$\leq \frac{\|\mathbf{A}\|_2^2}{\mu} \|\mathbf{u} - \mathbf{v}\|.$$

2. Since f is L-smooth, we know that f^* is $\frac{1}{L}$-strongly convex, but not necessarily differentiable (Point 4 of Proposition A.12). So

$$-\langle \hat{\nabla} d(\mathbf{u}) - \hat{\nabla} d(\mathbf{u}'), \mathbf{u} - \mathbf{u}' \rangle = \langle \mathbf{A} \hat{\nabla} f^*(\mathbf{A}^T \mathbf{u}) - \mathbf{A} \hat{\nabla} f^*(\mathbf{A}^T \mathbf{u}'), \mathbf{u} - \mathbf{u}' \rangle$$
$$= \langle \hat{\nabla} f^*(\mathbf{A}^T \mathbf{u}) - \hat{\nabla} f^*(\mathbf{A}^T \mathbf{u}'), \mathbf{A}^T \mathbf{u} - \mathbf{A}^T \mathbf{u}' \rangle$$
$$\overset{a}{\geq} \frac{1}{L} \|\mathbf{A}^T \mathbf{u} - \mathbf{A}^T \mathbf{u}'\|^2,$$

where $\hat{\nabla} d(\mathbf{u}) \in \partial d(\mathbf{u})$, $\hat{\nabla} f^*(\mathbf{u}) \in \partial f^*(\mathbf{u})$, and $\overset{a}{\geq}$ uses Proposition A.11. Let $\mathbf{A} = \mathbf{U} \mathbf{\Sigma} \mathbf{V}^T$ be the economic SVD (Definition A.1). Since $\mathbf{u}$ and $\mathbf{u}' \in Span(\mathbf{A})$, there exists $\mathbf{y}$ such that $\mathbf{u} - \mathbf{u}' = \mathbf{A}\mathbf{y} = \mathbf{U}\mathbf{\Sigma}\mathbf{V}^T\mathbf{y} = \mathbf{U}\mathbf{z}$. So we have $\mathbf{A}^T \mathbf{A} = \mathbf{V}\mathbf{\Sigma}^2\mathbf{V}^T$,

$\mathbf{AA}^T = \mathbf{U}\Sigma^2\mathbf{U}^T$ and

$$\begin{aligned}
\|\mathbf{A}^T\mathbf{u} - \mathbf{A}^T\mathbf{u}'\|^2 &= (\mathbf{u} - \mathbf{u}')^T\mathbf{AA}^T(\mathbf{u} - \mathbf{u}') \\
&= \mathbf{z}^T\mathbf{U}^T\mathbf{U}\Sigma^2\mathbf{U}^T\mathbf{U}\mathbf{z} \\
&= \mathbf{z}^T\Sigma^2\mathbf{z} \\
&\geq \sigma_r^2(\mathbf{A})\|\mathbf{z}\|^2 \\
&= \sigma_r^2(\mathbf{A})\|\mathbf{u} - \mathbf{u}'\|^2,
\end{aligned}$$

where we use $\|\mathbf{u} - \mathbf{u}'\|^2 = \mathbf{z}^T\mathbf{U}^T\mathbf{U}\mathbf{z} = \|\mathbf{z}\|^2$. $\qquad\square$

Now we introduce the following lemmas, which will be used in this section.

Lemma 3.2 *Suppose that $f(\mathbf{x})$ is convex and let $(\mathbf{x}^*, \boldsymbol{\lambda}^*)$ be a KKT point (Definition A.26) of problem (3.2), then we have $f(\mathbf{x}) - f(\mathbf{x}^*) + \langle\boldsymbol{\lambda}^*, \mathbf{Ax} - \mathbf{b}\rangle \geq 0, \forall\mathbf{x}$.*

Lemma 3.3 *Suppose that $f(\mathbf{x})$ is convex and let $(\mathbf{x}^*, \boldsymbol{\lambda}^*)$ be a KKT point of problem (3.2). If*

$$f(\mathbf{x}) - f(\mathbf{x}^*) + \langle\boldsymbol{\lambda}^*, \mathbf{Ax} - \mathbf{b}\rangle \leq \alpha_1,$$

$$\|\mathbf{Ax} - \mathbf{b}\| \leq \alpha_2,$$

then we have

$$-\|\boldsymbol{\lambda}^*\|\alpha_2 \leq f(\mathbf{x}) - f(\mathbf{x}^*) \leq \|\boldsymbol{\lambda}^*\|\alpha_2 + \alpha_1.$$

At last, we introduce the augmented Lagrangian function:

$$L_\beta(\mathbf{x}, \mathbf{u}) = f(\mathbf{x}) + \langle\mathbf{u}, \mathbf{Ax} - \mathbf{b}\rangle + \frac{\beta}{2}\|\mathbf{Ax} - \mathbf{b}\|^2,$$

and define

$$d_\beta(\mathbf{u}) = \min_{\mathbf{x}} L_\beta(\mathbf{x}, \mathbf{u}). \tag{3.3}$$

For any $\mathbf{u}$, we have $d(\mathbf{u}) \leq d_\beta(\mathbf{u})$. Moreover, for any $\mathbf{u}$, we have $d_\beta(\mathbf{u}) \leq f(\mathbf{x}^*)$. Since $d(\mathbf{u}^*) = f(\mathbf{x}^*)$, we know $d(\mathbf{u}^*) = d_\beta(\mathbf{u}^*) = f(\mathbf{x}^*)$. Further, $d_\beta(\mathbf{u})$ has a better smoothness property than $d(\mathbf{u})$.

Lemma 3.4 *Let $\mathcal{D}(\mathbf{u})$ denote the optimal solution set of* $\min_{\mathbf{x}} L_\beta(\mathbf{x}, \mathbf{u})$. *Then* $\mathbf{A}\mathbf{x}$ *is invariant over* $\mathcal{D}(\mathbf{u})$. *Moreover,* $d_\beta(\mathbf{u})$ *is differentiable and* $\nabla d_\beta(\mathbf{u}) = \mathbf{A}\mathbf{x}(\mathbf{u}) - \mathbf{b}$, *where* $\mathbf{x}(\mathbf{u}) \in \mathcal{D}(\mathbf{u})$ *is any minimizer of* $\min_{\mathbf{x}} L_\beta(\mathbf{x}, \mathbf{u})$. *We also have that* $d_\beta(\mathbf{u})$ *is* $\frac{1}{\beta}$-*smooth, i.e.,*

$$\|\nabla d_\beta(\mathbf{u}) - \nabla d_\beta(\mathbf{u}')\| \le \frac{1}{\beta}\|\mathbf{u} - \mathbf{u}'\|.$$

Proof Suppose that there exist $\mathbf{x}$ and $\mathbf{x}' \in \mathcal{D}(\mathbf{u})$ with $\mathbf{A}\mathbf{x} \ne \mathbf{A}\mathbf{x}'$. Then we have $d_\beta(\mathbf{u}) = L_\beta(\mathbf{x}, \mathbf{u}) = L_\beta(\mathbf{x}', \mathbf{u})$. Due to the convexity of $L_\beta(\mathbf{x}, \mathbf{u})$ with respect to $\mathbf{x}$, $\mathcal{D}(\mathbf{u})$ must be convex, implying $\overline{\mathbf{x}} = (\mathbf{x} + \mathbf{x}')/2 \in \mathcal{D}(\mathbf{u})$. By the convexity of f and strict convexity (Definition A.9) of $\|\cdot\|^2$, we have

$$\begin{aligned}
d_\beta(\mathbf{u}) &= \frac{1}{2}L_\beta(\mathbf{x}, \mathbf{u}) + \frac{1}{2}L_\beta(\mathbf{x}', \mathbf{u}) > f(\overline{\mathbf{x}}) + \langle \mathbf{A}\overline{\mathbf{x}} - \mathbf{b}, \mathbf{u} \rangle + \frac{\beta}{2}\|\mathbf{A}\overline{\mathbf{x}} - \mathbf{b}\|^2 \\
&= L_\beta(\overline{\mathbf{x}}, \mathbf{u}).
\end{aligned}$$

This contradicts the definition $d_\beta(\mathbf{u}) = \min_{\mathbf{x}} L_\beta(\mathbf{x}, \mathbf{u})$. Thus $\mathbf{A}\mathbf{x}$ is invariant over $\mathcal{D}(\mathbf{u})$. So $\partial d_\beta(\mathbf{u})$ is a singleton. By Danskin's theorem (Theorem A.1), we know $d_\beta(\mathbf{u})$ is differentiable and $\nabla d(\mathbf{u}) = \mathbf{A}\mathbf{x}(\mathbf{u}) - \mathbf{b}$, where $\mathbf{x}(\mathbf{u}) \in \mathcal{D}(\mathbf{u})$ is any minimizer of $\min_{\mathbf{x}} L_\beta(\mathbf{x}, \mathbf{u})$.

Let $\mathbf{x} = \operatorname{argmin}_{\mathbf{x}} L_\beta(\mathbf{x}, \mathbf{u})$ and $\mathbf{x}' = \operatorname{argmin}_{\mathbf{x}} L_\beta(\mathbf{x}, \mathbf{u}')$. Then we have

$$\mathbf{0} \in \partial f(\mathbf{x}) + \mathbf{A}^T \mathbf{u} + \beta \mathbf{A}^T(\mathbf{A}\mathbf{x} - \mathbf{b}),$$

$$\mathbf{0} \in \partial f(\mathbf{x}') + \mathbf{A}^T \mathbf{u}' + \beta \mathbf{A}^T(\mathbf{A}\mathbf{x}' - \mathbf{b}).$$

From the monotonicity of ∂f (Proposition A.11), we have

$$\left\langle -(\mathbf{A}^T \mathbf{u} + \beta \mathbf{A}^T(\mathbf{A}\mathbf{x} - \mathbf{b})) + (\mathbf{A}^T \mathbf{u}' + \beta \mathbf{A}^T(\mathbf{A}\mathbf{x}' - \mathbf{b})), \mathbf{x} - \mathbf{x}' \right\rangle \ge 0$$

$$\Rightarrow \langle \mathbf{u} - \mathbf{u}', \mathbf{A}\mathbf{x} - \mathbf{A}\mathbf{x}' \rangle + \beta \|\mathbf{A}\mathbf{x} - \mathbf{A}\mathbf{x}'\|^2 \le 0$$

$$\Rightarrow \beta \|\mathbf{A}\mathbf{x} - \mathbf{A}\mathbf{x}'\| \le \|\mathbf{u} - \mathbf{u}'\|.$$

So we have

$$\|\nabla d_\beta(\mathbf{u}) - \nabla d_\beta(\mathbf{u}')\| = \|\mathbf{A}\mathbf{x} - \mathbf{A}\mathbf{x}'\| \le \frac{1}{\beta}\|\mathbf{u} - \mathbf{u}'\|.$$

$\square$

3.2 Accelerated Penalty Method

The penalty method poses the constraint in problem (3.2) as a large penalty [17, 25, 28, 29, 33] and minimizes the following problem instead:

$$\min_{\mathbf{x}} f(\mathbf{x}) + \frac{\beta}{2}\|\mathbf{A}\mathbf{x} - \mathbf{b}\|^2. \tag{3.4}$$

Generally speaking, if β is of the order $\frac{1}{\epsilon}$ and

$$f(\mathbf{x}) + \frac{\beta}{2}\|\mathbf{A}\mathbf{x} - \mathbf{b}\|^2 \leq \min_{\mathbf{x}}\left(f(\mathbf{x}) + \frac{\beta}{2}\|\mathbf{A}\mathbf{x} - \mathbf{b}\|^2\right) + \epsilon,$$

then we can have $|f(\mathbf{x}) - f(\mathbf{x}^*)| \leq \epsilon$ and $\|\mathbf{A}\mathbf{x} - \mathbf{b}\| \leq \epsilon$ [17]. In fact, letting $(\mathbf{x}^*, \boldsymbol{\lambda}^*)$ be a KKT point of problem (3.2), we have

$$f(\mathbf{x}) + \frac{\beta}{2}\|\mathbf{A}\mathbf{x} - \mathbf{b}\|^2 \leq \min_{\mathbf{x}}\left(f(\mathbf{x}) + \frac{\beta}{2}\|\mathbf{A}\mathbf{x} - \mathbf{b}\|^2\right) + \epsilon \leq f(\mathbf{x}^*) + \epsilon,$$

$$f(\mathbf{x}^*) = f(\mathbf{x}^*) + \langle \boldsymbol{\lambda}^*, \mathbf{A}\mathbf{x}^* - \mathbf{b}\rangle \leq f(\mathbf{x}) + \langle \boldsymbol{\lambda}^*, \mathbf{A}\mathbf{x} - \mathbf{b}\rangle,$$

which leads to

$$f(\mathbf{x}) - f(\mathbf{x}^*) \leq \epsilon \text{ and } -\|\boldsymbol{\lambda}^*\|\|\mathbf{A}\mathbf{x} - \mathbf{b}\| \leq -\langle \boldsymbol{\lambda}^*, \mathbf{A}\mathbf{x} - \mathbf{b}\rangle \leq f(\mathbf{x}) - f(\mathbf{x}^*).$$

So

$$\frac{\beta}{2}\|\mathbf{A}\mathbf{x} - \mathbf{b}\|^2 - \|\boldsymbol{\lambda}^*\|\|\mathbf{A}\mathbf{x} - \mathbf{b}\| \leq \epsilon,$$

which further leads to

$$\|\mathbf{A}\mathbf{x} - \mathbf{b}\| \leq \frac{2\|\boldsymbol{\lambda}^*\|}{\beta} + \sqrt{\frac{2\epsilon}{\beta}} \leq \epsilon \text{ and } -\|\boldsymbol{\lambda}^*\|\epsilon \leq f(\mathbf{x}) - f(\mathbf{x}^*)$$

by letting $\beta = O\left(\frac{1}{\epsilon}\right)$.

Thus we can use the accelerated gradient methods described in the previous section to minimize the penalized problem (3.4). However, directly minimizing problem (3.4) with a large penalty makes the algorithm slow due to the ill-conditioning of $\frac{\beta}{2}\|\mathbf{A}\mathbf{x} - \mathbf{b}\|^2$ with a large β. To solve this problem, the continuation technique is often used [17], namely solving a sequence of problems (3.4) with increasing penalty parameters.

In this section, we follow [20] and introduce a little different strategy from the continuation technique. We increase the penalty parameter β at each iteration. In other words, we solve a sequence of subproblems in the original continuation technique with only *one* iteration and then immediately increase the penalty parameter. We adopt the acceleration technique discussed in the previous section and describe the algorithm in Algorithm 3.1, where θ_k, α_k, and η_k will be specified in Theorems 3.2 and 3.3.

Algorithm 3.1 Accelerated penalty method

Initialize $\mathbf{x}_0 = \mathbf{x}_{-1}$.
for $k = 0, 1, 2, 3, \cdots$ **do**
$\quad \mathbf{y}_k = \mathbf{x}_k + \frac{(\eta_k \theta_k - \mu)(1 - \theta_{k-1})}{(\eta_k - \mu)\theta_{k-1}}(\mathbf{x}_k - \mathbf{x}_{k-1}),$
$\quad \mathbf{x}_{k+1} = \mathbf{y}_k - \frac{1}{\eta_k}\left[\nabla f(\mathbf{y}_k) + \frac{\beta}{\alpha_k}\mathbf{A}^T(\mathbf{A}\mathbf{y}_k - \mathbf{b})\right].$
end for

We first give a general result in Theorem 3.1, which considers both the generally convex case and the strongly convex case. The following lemma gives some basic relations that will be used in the proof of Theorem 3.1.

Lemma 3.5 *Assume that sequences* $\{\alpha_k\}_{k=0}^{\infty}$ *and* $\{\theta_k\}_{k=0}^{\infty}$ *satisfy* $\frac{1-\theta_k}{\alpha_k} = \frac{1}{\alpha_{k-1}}$. *Define*

$$\overline{\boldsymbol{\lambda}}_{k+1} = \frac{\beta}{\alpha_k}(\mathbf{A}\mathbf{y}_k - \mathbf{b}),$$

$$\boldsymbol{\lambda}_{k+1} = \frac{\beta}{\alpha_k}(\mathbf{A}\mathbf{x}_{k+1} - \mathbf{b}),$$

$$\mathbf{w}_{k+1} = \frac{1}{\theta_k}\mathbf{x}_{k+1} - \frac{1-\theta_k}{\theta_k}\mathbf{x}_k. \tag{3.5}$$

Then we have

$$\boldsymbol{\lambda}_{k+1} - \boldsymbol{\lambda}_k = \frac{\beta}{\alpha_k}\left[\mathbf{A}\mathbf{x}_{k+1} - (1-\theta_k)\mathbf{A}\mathbf{x}_k - \theta_k\mathbf{b}\right], \tag{3.6}$$

$$\frac{\alpha_k}{2\beta}\|\boldsymbol{\lambda}_{k+1} - \overline{\boldsymbol{\lambda}}_{k+1}\|^2 \leq \frac{\beta\|\mathbf{A}^T\mathbf{A}\|_2}{2\alpha_k}\|\mathbf{x}_{k+1} - \mathbf{y}_k\|^2, \tag{3.7}$$

$$\mathbf{w}_k = \frac{\eta_k - \mu}{\eta_k\theta_k - \mu}\mathbf{y}_k - \frac{\eta_k(1-\theta_k)}{\eta_k\theta_k - \mu}\mathbf{x}_k. \tag{3.8}$$

Proof For the first relation, we have

$$\boldsymbol{\lambda}_{k+1} - \boldsymbol{\lambda}_k = \frac{\beta}{\alpha_k}(\mathbf{A}\mathbf{x}_{k+1} - \mathbf{b}) - \frac{\beta}{\alpha_{k-1}}(\mathbf{A}\mathbf{x}_k - \mathbf{b})$$

$$= \frac{\beta}{\alpha_k}(\mathbf{A}\mathbf{x}_{k+1} - \mathbf{b}) - \frac{\beta(1 - \theta_k)}{\alpha_k}(\mathbf{A}\mathbf{x}_k - \mathbf{b})$$

$$= \frac{\beta}{\alpha_k}[\mathbf{A}\mathbf{x}_{k+1} - (1 - \theta_k)\mathbf{A}\mathbf{x}_k - \theta_k\mathbf{b}].$$

For the second relation, we have

$$\frac{\alpha_k}{2\beta}\|\boldsymbol{\lambda}_{k+1} - \overline{\boldsymbol{\lambda}}_{k+1}\|^2 = \frac{\alpha_k}{2\beta}\left\|\frac{\beta}{\alpha_k}\mathbf{A}(\mathbf{x}_{k+1} - \mathbf{y}_k)\right\|^2 \le \frac{\beta\|\mathbf{A}^T\mathbf{A}\|_2}{2\alpha_k}\|\mathbf{x}_{k+1} - \mathbf{y}_k\|^2.$$

The third relation can be obtained from the definition of $\mathbf{y}_k$. $\square$

Now we give the main results in the following theorem.

Theorem 3.1 *Assume that $f(\mathbf{x})$ is L-smooth and μ-strongly convex. Let $\{\alpha_k\}_{k=0}^{\infty}$ be a decreasing sequence with $\frac{1}{\alpha_{-1}} = 0$ and $\alpha_k \ge 0$. Define θ_k and η_k as those satisfying $\frac{1-\theta_k}{\alpha_k} = \frac{1}{\alpha_{k-1}}$ and $\eta_k = L + \frac{\beta\|\mathbf{A}^T\mathbf{A}\|_2}{\alpha_k}$. Assume that the following two inequalities hold:*

$$\frac{\eta_{k-1}\theta_{k-1}^2}{2\alpha_{k-1}} \ge \frac{\eta_k\theta_k^2 - \mu\theta_k}{2\alpha_k}, \qquad \theta_k \ge \frac{\mu}{\eta_k}. \tag{3.9}$$

Then for Algorithm 3.1, we have

$$|f(\mathbf{x}_{K+1}) - f(\mathbf{x}^*)| \le O(\alpha_K), \quad \|\mathbf{A}\mathbf{x}_{K+1} - \mathbf{b}\| \le O(\alpha_K).$$

Proof From the second step, we have

$$\mathbf{0} = \nabla f(\mathbf{y}_k) + \mathbf{A}^T\overline{\boldsymbol{\lambda}}_{k+1} + \eta_k(\mathbf{x}_{k+1} - \mathbf{y}_k).$$

From the L-smoothness and the μ-strong convexity of f, we have

$$f(\mathbf{x}_{k+1}) \le f(\mathbf{y}_k) + \langle \nabla f(\mathbf{y}_k), \mathbf{x}_{k+1} - \mathbf{y}_k \rangle + \frac{L}{2}\|\mathbf{x}_{k+1} - \mathbf{y}_k\|^2$$

$$\le f(\mathbf{x}) - \frac{\mu}{2}\|\mathbf{x} - \mathbf{y}_k\|^2 + \langle \nabla f(\mathbf{y}_k), \mathbf{x}_{k+1} - \mathbf{x} \rangle + \frac{L}{2}\|\mathbf{x}_{k+1} - \mathbf{y}_k\|^2$$

$$= f(\mathbf{x}) - \frac{\mu}{2}\|\mathbf{x} - \mathbf{y}_k\|^2 + \left\langle \mathbf{A}^T\overline{\boldsymbol{\lambda}}_{k+1}, \mathbf{x} - \mathbf{x}_{k+1} \right\rangle$$

$$+ \eta_k\langle \mathbf{x}_{k+1} - \mathbf{y}_k, \mathbf{x} - \mathbf{x}_{k+1} \rangle + \frac{L}{2}\|\mathbf{x}_{k+1} - \mathbf{y}_k\|^2$$

$$= f(\mathbf{x}) + \left\langle \mathbf{A}^T\overline{\boldsymbol{\lambda}}_{k+1}, \mathbf{x} - \mathbf{x}_{k+1} \right\rangle + \eta_k \left\langle \mathbf{x}_{k+1} - \mathbf{y}_k, \mathbf{x} - \mathbf{y}_k \right\rangle$$

$$- \frac{\mu}{2}\|\mathbf{x} - \mathbf{y}_k\|^2 - \left(\frac{L}{2} + \frac{\beta\|\mathbf{A}^T\mathbf{A}\|_2}{\alpha_k} \right) \|\mathbf{x}_{k+1} - \mathbf{y}_k\|^2.$$

Letting $\mathbf{x} = \mathbf{x}_k$ and $\mathbf{x} = \mathbf{x}^*$, respectively, we obtain two inequalities. Multiplying the first inequality by $1 - \theta_k$ and the second by θ_k and adding them, we have

$$f(\mathbf{x}_{k+1}) - (1 - \theta_k)f(\mathbf{x}_k) - \theta_k f(\mathbf{x}^*)$$
$$\leq \left\langle \overline{\boldsymbol{\lambda}}_{k+1}, \theta_k\mathbf{A}\mathbf{x}^* + (1 - \theta_k)\mathbf{A}\mathbf{x}_k - \mathbf{A}\mathbf{x}_{k+1} \right\rangle$$
$$+ \eta_k \left\langle \mathbf{x}_{k+1} - \mathbf{y}_k, \theta_k\mathbf{x}^* + (1 - \theta_k)\mathbf{x}_k - \mathbf{y}_k \right\rangle$$
$$- \frac{\mu\theta_k}{2}\|\mathbf{x}^* - \mathbf{y}_k\|^2 - \left(\frac{L}{2} + \frac{\beta\|\mathbf{A}^T\mathbf{A}\|_2}{\alpha_k} \right) \|\mathbf{x}_{k+1} - \mathbf{y}_k\|^2.$$

Adding $\left\langle \boldsymbol{\lambda}^*, \mathbf{A}\mathbf{x}_{k+1} - (1 - \theta_k)\mathbf{A}\mathbf{x}_k - \theta_k\mathbf{A}\mathbf{x}^* \right\rangle$ to both sides and using $\mathbf{A}\mathbf{x}^* = \mathbf{b}$, we have

$$f(\mathbf{x}_{k+1}) - f(\mathbf{x}^*) + \left\langle \boldsymbol{\lambda}^*, \mathbf{A}\mathbf{x}_{k+1} - \mathbf{b} \right\rangle$$
$$- (1 - \theta_k)\left(f(\mathbf{x}_k) - f(\mathbf{x}^*) + \left\langle \boldsymbol{\lambda}^*, \mathbf{A}\mathbf{x}_k - \mathbf{b} \right\rangle \right)$$
$$\leq \left\langle \overline{\boldsymbol{\lambda}}_{k+1} - \boldsymbol{\lambda}^*, \theta_k\mathbf{A}\mathbf{x}^* + (1 - \theta_k)\mathbf{A}\mathbf{x}_k - \mathbf{A}\mathbf{x}_{k+1} \right\rangle$$
$$+ \eta_k \left\langle \mathbf{x}_{k+1} - \mathbf{y}_k, \theta_k\mathbf{x}^* + (1 - \theta_k)\mathbf{x}_k - \mathbf{y}_k \right\rangle$$
$$- \frac{\mu\theta_k}{2}\|\mathbf{x}^* - \mathbf{y}_k\|^2 - \left(\frac{L}{2} + \frac{\beta\|\mathbf{A}^T\mathbf{A}\|_2}{\alpha_k} \right) \|\mathbf{x}_{k+1} - \mathbf{y}_k\|^2$$
$$\overset{a}{=} \frac{\alpha_k}{\beta} \left\langle \overline{\boldsymbol{\lambda}}_{k+1} - \boldsymbol{\lambda}^*, \boldsymbol{\lambda}_k - \boldsymbol{\lambda}_{k+1} \right\rangle$$
$$+ \frac{\eta_k}{2} \left(\|\theta_k\mathbf{x}^* + (1 - \theta_k)\mathbf{x}_k - \mathbf{y}_k\|^2 - \|\theta_k\mathbf{x}^* + (1 - \theta_k)\mathbf{x}_k - \mathbf{x}_{k+1}\|^2 \right)$$
$$- \frac{\mu\theta_k}{2}\|\mathbf{x}^* - \mathbf{y}_k\|^2 - \frac{\beta\|\mathbf{A}^T\mathbf{A}\|_2}{2\alpha_k}\|\mathbf{y}_k - \mathbf{x}_{k+1}\|^2$$
$$\overset{b}{=} \frac{\alpha_k}{2\beta} \left(\|\boldsymbol{\lambda}_k - \boldsymbol{\lambda}^*\|^2 - \|\boldsymbol{\lambda}_{k+1} - \boldsymbol{\lambda}^*\|^2 - \|\overline{\boldsymbol{\lambda}}_{k+1} - \boldsymbol{\lambda}_k\|^2 + \|\boldsymbol{\lambda}_{k+1} - \overline{\boldsymbol{\lambda}}_{k+1}\|^2 \right)$$
$$+ \frac{\eta_k\theta_k^2}{2} \left(\left\| \mathbf{x}^* + \frac{1 - \theta_k}{\theta_k}\mathbf{x}_k - \frac{1}{\theta_k}\mathbf{y}_k \right\|^2 - \left\| \mathbf{x}^* + \frac{1 - \theta_k}{\theta_k}\mathbf{x}_k - \frac{1}{\theta_k}\mathbf{x}_{k+1} \right\|^2 \right)$$
$$- \frac{\mu\theta_k}{2}\|\mathbf{x}^* - \mathbf{y}_k\|^2 - \frac{\beta\|\mathbf{A}^T\mathbf{A}\|_2}{2\alpha_k}\|\mathbf{y}_k - \mathbf{x}_{k+1}\|^2$$

$$\leq \frac{\alpha_k}{2\beta} \left(\|\boldsymbol{\lambda}_k - \boldsymbol{\lambda}^*\|^2 - \|\boldsymbol{\lambda}_{k+1} - \boldsymbol{\lambda}^*\|^2 - \|\overline{\boldsymbol{\lambda}}_{k+1} - \boldsymbol{\lambda}_k\|^2 \right) - \frac{\mu\theta_k}{2} \|\mathbf{x}^* - \mathbf{y}_k\|^2$$

$$+ \frac{\eta_k \theta_k^2}{2} \left(\left\| \mathbf{x}^* + \frac{1-\theta_k}{\theta_k} \mathbf{x}_k - \frac{1}{\theta_k} \mathbf{y}_k \right\|^2 - \|\mathbf{w}_{k+1} - \mathbf{x}^*\|^2 \right),$$

where we use (3.6) and (A.1) in $\overset{a}{=}$, (A.3) in $\overset{b}{=}$, and (3.7) and (3.5) in $\overset{c}{\leq}$. Consider

$$\frac{\eta_k \theta_k^2}{2} \left\| \mathbf{x}^* + \frac{1-\theta_k}{\theta_k} \mathbf{x}_k - \frac{1}{\theta_k} \mathbf{y}_k \right\|^2$$

$$= \frac{\eta_k \theta_k^2}{2} \left\| \frac{\mu}{\eta_k \theta_k} (\mathbf{x}^* - \mathbf{y}_k) + \left(1 - \frac{\mu}{\eta_k \theta_k} \right) \right.$$

$$\left. \times \left(\mathbf{x}^* + \frac{\eta_k(1-\theta_k)}{\eta_k \theta_k - \mu} \mathbf{x}_k - \frac{\eta_k - \mu}{\eta_k \theta_k - \mu} \mathbf{y}_k \right) \right\|^2$$

$$\overset{a}{\leq} \frac{\mu\theta_k}{2} \|\mathbf{x}^* - \mathbf{y}_k\|^2 + \frac{\theta_k(\eta_k \theta_k - \mu)}{2} \left\| \mathbf{x}^* + \frac{\eta_k(1-\theta_k)}{\eta_k \theta_k - \mu} \mathbf{x}_k - \frac{\eta_k - \mu}{\eta_k \theta_k - \mu} \mathbf{y}_k \right\|^2$$

$$\overset{b}{=} \frac{\mu\theta_k}{2} \|\mathbf{x}^* - \mathbf{y}_k\|^2 + \frac{\theta_k(\eta_k \theta_k - \mu)}{2} \|\mathbf{w}_k - \mathbf{x}^*\|^2,$$

where $\overset{a}{\leq}$ uses the convexity of $\|\cdot\|^2$ and $\overset{b}{=}$ uses (3.8). Thus we have

$$f(\mathbf{x}_{k+1}) - f(\mathbf{x}^*) + \langle \boldsymbol{\lambda}^*, \mathbf{A}\mathbf{x}_{k+1} - \mathbf{b} \rangle$$

$$- (1 - \theta_k) \left(f(\mathbf{x}_k) - f(\mathbf{x}^*) + \langle \boldsymbol{\lambda}^*, \mathbf{A}\mathbf{x}_k - \mathbf{b} \rangle \right)$$

$$\leq \frac{\alpha_k}{2\beta} \left(\|\boldsymbol{\lambda}_k - \boldsymbol{\lambda}^*\|^2 - \|\boldsymbol{\lambda}_{k+1} - \boldsymbol{\lambda}^*\|^2 \right) + \frac{\theta_k(\eta_k \theta_k - \mu)}{2} \|\mathbf{w}_k - \mathbf{x}^*\|^2$$

$$- \frac{\eta_k \theta_k^2}{2} \|\mathbf{w}_{k+1} - \mathbf{x}^*\|^2.$$

Dividing both sides by α_k and using $\frac{1-\theta_k}{\alpha_k} = \frac{1}{\alpha_{k-1}}$ and (3.9), we have

$$\frac{1}{\alpha_k} \left(f(\mathbf{x}_{k+1}) - f(\mathbf{x}^*) + \langle \boldsymbol{\lambda}^*, \mathbf{A}\mathbf{x}_{k+1} - \mathbf{b} \rangle + \frac{\eta_k \theta_k^2}{2} \|\mathbf{w}_{k+1} - \mathbf{x}^*\|^2 \right)$$

$$+ \frac{1}{2\beta} \|\boldsymbol{\lambda}_{k+1} - \boldsymbol{\lambda}^*\|^2$$

$$\leq \frac{1}{\alpha_{k-1}} \left(f(\mathbf{x}_k) - f(\mathbf{x}^*) + \langle \boldsymbol{\lambda}^*, \mathbf{A}\mathbf{x}_k - \mathbf{b} \rangle + \frac{\eta_{k-1}\theta_{k-1}^2}{2} \|\mathbf{w}_k - \mathbf{x}^*\|^2 \right)$$

$$+ \frac{1}{2\beta} \|\boldsymbol{\lambda}_k - \boldsymbol{\lambda}^*\|^2.$$

So we have

$$
\frac{1}{\alpha_K} \left(f(\mathbf{x}_{K+1}) - f(\mathbf{x}^*) + \langle \boldsymbol{\lambda}^*, \mathbf{A}\mathbf{x}_{K+1} - \mathbf{b} \rangle + \frac{\eta_K \theta_K^2}{2} \| \mathbf{w}_{K+1} - \mathbf{x}^* \|^2 \right)
$$

$$
+ \frac{1}{2\beta} \| \boldsymbol{\lambda}_{K+1} - \boldsymbol{\lambda}^* \|^2 \leq \frac{1}{2\beta} \| \boldsymbol{\lambda}_0 - \boldsymbol{\lambda}^* \|^2,
$$

where we use $\frac{1}{\alpha_{-1}} = 0$. From Lemma 3.2, we have

$$
f(\mathbf{x}_{K+1}) - f(\mathbf{x}^*) + \langle \boldsymbol{\lambda}^*, \mathbf{A}\mathbf{x}_{K+1} - \mathbf{b} \rangle \leq \frac{\alpha_K \| \boldsymbol{\lambda}_0 - \boldsymbol{\lambda}^* \|^2}{2\beta},
$$

$$
\| \boldsymbol{\lambda}_{K+1} - \boldsymbol{\lambda}^* \| \leq \| \boldsymbol{\lambda}_0 - \boldsymbol{\lambda}^* \|.
$$

Since $\left\| \frac{\beta}{\alpha_K} (\mathbf{A}\mathbf{x}_{K+1} - \mathbf{b}) \right\| = \| \boldsymbol{\lambda}_{K+1} \| \leq \| \boldsymbol{\lambda}_{K+1} - \boldsymbol{\lambda}^* \| + \| \boldsymbol{\lambda}^* \| \leq \| \boldsymbol{\lambda}_0 - \boldsymbol{\lambda}^* \| + \| \boldsymbol{\lambda}^* \|$, we have

$$
\| \mathbf{A}\mathbf{x}_{K+1} - \mathbf{b} \| \leq \frac{\| \boldsymbol{\lambda}_0 - \boldsymbol{\lambda}^* \| + \| \boldsymbol{\lambda}^* \|}{\beta} \alpha_K.
$$

From Lemma 3.3, we can have the conclusion.   □

3.2.1 Generally Convex Objectives

We can specialize the value of α_k for the generally convex case and the strongly convex case and establish their convergence rates. We first consider the generally convex case and prove the $O\left(\frac{1}{K}\right)$ convergence rate.

Theorem 3.2 *Assume that $f(\mathbf{x})$ is L-smooth and convex. Let $\alpha_k = \theta_k = \frac{1}{k+1}$, then assumption (3.9) holds and we have*

$$
| f(\mathbf{x}_{K+1}) - f(\mathbf{x}^*) | \leq O(1/K) \text{ and } \| \mathbf{A}\mathbf{x}_{K+1} - \mathbf{b} \| \leq O(1/K).
$$

Proof If $\mu = 0$ and $\alpha_k = \theta_k$, then (3.9) reduces to $\eta_k \theta_k \leq \eta_{k-1} \theta_{k-1}$ and $\theta_k \geq 0$, which is true due to $0 \leq \theta_k < \theta_{k-1}$ and the definition of η_k. From $\frac{1 - \theta_k}{\alpha_k} = \frac{1}{\alpha_{k-1}}$ and $\frac{1}{\alpha_{-1}} = 0$ we have $\alpha_k = \frac{1}{k+1}$. So we have $\| \mathbf{A}\mathbf{x}_{K+1} - \mathbf{b} \| \leq O(1/K)$ and $| f(\mathbf{x}_{K+1}) - f(\mathbf{x}^*) | \leq O(1/K)$.   □

3.2.2 Strongly Convex Objectives

Then we consider the strongly convex case and give a faster $O\left(\frac{1}{K^2}\right)$ convergence rate.

Theorem 3.3 *Assume that $f(\mathbf{x})$ is L-smooth and μ-strongly convex. Let $\frac{1-\theta_k}{\theta_k^2} = \frac{1}{\theta_{k-1}^2}$, $\alpha_k = \theta_k^2$, $\theta_0 = 1$, and $\frac{\mu^2}{4L\|\mathbf{A}^T\mathbf{A}\|_2} \leq \beta \leq \frac{\mu}{\|\mathbf{A}^T\mathbf{A}\|_2}$, then assumption (3.9) holds and we have*

$$|f(\mathbf{x}_{K+1}) - f(\mathbf{x}^*)| \leq O\left(1/K^2\right) \text{ and } \|\mathbf{A}\mathbf{x}_{K+1} - \mathbf{b}\| \leq O\left(1/K^2\right).$$

Proof If $\mu > 0$ and $\alpha_k = \theta_k^2$, then (3.9) reduces to $\eta_k - \mu/\theta_k \leq \eta_{k-1}$ and $L\theta_k + \frac{\beta\|\mathbf{A}^T\mathbf{A}\|_2}{\theta_k} \geq \mu$. Consider $\eta_k - \mu/\theta_k - \eta_{k-1} = L + \frac{\beta\|\mathbf{A}^T\mathbf{A}\|_2}{\alpha_k} - \mu/\theta_k - \left(L + \frac{\beta\|\mathbf{A}^T\mathbf{A}\|_2}{\alpha_{k-1}}\right) = \beta\|\mathbf{A}^T\mathbf{A}\|_2\left(\frac{1}{\alpha_k} - \frac{1}{\alpha_{k-1}}\right) - \mu/\theta_k = \frac{\beta\|\mathbf{A}^T\mathbf{A}\|_2 - \mu}{\theta_k}$, where we use $\frac{1}{\alpha_k} - \frac{1}{\alpha_{k-1}} = \frac{\theta_k}{\alpha_k} = \frac{1}{\theta_k}$. So if $\beta \leq \frac{\mu}{\|\mathbf{A}^T\mathbf{A}\|_2}$, then $\eta_k - \mu/\theta_k \leq \eta_{k-1}$.

Then we consider $L\theta_k + \frac{\beta\|\mathbf{A}^T\mathbf{A}\|_2}{\theta_k} \geq \mu$. It holds if $\beta \geq \frac{\theta_k\mu - L\theta_k^2}{\|\mathbf{A}^T\mathbf{A}\|_2}$. Since $\theta\mu - L\theta^2 \leq \frac{\mu^2}{4L}, \forall\theta$, we only need $\beta \geq \frac{\mu^2}{4L\|\mathbf{A}^T\mathbf{A}\|_2}$. So we finally get the condition $\frac{\mu^2}{4L\|\mathbf{A}^T\mathbf{A}\|_2} \leq \beta \leq \frac{\mu}{\|\mathbf{A}^T\mathbf{A}\|_2}$. Since $\theta_0 = 1$ and $\frac{1-\theta_k}{\theta_k^2} = \frac{1}{\theta_{k-1}^2}$, from Lemma 2.3 we can easily have $\theta_k \leq \frac{2}{k+2}$ and $\alpha_k \leq \frac{4}{(k+2)^2}$. Thus we have $\|\mathbf{A}\mathbf{x}_{K+1} - \mathbf{b}\| \leq O\left(1/K^2\right)$ and $|f(\mathbf{x}_{K+1}) - f(\mathbf{x}^*)| \leq O\left(1/K^2\right)$. $\qquad\square$

3.3 Accelerated Lagrange Multiplier Method

In this section, we consider the general problem (3.1) and use the methods described in the previous section to maximize its Lagrange dual (A.13). When $f(\mathbf{x})$ is μ-strongly convex and each g_i has bounded subgradient, due to Danskin's theorem [1] (Theorem A.1) we know that $d(\mathbf{u}, \mathbf{v})$ is convex, differentiable (see also Lemma 3.4) and

$$\nabla d(\mathbf{u}, \mathbf{v}) = \left[\left(\mathbf{A}\mathbf{x}^*(\mathbf{u}, \mathbf{v}) - \mathbf{b}\right)^T, g_1(\mathbf{x}^*(\mathbf{u}, \mathbf{v})), \cdots, g_p(\mathbf{x}^*(\mathbf{u}, \mathbf{v}))\right]^T, \quad (3.10)$$

where $\mathbf{x}^*(\mathbf{u}, \mathbf{v}) = \mathrm{argmin}_{\mathbf{x}} L(\mathbf{x}, \mathbf{u}, \mathbf{v})$. Then

$$d(\mathbf{u}, \mathbf{v}) = f(\mathbf{x}^*(\mathbf{u}, \mathbf{v})) + \langle\mathbf{u}, \mathbf{A}\mathbf{x}^*(\mathbf{u}, \mathbf{v}) - \mathbf{b}\rangle + \sum_{i=1}^{p} v_i g_i(\mathbf{x}^*(\mathbf{u}, \mathbf{v}))$$

$$= f(\mathbf{x}^*(\mathbf{u}, \mathbf{v})) + \langle\boldsymbol{\lambda}, \nabla d(\mathbf{u}, \mathbf{v})\rangle, \quad (3.11)$$

Moreover, from Proposition 3.3 in [23] we have the Lipschitz smoothness condition of $d(\mathbf{u}, \mathbf{v})$:

$$\|\nabla d(\mathbf{u}, \mathbf{v}) - \nabla d(\mathbf{u}', \mathbf{v}')\| \leq L \|(\mathbf{u}, \mathbf{v}) - (\mathbf{u}', \mathbf{v}')\|, \forall (\mathbf{u}, \mathbf{v}), (\mathbf{u}', \mathbf{v}') \in \mathcal{D},$$

where

$$L = \frac{\sqrt{p+1} \max\{\|\mathbf{A}\|_2, \max_i L_{g_i}\}}{\mu} \sqrt{\|\mathbf{A}\|_2^2 + \sum_{i=1}^{m} L_{g_i}^2},$$

in which L_{g_i} is an upper bound of $\|\partial g_i\|$. Thus we can use the gradient ascent or accelerated gradient ascent to maximize $d(\mathbf{u}, \mathbf{v})$. The Lagrange multiplier method solves problem (3.1) via maximizing its dual function using the gradient ascent. It consists of the following step at each iteration:

$$\boldsymbol{\lambda}_{k+1} = \text{Proj}_{\boldsymbol{\lambda} \in \mathcal{D}} (\boldsymbol{\lambda}_k + \beta \nabla d(\boldsymbol{\lambda}_k)),$$

where $\boldsymbol{\lambda} = (\mathbf{u}, \mathbf{v})$ and Proj is the projection operator. Instead of the gradient ascent, [23] used the accelerated gradient ascent to solve the dual problem, which leads to the accelerated Lagrange multiplier method. We describe it in Algorithm 3.2, where we use Algorithm 2.3 rather than Algorithm 2.2 in the dual space since we need to ensure $\boldsymbol{v}_k$, $\boldsymbol{\mu}_k$, and $\boldsymbol{\lambda}_k \in \mathcal{D}$. Algorithm 2.2 cannot guarantee this.

Algorithm 3.2 Accelerated Lagrange multiplier method

Initialize $\boldsymbol{v}_0 = \boldsymbol{\mu}_0 = \boldsymbol{\lambda}_0$.
for $k = 0, 1, 2, 3, \cdots$ **do**
$\quad \boldsymbol{v}_k = (1 - \theta_k)\boldsymbol{\lambda}_k + \theta_k \boldsymbol{\mu}_k,$
$\quad \boldsymbol{\mu}_{k+1} = \text{Proj}_{\boldsymbol{\mu} \in \mathcal{D}} \left(\boldsymbol{\mu}_k + \frac{1}{\theta_k L} \nabla d(\boldsymbol{v}_k) \right),$
$\quad \boldsymbol{\lambda}_{k+1} = (1 - \theta_k)\boldsymbol{\lambda}_k + \theta_k \boldsymbol{\mu}_{k+1}.$
end for

From Theorem 2.3, we have the following convergence rate theorem in the dual space.

Theorem 3.4 *Assume that $f(\mathbf{x})$ is μ-strongly convex and $\|\partial g_i\| \leq L_{g_i}$. Let $\theta_0 = 1$ and $\theta_k = \frac{\sqrt{\theta_{k-1}^4 + 4\theta_{k-1}^2} - \theta_{k-1}^2}{2}$. Then for Algorithm 3.2, with Algorithm 2.3 used in the dual space, we have*

$$-d(\boldsymbol{\lambda}_{K+1}) + d(\boldsymbol{\lambda}^*) \leq \frac{L}{(K+2)^2} \|\boldsymbol{\lambda}_0 - \boldsymbol{\lambda}^*\|^2.$$

3.3.1 Recovering the Primal Solution

It is not satisfactory to establish the iteration complexity only in the dual space. We should recover the primal solutions from the dual iterates and need to estimate how quickly the primal solutions converge. We describe the main results studied in [23], which answers this question.

Lemma 3.6 *For any* $\mathbf{x} = \text{argmin}_{\mathbf{x}} L(\mathbf{x}, \mathbf{u}, \mathbf{v})$*, we have*

$$\sqrt{\|\mathbf{Ax} - \mathbf{b}\|^2 + \sum_{i=1}^{p} (\max\{0, g_i(\mathbf{x})\})^2} \leq \sqrt{\frac{2}{L} \left[d(\boldsymbol{\lambda}^*) - d(\boldsymbol{\lambda}) \right]},$$

$$f(\mathbf{x}^*) - f(\mathbf{x}) \leq \|\boldsymbol{\lambda}^*\| \sqrt{\frac{2(p+m)}{L} \left[d(\boldsymbol{\lambda}^*) - d(\boldsymbol{\lambda}) \right]},$$

$$f(\mathbf{x}) - f(\mathbf{x}^*) \leq 2[d(\boldsymbol{\lambda}^*) - d(\boldsymbol{\lambda})] + \|\boldsymbol{\lambda}\|_{\infty} \sqrt{2L(p+m)[d(\boldsymbol{\lambda}^*) - d(\boldsymbol{\lambda})]}.$$

Proof Let $A(\boldsymbol{\lambda}) = \{i > m : \lambda_i + \frac{1}{L}\nabla_i d(\boldsymbol{\lambda}) < 0\}$ and $I(\boldsymbol{\lambda}) = \{1, 2, \cdots, m + p\}\backslash A(\boldsymbol{\lambda})$, which means that the projection onto $\mathcal{D}$ is active and inactive, respectively. Then

$$d(\boldsymbol{\lambda}^*) - d(\boldsymbol{\lambda}) \geq d\left(\text{Proj}_{\mathcal{D}} \left(\boldsymbol{\lambda} + \frac{1}{L}\nabla d(\boldsymbol{\lambda}) \right) \right) - d(\boldsymbol{\lambda})$$

$$\overset{a}{\geq} \left\langle \nabla d(\boldsymbol{\lambda}), \text{Proj}_{\mathcal{D}} \left(\boldsymbol{\lambda} + \frac{1}{L}\nabla d(\boldsymbol{\lambda}) \right) - \boldsymbol{\lambda} \right\rangle$$

$$\quad - \frac{L}{2} \left\| \text{Proj}_{\mathcal{D}} \left(\boldsymbol{\lambda} + \frac{1}{L}\nabla d(\boldsymbol{\lambda}) \right) - \boldsymbol{\lambda} \right\|^2$$

$$\overset{b}{=} \sum_{i \in A(\boldsymbol{\lambda})} \left(-\langle \nabla_i d(\boldsymbol{\lambda}), \lambda_i \rangle - \frac{L}{2}\lambda_i^2 \right) + \sum_{i \in I(\boldsymbol{\lambda})} \frac{1}{2L}\|\nabla_i d(\boldsymbol{\lambda})\|^2$$

$$\overset{c}{\geq} \sum_{i \in A(\boldsymbol{\lambda})} \left(-\frac{1}{2}\langle \nabla_i d(\boldsymbol{\lambda}), \lambda_i \rangle \right) + \sum_{i \in I(\boldsymbol{\lambda})} \frac{1}{2L}\|\nabla_i d(\boldsymbol{\lambda})\|^2 \tag{3.12a}$$

$$\overset{d}{\geq} \sum_{i \in I(\boldsymbol{\lambda})} \frac{1}{2L}\|\nabla_i d(\boldsymbol{\lambda})\|^2 \tag{3.12b}$$

$$\overset{e}{=} \frac{1}{2L}\|\mathbf{Ax} - \mathbf{b}\|^2 + \frac{1}{2L} \sum_{i>m, i \in I(\boldsymbol{\lambda})} (g_{i-m}(\mathbf{x}))^2$$

$$\overset{f}{\geq} \frac{1}{2L}\|\mathbf{Ax} - \mathbf{b}\|^2 + \frac{1}{2L} \sum_{i=1}^{p} (\max\{0, g_i(\mathbf{x})\})^2,$$

where we use the L-smoothness of the concave function $d(\lambda)$ in $\overset{a}{\geq}$, the fact that $\mathrm{Proj}_{\mathcal{D}}\left(\lambda_i + \frac{1}{L}\nabla_i d(\lambda)\right) = 0$ if $i \in A(\lambda)$ and $\mathrm{Proj}_{\mathcal{D}}\left(\lambda_i + \frac{1}{L}\nabla_i d(\lambda)\right) = \lambda_i + \frac{1}{L}\nabla_i d(\lambda)$ if $i \in I(\lambda)$ in $\overset{b}{=}$, the fact that $\lambda_i^2 \leq \left\langle \lambda_i, -\frac{1}{L}\nabla_i d(\lambda)\right\rangle$ for $i \in A(\lambda)$ due to $\left\langle \lambda_i, \lambda_i + \frac{1}{L}\nabla_i d(\lambda)\right\rangle \leq 0$ in $\overset{c}{\geq}$, $\langle \nabla_i d(\lambda), \lambda_i\rangle \leq -L\lambda_i^2 \leq 0$ for $i \in A(\lambda)$ in $\overset{d}{\geq}$, (3.10) in $\overset{e}{=}$, and $g_{i-m}(\mathbf{x}) \geq 0 \Rightarrow i \in I(\lambda)$ in $\overset{f}{\geq}$. Since

$$\sum_{i \in I(\lambda)} \|\nabla_i d(\lambda)\|^2 \geq \frac{1}{|I(\lambda)|}\left(\sum_{i \in I(\lambda)} |\nabla_i d(\lambda)|\right)^2$$

$$\geq \frac{1}{p+m}\left(\frac{1}{\|\lambda\|_\infty}\sum_{i \in I(\lambda)} \langle \nabla_i d(\lambda), \lambda_i\rangle\right)^2,$$

we have

$$\sum_{i \in I(\lambda)} -\langle \nabla_i d(\lambda), \lambda_i\rangle \leq \|\lambda\|_\infty\sqrt{2L(p+m)[d(\lambda^*) - d(\lambda)]}$$

from (3.12b) and

$$\sum_{i \in A(\lambda)} -\langle \nabla_i d(\lambda), \lambda_i\rangle \leq 2[d(\lambda^*) - d(\lambda)]$$

from (3.12a). So by (3.11) and the strong duality, we have

$$f(\mathbf{x}) - f(\mathbf{x}^*) = -\langle \lambda, \nabla d(\lambda)\rangle + d(\lambda) - d(\lambda^*)$$

$$\leq -\langle \lambda, \nabla d(\lambda)\rangle$$

$$\leq 2[d(\lambda^*) - d(\lambda)] + \|\lambda\|_\infty\sqrt{2L(p+m)[d(\lambda^*) - d(\lambda)]}.$$

On the other hand, we have

$$f(\mathbf{x}^*) \overset{a}{=} f(\mathbf{x}^*) + \langle \mathbf{u}^*, \mathbf{A}\mathbf{x}^* - \mathbf{b}\rangle + \sum_{i=1}^{p} v_i^* g_i(\mathbf{x}^*)$$

$$\overset{b}{\leq} f(\mathbf{x}) + \langle \mathbf{u}^*, \mathbf{A}\mathbf{x} - \mathbf{b}\rangle + \sum_{i=1}^{p} v_i^* g_i(\mathbf{x})$$

$$\overset{c}{\leq} f(\mathbf{x}) + \langle \mathbf{u}^*, \mathbf{Ax} - \mathbf{b} \rangle + \sum_{i=1}^{p} v_i^* \max\{0, g_i(\mathbf{x})\}$$

$$\leq f(\mathbf{x}) + \sqrt{p+m}\|\boldsymbol{\lambda}^*\| \sqrt{\|\mathbf{Ax} - \mathbf{b}\|^2 + \sum_{i=1}^{p} (\max\{0, g_i(\mathbf{x})\})^2},$$

where we use the KKT condition (Definition A.26) in $\overset{a}{=}$,

$$\mathbf{x}^* = \underset{\mathbf{x}}{\operatorname{argmin}} \left(f(\mathbf{x}) + \langle \mathbf{u}^*, \mathbf{Ax} - \mathbf{b} \rangle + \sum_{i=1}^{m} v_i^* g_i(\mathbf{x}) \right)$$

in $\overset{b}{\leq}$, and $v_i^* \geq \mathbf{0}$ in $\overset{c}{\leq}$.   □

From Lemma 3.6, we can see that when the algorithm in the dual space has an $O\left(\frac{1}{K^2}\right)$ convergence rate, it only has an $O\left(\frac{1}{K}\right)$ convergence rate in the primal space. The rate can be improved in the primal space via averaging the primal solutions appropriately, e.g., see [18, 26, 27, 31, 36]. We omit the details.

3.3.2 Accelerated Augmented Lagrange Multiplier Method

In Sect. 3.3, we use the accelerated gradient ascent to solve the dual problem. In this section, we consider the objective (3.3) induced by the augmented Lagrangian function, i.e., using the accelerated gradient ascent to maximize $d_\beta(\mathbf{u})$. Lemma 3.4 shows that $d_\beta(\mathbf{u})$ is smooth, no matter whether f is strongly convex or not. This is the advantage of the augmented Lagrange multiplier method over the Lagrange multiplier method. Specially, when applying Algorithm 2.6 in the dual space, we have the following iterations:

$$\begin{aligned}
\widetilde{\boldsymbol{\lambda}}_k &= \boldsymbol{\lambda}_k + \frac{\theta_k(1 - \theta_{k-1})}{\theta_{k-1}}(\boldsymbol{\lambda}_k - \boldsymbol{\lambda}_{k-1}), \\
\boldsymbol{\lambda}_{k+1} &= \widetilde{\boldsymbol{\lambda}}_k + \beta \widetilde{\nabla} d_\beta(\widetilde{\boldsymbol{\lambda}}_k),
\end{aligned} \tag{3.13}$$

where $\widetilde{\nabla} d_\beta(\widetilde{\boldsymbol{\lambda}}_k)$ is an error corrupted gradient of $\nabla d_\beta(\widetilde{\boldsymbol{\lambda}}_k)$. From Theorem 2.6 we need

$$\left\| \nabla d_\beta(\widetilde{\boldsymbol{\lambda}}_k) - \widetilde{\nabla} d_\beta(\widetilde{\boldsymbol{\lambda}}_k) \right\| \leq \frac{1}{(k+1)^{2+\delta}} \sqrt{\frac{1}{2\beta}}$$

for the second step. Define

$$\mathbf{x}_{k+1}^* = \operatorname*{argmin}_{\mathbf{x}} \left(f(\mathbf{x}) + \langle \widetilde{\boldsymbol{\lambda}}_k, \mathbf{Ax} - \mathbf{b} \rangle + \frac{\beta}{2} \|\mathbf{Ax} - \mathbf{b}\|^2 \right).$$

Then we have $\nabla d_\beta(\widetilde{\boldsymbol{\lambda}}_k) = \mathbf{Ax}_{k+1}^* - \mathbf{b}$. Define $\widetilde{\nabla} d_\beta(\widetilde{\boldsymbol{\lambda}}_k) = \mathbf{Ax}_{k+1} - \mathbf{b}$ for some $\mathbf{x}_{k+1}$, then we only need

$$\left\| \mathbf{Ax}_{k+1}^* - \mathbf{Ax}_{k+1} \right\|^2 \le \frac{1}{(k+1)^{2+\delta}} \sqrt{\frac{1}{2\beta}}. \tag{3.14}$$

In conclusion, iteration (3.13) leads to the accelerated augmented Lagrange multiplier method [11], which is described in Algorithm 3.3. From Theorem 2.6, we can give the convergence rate of Algorithm 3.3 in the dual space directly. The convergence rate in the primal space can be recovered by Lemma 3.6.

Algorithm 3.3 Accelerated augmented Lagrange multiplier method

---
Initialize $\mathbf{x}_0 = \mathbf{x}_{-1}$ and $\theta_0 = 1$.
for $k = 0, 1, 2, 3, \cdots$ **do**
　　$\widetilde{\boldsymbol{\lambda}}_k = \boldsymbol{\lambda}_k + \frac{\theta_k(1 - \theta_{k-1})}{\theta_{k-1}} (\boldsymbol{\lambda}_k - \boldsymbol{\lambda}_{k-1})$,
　　$\mathbf{x}_{k+1} \approx \operatorname*{argmin}_{\mathbf{x}} \left(f(\mathbf{x}) + \langle \widetilde{\boldsymbol{\lambda}}_k, \mathbf{Ax} - \mathbf{b} \rangle + \frac{\beta}{2} \|\mathbf{Ax} - \mathbf{b}\|^2 \right)$,
　　$\boldsymbol{\lambda}_{k+1} = \widetilde{\boldsymbol{\lambda}}_k + \beta(\mathbf{Ax}_{k+1} - \mathbf{b})$.
end for

Theorem 3.5 *Suppose that $f(\mathbf{x})$ is convex. Let $\theta_0 = 1$, $\theta_{k+1} = \frac{\sqrt{\theta_k^4 + 4\theta_k^2} - \theta_k^2}{2}$, and (3.14) be satisfied, where δ can be any small positive constant. Then for Algorithm 3.3, we have*

$$d_\beta(\boldsymbol{\lambda}^*) - d_\beta(\boldsymbol{\lambda}_{K+1}) \le \frac{4}{(K+2)^2} \left(\frac{1}{\beta} \|\boldsymbol{\lambda}_0 - \boldsymbol{\lambda}^*\|^2 + \frac{2}{1+2\delta} + \frac{18}{\delta^2} \right).$$

3.4　Alternating Direction Method of Multiplier and Its Non-ergodic Accelerated Variant

In this section, we consider the following convex problem with a linear constraint and a separable objective function, which is the sum of two functions with decoupled variables:

$$\min_{\mathbf{x}, \mathbf{y}} f(\mathbf{x}) + g(\mathbf{y}), \quad s.t. \quad \mathbf{Ax} + \mathbf{By} = \mathbf{b}. \tag{3.15}$$

Introduce the augmented Lagrangian function

$$L_\beta(\mathbf{x}, \mathbf{y}, \lambda) = f(\mathbf{x}) + g(\mathbf{y}) + \langle \mathbf{Ax} + \mathbf{By} - \mathbf{b}, \lambda \rangle + \frac{\beta}{2} \| \mathbf{Ax} + \mathbf{By} - \mathbf{b} \|^2.$$

The alternating direction method of multiplier (ADMM) is a popular method to solve problem (3.15) and has broad applications, see, e.g., [2, 7, 21, 22] and references therein. It alternately updates $\mathbf{x}$, $\mathbf{y}$, and λ and we describe it in Algorithm 3.4. When f and g are not simple and $\mathbf{A}$ and $\mathbf{B}$ are non-unitary, the cost of solving the subproblems may be high. Thus the linearized ADMM is proposed by linearizing the augmentation term $\| \mathbf{Ax} + \mathbf{By} - \mathbf{b} \|^2$ and the complex f and g [14, 22, 34, 37] such that the subproblems may even have closed form solutions. For simplicity, we only consider the original ADMM.

Algorithm 3.4 Alternating direction method of multiplier (ADMM)

for $k = 0, 1, 2, 3, \cdots$ do
  $\mathbf{x}_{k+1} = \text{argmin}_\mathbf{x} \, L_{\beta_k}(\mathbf{x}, \mathbf{y}_k, \lambda_k),$
  $\mathbf{y}_{k+1} = \text{argmin}_\mathbf{y} \, L_{\beta_k}(\mathbf{x}_{k+1}, \mathbf{y}, \lambda_k),$
  $\lambda_{k+1} = \lambda_k + \beta_k(\mathbf{Ax}_{k+1} + \mathbf{By}_{k+1} - \mathbf{b}).$
end for

In this section, we focus on the convergence rate analysis of ADMM for several scenarios.[1] We first give several useful lemmas. The following one measures the optimality condition of the first two steps of ADMM and the KKT condition.

Lemma 3.7 *For Algorithm 3.4, we have*

$$\mathbf{0} \in \partial f(\mathbf{x}_{k+1}) + \mathbf{A}^T \lambda_k + \beta_k \mathbf{A}^T (\mathbf{Ax}_{k+1} + \mathbf{By}_k - \mathbf{b}),$$

$$\mathbf{0} \in \partial g(\mathbf{y}_{k+1}) + \mathbf{B}^T \lambda_k + \beta_k \mathbf{B}^T (\mathbf{Ax}_{k+1} + \mathbf{By}_{k+1} - \mathbf{b}), \quad (3.16)$$

$$\lambda_{k+1} - \lambda_k = \beta_k(\mathbf{Ax}_{k+1} + \mathbf{By}_{k+1} - \mathbf{b}), \quad (3.17)$$

$$\mathbf{0} \in \partial f(\mathbf{x}^*) + \mathbf{A}^T \lambda^*,$$

$$\mathbf{0} \in \partial g(\mathbf{y}^*) + \mathbf{B}^T \lambda^*, \quad (3.18)$$

$$\mathbf{Ax}^* + \mathbf{By}^* = \mathbf{b}. \quad (3.19)$$

Define two variables:

$$\hat{\nabla} f(\mathbf{x}_{k+1}) = -\mathbf{A}^T \lambda_k - \beta_k \mathbf{A}^T (\mathbf{Ax}_{k+1} + \mathbf{By}_k - \mathbf{b}),$$

$$\hat{\nabla} g(\mathbf{y}_{k+1}) = -\mathbf{B}^T \lambda_k - \beta_k \mathbf{B}^T (\mathbf{Ax}_{k+1} + \mathbf{By}_{k+1} - \mathbf{b}).$$

---

[1]The four cases in Sects. 3.4.1–3.4.4 assume different conditions. They are not accelerated and cannot compare with each other. However, the convergence rate in Sect. 3.4.5 is truly accelerated.

Then we have $\hat{\nabla} f(\mathbf{x}_{k+1}) \in \partial f(\mathbf{x}_{k+1})$ and $\hat{\nabla} g(\mathbf{y}_{k+1}) \in \partial g(\mathbf{y}_{k+1})$, and they further lead to the following lemma.

Lemma 3.8 *For Algorithm 3.4, we have*

$$\left\langle \hat{\nabla} g(\mathbf{y}_{k+1}), \mathbf{y}_{k+1} - \mathbf{y} \right\rangle = -\langle \boldsymbol{\lambda}_{k+1}, \mathbf{B}\mathbf{y}_{k+1} - \mathbf{B}\mathbf{y} \rangle \tag{3.20}$$

and

$$\left\langle \hat{\nabla} f(\mathbf{x}_{k+1}), \mathbf{x}_{k+1} - \mathbf{x} \right\rangle + \left\langle \hat{\nabla} g(\mathbf{y}_{k+1}), \mathbf{y}_{k+1} - \mathbf{y} \right\rangle$$
$$= -\langle \boldsymbol{\lambda}_{k+1}, \mathbf{A}\mathbf{x}_{k+1} + \mathbf{B}\mathbf{y}_{k+1} - \mathbf{A}\mathbf{x} - \mathbf{B}\mathbf{y} \rangle + \beta_k \langle \mathbf{B}\mathbf{y}_{k+1} - \mathbf{B}\mathbf{y}_k, \mathbf{A}\mathbf{x}_{k+1} - \mathbf{A}\mathbf{x} \rangle . \tag{3.21}$$

Proof From (3.17) we also have

$$\left\langle \hat{\nabla} f(\mathbf{x}_{k+1}), \mathbf{x}_{k+1} - \mathbf{x} \right\rangle$$
$$= -\left\langle \mathbf{A}^T \boldsymbol{\lambda}_k + \beta_k \mathbf{A}^T (\mathbf{A}\mathbf{x}_{k+1} + \mathbf{B}\mathbf{y}_k - \mathbf{b}), \mathbf{x}_{k+1} - \mathbf{x} \right\rangle$$
$$= -\langle \boldsymbol{\lambda}_{k+1}, \mathbf{A}\mathbf{x}_{k+1} - \mathbf{A}\mathbf{x} \rangle + \beta_k \langle \mathbf{B}\mathbf{y}_{k+1} - \mathbf{B}\mathbf{y}_k, \mathbf{A}\mathbf{x}_{k+1} - \mathbf{A}\mathbf{x} \rangle$$

and

$$\left\langle \hat{\nabla} g(\mathbf{y}_{k+1}), \mathbf{y}_{k+1} - \mathbf{y} \right\rangle = -\langle \boldsymbol{\lambda}_{k+1}, \mathbf{B}\mathbf{y}_{k+1} - \mathbf{B}\mathbf{y} \rangle .$$

Adding them together, we can have (3.21). $\square$

The following lemma can be used to prove the monotonicity of ADMM.

Lemma 3.9 *For Algorithm 3.4, we have*

$$\langle \boldsymbol{\lambda}_{k+1} - \boldsymbol{\lambda}_k, \mathbf{B}\mathbf{y}_{k+1} - \mathbf{B}\mathbf{y}_k \rangle \leq 0.$$

Proof (3.20) gives

$$\left\langle \hat{\nabla} g(\mathbf{y}_k), \mathbf{y}_k - \mathbf{y} \right\rangle + \langle \boldsymbol{\lambda}_k, \mathbf{B}\mathbf{y}_k - \mathbf{B}\mathbf{y} \rangle = 0. \tag{3.22}$$

Letting $\mathbf{y} = \mathbf{y}_k$ in (3.20) and $\mathbf{y} = \mathbf{y}_{k+1}$ in (3.22) and adding them together, we have

$$\left\langle \hat{\nabla} g(\mathbf{y}_{k+1}) - \hat{\nabla} g(\mathbf{y}_k), \mathbf{y}_{k+1} - \mathbf{y}_k \right\rangle + \langle \boldsymbol{\lambda}_{k+1} - \boldsymbol{\lambda}_k, \mathbf{B}\mathbf{y}_{k+1} - \mathbf{B}\mathbf{y}_k \rangle = 0.$$

Using the monotonicity of ∂g (Proposition A.11), we have the conclusion. $\square$

Based on Lemma 3.9, we can give the monotonicity in the following lemma.

Lemma 3.10 *Let $\beta_k = \beta, \forall k$. For Algorithm 3.4, we have*

$$\frac{1}{2\beta}\|\lambda_{k+1} - \lambda_k\|^2 + \frac{\beta}{2}\|\mathbf{B}\mathbf{y}_{k+1} - \mathbf{B}\mathbf{y}_k\|^2 \leq \frac{1}{2\beta}\|\lambda_k - \lambda_{k-1}\|^2$$
$$+ \frac{\beta}{2}\|\mathbf{B}\mathbf{y}_k - \mathbf{B}\mathbf{y}_{k-1}\|^2.$$

Proof (3.21) gives

$$\left\langle \hat{\nabla} f(\mathbf{x}_k), \mathbf{x}_k - \mathbf{x} \right\rangle + \left\langle \hat{\nabla} g(\mathbf{y}_k), \mathbf{y}_k - \mathbf{y} \right\rangle$$
$$= -\langle \lambda_k, \mathbf{A}\mathbf{x}_k + \mathbf{B}\mathbf{y}_k - \mathbf{A}\mathbf{x} - \mathbf{B}\mathbf{y}\rangle + \beta\langle \mathbf{B}\mathbf{y}_k - \mathbf{B}\mathbf{y}_{k-1}, \mathbf{A}\mathbf{x}_k - \mathbf{A}\mathbf{x}\rangle. \quad (3.23)$$

Letting $(\mathbf{x}, \mathbf{y}, \lambda) = (\mathbf{x}_k, \mathbf{y}_k, \lambda_k)$ in (3.21) and $(\mathbf{x}, \mathbf{y}, \lambda) = (\mathbf{x}_{k+1}, \mathbf{y}_{k+1}, \lambda_{k_1})$ in (3.23), adding them together, and using (3.17), we have

$$\left\langle \hat{\nabla} f(\mathbf{x}_{k+1}) - \hat{\nabla} f(\mathbf{x}_k), \mathbf{x}_{k+1} - \mathbf{x}_k \right\rangle + \left\langle \hat{\nabla} g(\mathbf{y}_{k+1}) - \hat{\nabla} g(\mathbf{y}_k), \mathbf{y}_{k+1} - \mathbf{y}_k \right\rangle$$
$$= -\langle \lambda_{k+1} - \lambda_k, \mathbf{A}\mathbf{x}_{k+1} + \mathbf{B}\mathbf{y}_{k+1} - \mathbf{A}\mathbf{x}_k - \mathbf{B}\mathbf{y}_k\rangle$$
$$+ \beta\langle \mathbf{B}\mathbf{y}_{k+1} - \mathbf{B}\mathbf{y}_k - (\mathbf{B}\mathbf{y}_k - \mathbf{B}\mathbf{y}_{k-1}), \mathbf{A}\mathbf{x}_{k+1} - \mathbf{A}\mathbf{x}_k\rangle$$
$$= -\frac{1}{\beta}\langle \lambda_{k+1} - \lambda_k, \lambda_{k+1} - \lambda_k - (\lambda_k - \lambda_{k-1})\rangle$$
$$+ \langle \mathbf{B}\mathbf{y}_{k+1} - \mathbf{B}\mathbf{y}_k - (\mathbf{B}\mathbf{y}_k - \mathbf{B}\mathbf{y}_{k-1}), \lambda_{k+1}$$
$$- \lambda_k - \beta\mathbf{B}\mathbf{y}_{k+1} - (\lambda_k - \lambda_{k-1} - \beta\mathbf{B}\mathbf{y}_k)\rangle$$
$$\overset{a}{=} \frac{1}{2\beta}\left[\|\lambda_k - \lambda_{k-1}\|^2 - \|\lambda_{k+1} - \lambda_k\|^2 - \|\lambda_{k+1} - \lambda_k - (\lambda_k - \lambda_{k-1})\|^2\right]$$
$$+ \frac{\beta}{2}\left[\|\mathbf{B}\mathbf{y}_k - \mathbf{B}\mathbf{y}_{k-1}\|^2 - \|\mathbf{B}\mathbf{y}_{k+1} - \mathbf{B}\mathbf{y}_k\|^2\right.$$
$$\left. - \|\mathbf{B}\mathbf{y}_{k+1} - \mathbf{B}\mathbf{y}_k - (\mathbf{B}\mathbf{y}_k - \mathbf{B}\mathbf{y}_{k-1})\|^2\right]$$
$$+ \langle \mathbf{B}\mathbf{y}_{k+1} - \mathbf{B}\mathbf{y}_k - (\mathbf{B}\mathbf{y}_k - \mathbf{B}\mathbf{y}_{k-1}), \lambda_{k+1} - \lambda_k - (\lambda_k - \lambda_{k-1})\rangle$$
$$= \frac{1}{2\beta}\left(\|\lambda_k - \lambda_{k-1}\|^2 - \|\lambda_{k+1} - \lambda_k\|^2\right)$$
$$+ \frac{\beta}{2}\left(\|\mathbf{B}\mathbf{y}_k - \mathbf{B}\mathbf{y}_{k-1}\|^2 - \|\mathbf{B}\mathbf{y}_{k+1} - \mathbf{B}\mathbf{y}_k\|^2\right)$$
$$- \left[\frac{1}{2\beta}\|\lambda_{k+1} - \lambda_k - (\lambda_k - \lambda_{k-1})\|^2\right.$$

$$+ \frac{\beta}{2} \|\mathbf{By}_{k+1} - \mathbf{By}_k - (\mathbf{By}_k - \mathbf{By}_{k-1})\|^2$$

$$- \langle \mathbf{By}_{k+1} - \mathbf{By}_k - (\mathbf{By}_k - \mathbf{By}_{k-1}), \boldsymbol{\lambda}_{k+1} - \boldsymbol{\lambda}_k - (\boldsymbol{\lambda}_k - \boldsymbol{\lambda}_{k-1}) \rangle \Big]$$

$$\leq \frac{1}{2\beta} \left(\|\boldsymbol{\lambda}_k - \boldsymbol{\lambda}_{k-1}\|^2 - \|\boldsymbol{\lambda}_{k+1} - \boldsymbol{\lambda}_k\|^2 \right)$$

$$+ \frac{\beta}{2} \left(\|\mathbf{By}_k - \mathbf{By}_{k-1}\|^2 - \|\mathbf{By}_{k+1} - \mathbf{By}_k\|^2 \right),$$

where $\overset{a}{=}$ uses (A.1). Using the monotonicity of ∂f and ∂g (Proposition A.11), we have the conclusion.   □

Lemma 3.8 leads to the following lemma, which further leads to Lemma 3.12. The convergence rate of ADMM can be immediately obtained from Lemma 3.12.

Lemma 3.11 *For Algorithm 3.4, we have*

$$\left\langle \hat{\nabla} f(\mathbf{x}_{k+1}), \mathbf{x}_{k+1} - \mathbf{x}^* \right\rangle + \left\langle \hat{\nabla} g(\mathbf{y}_{k+1}), \mathbf{y}_{k+1} - \mathbf{y}^* \right\rangle + \left\langle \boldsymbol{\lambda}^*, \mathbf{Ax}_{k+1} + \mathbf{By}_{k+1} - \mathbf{b} \right\rangle$$

$$\leq \frac{1}{2\beta_k} \|\boldsymbol{\lambda}_k - \boldsymbol{\lambda}^*\|^2 - \frac{1}{2\beta_k} \|\boldsymbol{\lambda}_{k+1} - \boldsymbol{\lambda}^*\|^2 - \frac{1}{2\beta_k} \|\boldsymbol{\lambda}_{k+1} - \boldsymbol{\lambda}_k\|^2$$

$$+ \frac{\beta_k}{2} \|\mathbf{By}_k - \mathbf{By}^*\|^2 - \frac{\beta_k}{2} \|\mathbf{By}_{k+1} - \mathbf{By}^*\|^2 - \frac{\beta_k}{2} \|\mathbf{By}_{k+1} - \mathbf{By}_k\|^2.$$

Proof Letting $(\mathbf{x}, \mathbf{y}, \boldsymbol{\lambda}) = (\mathbf{x}^*, \mathbf{y}^*, \boldsymbol{\lambda}^*)$ in (3.21), adding $\langle \boldsymbol{\lambda}^*, \mathbf{Ax}_{k+1} + \mathbf{By}_{k+1} - \mathbf{b} \rangle$ to both sides, and using (3.17), (3.19), and the identities in Lemma A.1, we can have

$$\left\langle \hat{\nabla} f(\mathbf{x}_{k+1}), \mathbf{x}_{k+1} - \mathbf{x}^* \right\rangle + \left\langle \hat{\nabla} g(\mathbf{y}_{k+1}), \mathbf{y}_{k+1} - \mathbf{y}^* \right\rangle + \left\langle \boldsymbol{\lambda}^*, \mathbf{Ax}_{k+1} + \mathbf{By}_{k+1} - \mathbf{b} \right\rangle$$

$$= - \left\langle \boldsymbol{\lambda}_{k+1} - \boldsymbol{\lambda}^*, \mathbf{Ax}_{k+1} + \mathbf{By}_{k+1} - \mathbf{b} \right\rangle + \beta_k \left\langle \mathbf{By}_{k+1} - \mathbf{By}_k, \mathbf{Ax}_{k+1} - \mathbf{Ax}^* \right\rangle$$

$$= - \frac{1}{\beta_k} \left\langle \boldsymbol{\lambda}_{k+1} - \boldsymbol{\lambda}^*, \boldsymbol{\lambda}_{k+1} - \boldsymbol{\lambda}_k \right\rangle + \left\langle \mathbf{By}_{k+1} - \mathbf{By}_k, \boldsymbol{\lambda}_{k+1} - \boldsymbol{\lambda}_k \right\rangle$$

$$- \beta_k \left\langle \mathbf{By}_{k+1} - \mathbf{By}_k, \mathbf{By}_{k+1} - \mathbf{By}^* \right\rangle$$

$$\overset{a}{=} \frac{1}{2\beta_k} \|\boldsymbol{\lambda}_k - \boldsymbol{\lambda}^*\|^2 - \frac{1}{2\beta_k} \|\boldsymbol{\lambda}_{k+1} - \boldsymbol{\lambda}^*\|^2 - \frac{1}{2\beta_k} \|\boldsymbol{\lambda}_{k+1} - \boldsymbol{\lambda}_k\|^2$$

$$+ \frac{\beta_k}{2} \|\mathbf{By}_k - \mathbf{By}^*\|^2 - \frac{\beta_k}{2} \|\mathbf{By}_{k+1} - \mathbf{By}^*\|^2 - \frac{\beta_k}{2} \|\mathbf{By}_{k+1} - \mathbf{By}_k\|^2$$

$$+ \left\langle \mathbf{By}_{k+1} - \mathbf{By}_k, \boldsymbol{\lambda}_{k+1} - \boldsymbol{\lambda}_k \right\rangle, \tag{3.24}$$

where $\overset{a}{=}$ uses (A.1). From Lemma 3.9, we can have the conclusion.   □

Lemma 3.12 *Suppose that $f(\mathbf{x})$ and $g(\mathbf{y})$ are convex. Then for Algorithm 3.4, we have*

$$f(\mathbf{x}_{k+1}) + g(\mathbf{y}_{k+1}) - f(\mathbf{x}^*) - g(\mathbf{y}^*) + \langle \boldsymbol{\lambda}^*, \mathbf{A}\mathbf{x}_{k+1} + \mathbf{B}\mathbf{y}_{k+1} - \mathbf{b} \rangle$$

$$\leq \frac{1}{2\beta_k} \|\boldsymbol{\lambda}_k - \boldsymbol{\lambda}^*\|^2 - \frac{1}{2\beta_k} \|\boldsymbol{\lambda}_{k+1} - \boldsymbol{\lambda}^*\|^2 + \frac{\beta_k}{2} \|\mathbf{B}\mathbf{y}_k - \mathbf{B}\mathbf{y}^*\|^2$$

$$- \frac{\beta_k}{2} \|\mathbf{B}\mathbf{y}_{k+1} - \mathbf{B}\mathbf{y}^*\|^2. \tag{3.25}$$

If we further assume that $g(\mathbf{y})$ is μ-strongly convex, then we have

$$f(\mathbf{x}_{k+1}) + g(\mathbf{y}_{k+1}) - f(\mathbf{x}^*) - g(\mathbf{y}^*) + \langle \boldsymbol{\lambda}^*, \mathbf{A}\mathbf{x}_{k+1} + \mathbf{B}\mathbf{y}_{k+1} - \mathbf{b} \rangle$$

$$\leq \frac{1}{2\beta_k} \|\boldsymbol{\lambda}_k - \boldsymbol{\lambda}^*\|^2 - \frac{1}{2\beta_k} \|\boldsymbol{\lambda}_{k+1} - \boldsymbol{\lambda}^*\|^2$$

$$+ \frac{\beta_k}{2} \|\mathbf{B}\mathbf{y}_k - \mathbf{B}\mathbf{y}^*\|^2 - \frac{\beta_k}{2} \|\mathbf{B}\mathbf{y}_{k+1} - \mathbf{B}\mathbf{y}^*\|^2 - \frac{\mu}{2} \|\mathbf{y}_{k+1} - \mathbf{y}^*\|^2. \tag{3.26}$$

If we further assume that $g(\mathbf{y})$ is L-smooth, then we have

$$f(\mathbf{x}_{k+1}) + g(\mathbf{y}_{k+1}) - f(\mathbf{x}^*) - g(\mathbf{y}^*) + \langle \boldsymbol{\lambda}^*, \mathbf{A}\mathbf{x}_{k+1} + \mathbf{B}\mathbf{y}_{k+1} - \mathbf{b} \rangle$$

$$\leq \frac{1}{2\beta_k} \|\boldsymbol{\lambda}_k - \boldsymbol{\lambda}^*\|^2 - \frac{1}{2\beta_k} \|\boldsymbol{\lambda}_{k+1} - \boldsymbol{\lambda}^*\|^2$$

$$+ \frac{\beta_k}{2} \|\mathbf{B}\mathbf{y}_k - \mathbf{B}\mathbf{y}^*\|^2 - \frac{\beta_k}{2} \|\mathbf{B}\mathbf{y}_{k+1} - \mathbf{B}\mathbf{y}^*\|^2 - \frac{1}{2L} \|\nabla g(\mathbf{y}_{k+1}) - \nabla g(\mathbf{y}^*)\|^2. \tag{3.27}$$

Proof We use Lemma 3.11 to prove these conclusions. From the convexity of $f(\mathbf{x})$ and $g(\mathbf{y})$, we have

$$f(\mathbf{x}_{k+1}) + g(\mathbf{y}_{k+1}) - f(\mathbf{x}^*) - g(\mathbf{y}^*) + \langle \boldsymbol{\lambda}^*, \mathbf{A}\mathbf{x}_{k+1} + \mathbf{B}\mathbf{y}_{k+1} - \mathbf{b} \rangle$$

$$\overset{a}{\leq} \langle \hat{\nabla} f(\mathbf{x}_{k+1}), \mathbf{x}_{k+1} - \mathbf{x}^* \rangle + \langle \hat{\nabla} g(\mathbf{y}_{k+1}), \mathbf{y}_{k+1} - \mathbf{y}^* \rangle$$

$$+ \langle \boldsymbol{\lambda}^*, \mathbf{A}\mathbf{x}_{k+1} + \mathbf{B}\mathbf{y}_{k+1} - \mathbf{b} \rangle$$

$$\leq \frac{1}{2\beta_k} \|\boldsymbol{\lambda}_k - \boldsymbol{\lambda}^*\|^2 - \frac{1}{2\beta_k} \|\boldsymbol{\lambda}_{k+1} - \boldsymbol{\lambda}^*\|^2 + \frac{\beta_k}{2} \|\mathbf{B}\mathbf{y}_k - \mathbf{B}\mathbf{y}^*\|^2$$

$$- \frac{\beta_k}{2} \|\mathbf{B}\mathbf{y}_{k+1} - \mathbf{B}\mathbf{y}^*\|^2.$$

When $g(\mathbf{y})$ is strongly convex, we can have an extra $\frac{\mu}{2}\|\mathbf{y}_{k+1} - \mathbf{y}^*\|^2$ in the left-hand side of $\overset{a}{\leq}$ thus have (3.26). When $g(\mathbf{y})$ is L-smooth, from (A.7) we can have an extra $\frac{1}{2L}\|\nabla g(\mathbf{y}_{k+1}) - \nabla g(\mathbf{y}^*)\|^2$ in the left-hand side of $\overset{a}{\leq}$ thus have (3.27).   □

Based on the above properties, we can prove the convergence rate of ADMM. We consider four scenarios in the following sections, respectively. We also discuss the non-ergodic convergence rate of ADMM.

3.4.1 Generally Convex and Nonsmooth Case

We first consider the scenario that f and g are both generally convex and nonsmooth. In this case, [12] proved the ergodic $O\left(\frac{1}{K}\right)$ convergence rate. The following theorem summarizes the result.

Theorem 3.6 *Suppose that $f(\mathbf{x})$ and $g(\mathbf{y})$ are convex. Let $\beta_k = \beta, \forall k$. Then for Algorithm 3.4, we have*

$$|f(\hat{\mathbf{x}}_{K+1}) + g(\hat{\mathbf{y}}_{K+1}) - f(\mathbf{x}^*) - g(\mathbf{y}^*)|$$
$$\leq \frac{1}{K+1}\left(\frac{1}{2\beta}\|\boldsymbol{\lambda}_0 - \boldsymbol{\lambda}^*\|^2 + \frac{\beta}{2}\|\mathbf{By}_0 - \mathbf{By}^*\|^2\right)$$
$$+ \frac{\|\boldsymbol{\lambda}^*\|}{K+1}\left(\frac{2}{\beta}\|\boldsymbol{\lambda}_0 - \boldsymbol{\lambda}^*\| + \|\mathbf{By}_0 - \mathbf{By}^*\|\right),$$

$$\|\mathbf{A}\hat{\mathbf{x}}_{K+1} + \mathbf{B}\hat{\mathbf{y}}_{K+1} - \mathbf{b}\| \leq \frac{1}{K+1}\left(\frac{2}{\beta}\|\boldsymbol{\lambda}_0 - \boldsymbol{\lambda}^*\| + \|\mathbf{By}_0 - \mathbf{By}^*\|\right),$$

where $\hat{\mathbf{x}}_{K+1} = \frac{1}{K+1}\sum_{k=1}^{K+1}\mathbf{x}_k$ and $\hat{\mathbf{y}}_{K+1} = \frac{1}{K+1}\sum_{k=1}^{K+1}\mathbf{y}_k$.

Proof Summing (3.25) over $k = 0, 1, \cdots, K$, dividing both sides with $K + 1$, and using the convexity of $f(\mathbf{x})$ and $g(\mathbf{y})$, we have

$$f(\hat{\mathbf{x}}_{K+1}) + g(\hat{\mathbf{y}}_{K+1}) - f(\mathbf{x}^*) - g(\mathbf{y}^*) + \langle\boldsymbol{\lambda}^*, \mathbf{A}\hat{\mathbf{x}}_{K+1} + \mathbf{B}\hat{\mathbf{y}}_{K+1} - \mathbf{b}\rangle$$
$$\leq \frac{1}{K+1}\left(\frac{1}{2\beta}\|\boldsymbol{\lambda}_0 - \boldsymbol{\lambda}^*\|^2 - \frac{1}{2\beta}\|\boldsymbol{\lambda}_{K+1} - \boldsymbol{\lambda}^*\|^2 + \frac{\beta}{2}\|\mathbf{By}_0 - \mathbf{By}^*\|^2\right).$$

From Lemma 3.2, we know that the left-hand side of the above inequality is non-negative, which leads to $\|\boldsymbol{\lambda}_{K+1} - \boldsymbol{\lambda}^*\|^2 \leq \|\boldsymbol{\lambda}_0 - \boldsymbol{\lambda}^*\|^2 + \beta^2\|\mathbf{By}_0 - \mathbf{By}^*\|^2$.

From (3.17), we have

$$
\begin{aligned}
\|\mathbf{A}\hat{\mathbf{x}}_{K+1} + \mathbf{B}\hat{\mathbf{y}}_{K+1} - \mathbf{b}\| &= \frac{1}{\beta(K+1)} \left\| \sum_{k=0}^{K} (\boldsymbol{\lambda}_{k+1} - \boldsymbol{\lambda}_k) \right\| \\
&= \frac{1}{\beta(K+1)} \|\boldsymbol{\lambda}_{K+1} - \boldsymbol{\lambda}_0\| \\
&\leq \frac{1}{\beta(K+1)} \left(2\|\boldsymbol{\lambda}_0 - \boldsymbol{\lambda}^*\| + \beta\|\mathbf{B}\mathbf{y}_0 - \mathbf{B}\mathbf{y}^*\| \right).
\end{aligned}
$$

From Lemma 3.3, we can have the conclusion. □

Ouyang et al. [30] and Lu et al. [24] also studied the accelerated ADMM when the objectives are composed of an L-smooth part and a nonsmooth part. They can partially accelerate ADMM with a better dependence on L, i.e., $O\left(\frac{L}{K^2} + \frac{1}{K}\right)$. However, their entire complexity remains $O\left(\frac{1}{K}\right)$. We omit the details.

3.4.2 Strongly Convex and Nonsmooth Case

When g is strongly convex, the convergence rate can be improved to $O\left(\frac{1}{K^2}\right)$ via setting increasing β_k, e.g., see [38]. Moreover, we do not use the acceleration techniques described for the unconstrained problem, e.g., the extrapolation technique and the using of multiple sequences. Via comparing the proofs with the previous scenario, the only difference is that we use the increasing penalty parameters here.[2]

Theorem 3.7 *Assume that $f(\mathbf{x})$ is convex and $g(\mathbf{y})$ is μ-strongly convex.* *Let* $\beta_{k+1}^2 \leq \beta_k^2 + \frac{\mu}{\|\mathbf{B}\|_2^2}\beta_k$, $\hat{\mathbf{x}}_K = \left(\sum_{k=0}^{K} \beta_k\right)^{-1} \sum_{k=0}^{K} \beta_k \mathbf{x}_{k+1}$, *and* $\hat{\mathbf{y}}_K = \left(\sum_{k=0}^{K} \beta_k\right)^{-1} \sum_{k=0}^{K} \beta_k \mathbf{y}_{k+1}$, *then we have*

$$
\begin{aligned}
&|f(\hat{\mathbf{x}}_{K+1}) + g(\hat{\mathbf{y}}_{K+1}) - f(\mathbf{x}^*) - g(\mathbf{y}^*)| \\
&\leq \frac{1}{\sum_{k=0}^{K} \beta_k} \left(\frac{1}{2}\|\boldsymbol{\lambda}_0 - \boldsymbol{\lambda}^*\|^2 + \frac{\beta_0^2}{2}\|\mathbf{B}\mathbf{y}_0 - \mathbf{B}\mathbf{y}^*\|^2 \right)
\end{aligned}
$$

---

[2]In fact, the faster rate is due to the stronger assumption, i.e., the strong convexity of g, rather than the acceleration technique.

$$+ \frac{\|\boldsymbol{\lambda}^*\|}{\sum_{k=0}^{K} \beta_k} \left(2\|\boldsymbol{\lambda}_0 - \boldsymbol{\lambda}^*\| + \beta_0 \|\mathbf{B}\mathbf{y}_0 - \mathbf{B}\mathbf{y}^*\| \right),$$

$$\|\mathbf{A}\hat{\mathbf{x}}_{K+1} + \mathbf{B}\hat{\mathbf{y}}_{K+1} - \mathbf{b}\| \leq \frac{1}{\sum_{k=0}^{K} \beta_k} \left(2\|\boldsymbol{\lambda}_0 - \boldsymbol{\lambda}^*\| + \beta_0 \|\mathbf{B}\mathbf{y}_0 - \mathbf{B}\mathbf{y}^*\| \right).$$

Proof Multiplying both sides of (3.26) by β_k and using $\|\mathbf{y}_{k+1} - \mathbf{y}^*\|^2 \geq \frac{1}{\|\mathbf{B}\|_2^2} \|\mathbf{B}\mathbf{y}_{k+1} - \mathbf{B}\mathbf{y}^*\|^2$, we have

$$\beta_k \left(f(\mathbf{x}_{k+1}) + g(\mathbf{y}_{k+1}) - f(\mathbf{x}^*) - g(\mathbf{y}^*) + \langle \boldsymbol{\lambda}^*, \mathbf{A}\mathbf{x}_{k+1} + \mathbf{B}\mathbf{y}_{k+1} - \mathbf{b} \rangle \right)$$

$$\leq \frac{1}{2} \|\boldsymbol{\lambda}_k - \boldsymbol{\lambda}^*\|^2 - \frac{1}{2} \|\boldsymbol{\lambda}_{k+1} - \boldsymbol{\lambda}^*\|^2 + \frac{\beta_k^2}{2} \|\mathbf{B}\mathbf{y}_k - \mathbf{B}\mathbf{y}^*\|^2$$

$$- \left(\frac{\beta_k^2}{2} + \frac{\mu \beta_k}{2\|\mathbf{B}\|_2^2} \right) \|\mathbf{B}\mathbf{y}_{k+1} - \mathbf{B}\mathbf{y}^*\|^2$$

$$\leq \frac{1}{2} \|\boldsymbol{\lambda}_k - \boldsymbol{\lambda}^*\|^2 - \frac{1}{2} \|\boldsymbol{\lambda}_{k+1} - \boldsymbol{\lambda}^*\|^2 + \frac{\beta_k^2}{2} \|\mathbf{B}\mathbf{y}_k - \mathbf{B}\mathbf{y}^*\|^2$$

$$- \frac{\beta_{k+1}^2}{2} \|\mathbf{B}\mathbf{y}_{k+1} - \mathbf{B}\mathbf{y}^*\|^2.$$

Summing over $k = 0, 1, \cdots, K$, and dividing both sides with $\sum_{k=0}^{K} \beta_k$, we have

$$f(\hat{\mathbf{x}}_K) + g(\hat{\mathbf{y}}_K) - f(\mathbf{x}^*) - g(\mathbf{y}^*) + \langle \boldsymbol{\lambda}^*, \mathbf{A}\hat{\mathbf{x}}_K + \mathbf{B}\hat{\mathbf{y}}_K - \mathbf{b} \rangle$$

$$\leq \frac{1}{\sum_{k=0}^{K} \beta_k} \left(\frac{1}{2} \|\boldsymbol{\lambda}_0 - \boldsymbol{\lambda}^*\|^2 - \frac{1}{2} \|\boldsymbol{\lambda}_{K+1} - \boldsymbol{\lambda}^*\|^2 + \frac{\beta_0^2}{2} \|\mathbf{B}\mathbf{y}_0 - \mathbf{B}\mathbf{y}^*\|^2 \right).$$

Similar to the proof of Theorem 3.6, we have

$$\|\mathbf{A}\hat{\mathbf{x}}_{K+1} + \mathbf{B}\hat{\mathbf{y}}_{K+1} - \mathbf{b}\| = \frac{1}{\sum_{k=0}^{K} \beta_k} \left\| \sum_{k=0}^{K} (\boldsymbol{\lambda}_{k+1} - \boldsymbol{\lambda}_k) \right\|$$

$$= \frac{1}{\sum_{k=0}^{K} \beta_k} \|\boldsymbol{\lambda}_{K+1} - \boldsymbol{\lambda}_0\|$$

$$\leq \frac{1}{\sum_{k=0}^{K} \beta_k} \left(2\|\boldsymbol{\lambda}_0 - \boldsymbol{\lambda}^*\| + \beta_0 \|\mathbf{B}\mathbf{y}_0 - \mathbf{B}\mathbf{y}^*\| \right).$$

Similar to the induction of Theorem 3.6, we can have the conclusion.   □

By using the following lemma, we see that the convergence rate is $O\left(\frac{1}{K^2} \right)$.

Lemma 3.13 *Let* $\beta_k = \frac{\mu(k+1)}{3\|\mathbf{B}\|_2^2}$. *Then* $\{\beta_k\}$ *satisfy* $\beta_{k+1}^2 \leq \beta_k^2 + \frac{\mu}{\|\mathbf{B}\|_2^2}\beta_k$ *and*
$\frac{1}{\sum_{k=0}^K \beta_k} \leq \frac{6\|\mathbf{B}\|_2^2}{\mu(K+1)^2}$.

3.4.3 Generally Convex and Smooth Case

Now we consider the scenario that g is smooth and both f and g are generally convex. We describe the results in [35] in the following theorem.

Theorem 3.8 *Assume that* $f(\mathbf{x})$ *and* $g(\mathbf{y})$ *are convex and* $g(\mathbf{y})$ *is* L-*smooth.*
Let $\frac{1}{2\beta_k^2} + \frac{\sigma^2}{2L\beta_k} \geq \frac{1}{2\beta_{k+1}^2}$, $\hat{\mathbf{x}}_K = \left(\sum_{k=0}^K \frac{1}{\beta_k}\right)^{-1} \sum_{k=0}^K \frac{1}{\beta_k}\mathbf{x}_{k+1}$, *and* $\hat{\mathbf{y}}_K =$
$\left(\sum_{k=0}^K \frac{1}{\beta_k}\right)^{-1} \sum_{k=0}^K \frac{1}{\beta_k}\mathbf{y}_{k+1}$, *where* $\sigma = \sigma_{\min}(\mathbf{B})$, *then we have*

$$f(\hat{\mathbf{x}}_{K+1}) + g(\hat{\mathbf{y}}_{K+1}) - f(\mathbf{x}^*) - g(\mathbf{y}^*) + \langle \boldsymbol{\lambda}^*, \mathbf{A}\hat{\mathbf{x}}_{K+1} + \mathbf{B}\hat{\mathbf{y}}_{K+1} - \mathbf{b}\rangle$$

$$\leq \frac{1}{\sum_{k=0}^K \frac{1}{\beta_k}} \left(\frac{1}{2\beta_0^2}\|\boldsymbol{\lambda}_0 - \boldsymbol{\lambda}^*\|^2 + \frac{1}{2}\|\mathbf{B}\mathbf{y}_0 - \mathbf{B}\mathbf{y}^*\|^2\right),$$

$$\|\boldsymbol{\lambda}_{K+1} - \boldsymbol{\lambda}^*\|^2 \leq \beta_{K+1}^2 \left(\frac{1}{\beta_0^2}\|\boldsymbol{\lambda}_0 - \boldsymbol{\lambda}^*\|^2 + \|\mathbf{B}\mathbf{y}_0 - \mathbf{B}\mathbf{y}^*\|^2\right).$$

Proof From $\|\mathbf{B}^T\boldsymbol{\lambda}\| \geq \sigma\|\boldsymbol{\lambda}\|$, where $\sigma = \sigma_{\min}(\mathbf{B})$, (3.17), (3.16), and (3.18), we have

$$\frac{\sigma^2}{2L}\|\boldsymbol{\lambda}_{k+1} - \boldsymbol{\lambda}^*\|^2 \leq \frac{1}{2L}\|\mathbf{B}^T(\boldsymbol{\lambda}_{k+1} - \boldsymbol{\lambda}^*)\|^2$$

$$= \frac{1}{2L}\|\mathbf{B}^T(\boldsymbol{\lambda}_k - \boldsymbol{\lambda}^*) + \beta\mathbf{B}^T(\mathbf{A}\mathbf{x}_{k+1} + \mathbf{B}\mathbf{y}_{k+1} - \mathbf{b})\|^2$$

$$= \frac{1}{2L}\|\nabla g(\mathbf{y}_{k+1}) + \mathbf{B}^T\boldsymbol{\lambda}^*\|^2$$

$$= \frac{1}{2L}\|\nabla g(\mathbf{y}_{k+1}) - \nabla g(\mathbf{y}^*)\|^2. \tag{3.28}$$

Dividing both sides of (3.27) by β_k and using (3.28) and $\frac{1}{2\beta_k^2} + \frac{\sigma^2}{2L\beta_k} \geq \frac{1}{2\beta_{k+1}^2}$, we have

$$\frac{1}{\beta_k}\left(f(\mathbf{x}_{k+1}) + g(\mathbf{y}_{k+1}) - f(\mathbf{x}^*) - g(\mathbf{y}^*) + \langle \boldsymbol{\lambda}^*, \mathbf{A}\mathbf{x}_{k+1} + \mathbf{B}\mathbf{y}_{k+1} - \mathbf{b}\rangle\right)$$

$$\leq \frac{1}{2\beta_k^2}\|\boldsymbol{\lambda}_k - \boldsymbol{\lambda}^*\|^2 - \left(\frac{1}{2\beta_k^2} + \frac{\sigma^2}{2L\beta_k}\right)\|\boldsymbol{\lambda}_{k+1} - \boldsymbol{\lambda}^*\|^2$$

$$+ \frac{1}{2}\|\mathbf{B}\mathbf{y}_k - \mathbf{B}\mathbf{y}^*\|^2 - \frac{1}{2}\|\mathbf{B}\mathbf{y}_{k+1} - \mathbf{B}\mathbf{y}^*\|^2$$

$$\leq \frac{1}{2\beta_k^2}\|\boldsymbol{\lambda}_k - \boldsymbol{\lambda}^*\|^2 - \frac{1}{2\beta_{k+1}^2}\|\boldsymbol{\lambda}_{k+1} - \boldsymbol{\lambda}^*\|^2 + \frac{1}{2}\|\mathbf{B}\mathbf{y}_k - \mathbf{B}\mathbf{y}^*\|^2$$

$$- \frac{1}{2}\|\mathbf{B}\mathbf{y}_{k+1} - \mathbf{B}\mathbf{y}^*\|^2.$$

Summing over $k = 0, 1, \cdots, K$ and dividing both sides with $\sum_{k=0}^{K}\frac{1}{\beta_k}$, we have

$$f(\hat{\mathbf{x}}_{K+1}) + g(\hat{\mathbf{y}}_{K+1}) - f(\mathbf{x}^*) - g(\mathbf{y}^*) + \langle\boldsymbol{\lambda}^*, \mathbf{A}\hat{\mathbf{x}}_{K+1} + \mathbf{B}\hat{\mathbf{y}}_{K+1} - \mathbf{b}\rangle$$

$$\leq \frac{1}{\sum_{k=0}^{K}\frac{1}{\beta_k}}\left(\frac{1}{2\beta_0^2}\|\boldsymbol{\lambda}_0 - \boldsymbol{\lambda}^*\|^2 - \frac{1}{2\beta_{K+1}^2}\|\boldsymbol{\lambda}_{K+1} - \boldsymbol{\lambda}^*\|^2 + \frac{1}{2}\|\mathbf{B}\mathbf{y}_0 - \mathbf{B}\mathbf{y}^*\|^2\right).$$

The proof is complete. $\square$

By using the following lemma, we see that the convergence rate is $O\left(\frac{1}{K^2}\right)$ when **B** is of full row rank.

Lemma 3.14 *Suppose* $\sigma > 0$ *and let* $\frac{1}{\beta_k} = \frac{\sigma^2(k+1)}{3L}$. *Then* $\{\beta_k\}$ *satisfy* $\frac{1}{2\beta_k^2} + \frac{\sigma^2}{2L\beta_k} \geq \frac{1}{2\beta_{k+1}^2}$ *and* $\frac{1}{\sum_{k=0}^{K}\frac{1}{\beta_k}} \leq \frac{6L}{\sigma^2(K+1)^2}$.

Different from the above two scenarios, Theorem 3.8 does not measure the distance to the optimal objective value and the violation of the constraint. In fact, as claimed in the following remark, we may not prove a faster rate for the violation of the constraint. Although the convergence is proven in [35], the convergence rate is very slow, rather than "accelerated." We involve this scenario only for providing a complete comparison with other scenarios.

Remark 3.1 Different from Theorem 3.7, we have

$$\|\mathbf{A}\hat{\mathbf{x}}_{K+1} + \mathbf{B}\hat{\mathbf{y}}_{K+1} - \mathbf{b}\| = \left\| \frac{\sum_{k=0}^{K} \frac{1}{\beta_k}(\mathbf{A}\mathbf{x}_{k+1} + \mathbf{B}\mathbf{y}_{k+1} - \mathbf{b})}{\sum_{k=0}^{K} \frac{1}{\beta_k}} \right\|$$

$$= \frac{1}{\sum_{k=0}^{K} \frac{1}{\beta_k}} \left\| \sum_{k=0}^{K} \frac{\lambda_{k+1} - \lambda_k}{\beta_k^2} \right\|$$

$$\leq \frac{1}{\sum_{k=0}^{K} \frac{1}{\beta_k}} \sum_{k=0}^{K} \frac{\|\lambda_k - \lambda^*\| + \|\lambda_{k+1} - \lambda^*\|}{\beta_k^2}$$

$$\lesssim \frac{1}{\sum_{k=0}^{K} \frac{1}{\beta_k}} \sum_{k=0}^{K} \frac{2C}{\beta_k} = 2C,$$

where we let $C = \sqrt{\frac{1}{\beta_0^2}\|\lambda_0 - \lambda^*\|^2 + \|\mathbf{B}\mathbf{y}_0 - \mathbf{B}\mathbf{y}^*\|^2}$. The above suggests that $\|\mathbf{A}\hat{\mathbf{x}}_{K+1} + \mathbf{B}\hat{\mathbf{y}}_{K+1} - \mathbf{b}\|$ may not be decreasing. Similarly, we also have $\|\mathbf{A}\mathbf{x}_{K+1} + \mathbf{B}\mathbf{y}_{K+1} - \mathbf{b}\| = \frac{\|\lambda_K - \lambda_{K+1}\|}{\beta_K} \lesssim 2C$. The reason is that we use decreasing penalty parameters $\{\beta_k\}$.

3.4.4 Strongly Convex and Smooth Case

At last, we discuss the scenario that g is both strongly convex and smooth. In this case, the faster convergence rate can be proven via carefully choosing the penalty parameter [10]. Similar to the previous scenarios, we also do not use the acceleration techniques described in the previous section.

Theorem 3.9 *Assume that $f(\mathbf{x})$ is convex and $g(\mathbf{y})$ is μ-strongly convex and L-smooth. Assume that $\|\mathbf{B}^T\lambda\| \geq \sigma\|\lambda\|$, $\forall\lambda$, where $\sigma = \sigma_{\min}(\mathbf{B}) > 0$. Let $\beta_k = \beta = \frac{\sqrt{\mu L}}{\sigma\|\mathbf{B}\|_2}$, then we have*

$$\frac{1}{2\beta}\|\lambda_{k+1} - \lambda^*\|^2 + \frac{\beta}{2}\|\mathbf{B}\mathbf{y}_{k+1} - \mathbf{B}\mathbf{y}^*\|^2$$

$$\leq \frac{1}{1 + \frac{1}{2}\sqrt{\frac{\mu}{L}}\frac{\sigma}{\|\mathbf{B}\|_2}} \left(\frac{1}{2\beta}\|\lambda_k - \lambda^*\|^2 + \frac{\beta}{2}\|\mathbf{B}\mathbf{y}_k - \mathbf{B}\mathbf{y}^*\|^2 \right).$$

Proof From (3.26)–(3.28), and Lemma 3.2, we have

$$\frac{\mu}{2\|\mathbf{B}\|_2^2}\|\mathbf{B}\mathbf{y}_{k+1} - \mathbf{B}\mathbf{y}^*\|^2 \leq \frac{1}{2\beta}\|\boldsymbol{\lambda}_k - \boldsymbol{\lambda}^*\|^2 - \frac{1}{2\beta}\|\boldsymbol{\lambda}_{k+1} - \boldsymbol{\lambda}^*\|^2$$

$$+\frac{\beta}{2}\|\mathbf{B}\mathbf{y}_k - \mathbf{B}\mathbf{y}^*\|^2 - \frac{\beta}{2}\|\mathbf{B}\mathbf{y}_{k+1} - \mathbf{B}\mathbf{y}^*\|^2 \quad (3.29)$$

and

$$\frac{\sigma^2}{2L}\|\boldsymbol{\lambda}_{k+1} - \boldsymbol{\lambda}^*\|^2 \leq \frac{1}{2\beta}\|\boldsymbol{\lambda}_k - \boldsymbol{\lambda}^*\|^2 - \frac{1}{2\beta}\|\boldsymbol{\lambda}_{k+1} - \boldsymbol{\lambda}^*\|^2$$

$$+\frac{\beta}{2}\|\mathbf{B}\mathbf{y}_k - \mathbf{B}\mathbf{y}^*\|^2 - \frac{\beta}{2}\|\mathbf{B}\mathbf{y}_{k+1} - \mathbf{B}\mathbf{y}^*\|^2. \quad (3.30)$$

Multiplying (3.30) by t, multiplying (3.29) by $1 - t$, adding them together, and rearranging the terms, we have

$$\left(\frac{\sigma^2 t}{2L} + \frac{1}{2\beta}\right)\|\boldsymbol{\lambda}_{k+1} - \boldsymbol{\lambda}^*\|^2 + \left[\frac{\beta}{2} + \frac{\mu(1-t)}{2\|\mathbf{B}\|_2^2}\right]\|\mathbf{B}\mathbf{y}_{k+1} - \mathbf{B}\mathbf{y}^*\|^2$$

$$\leq \frac{1}{2\beta}\|\boldsymbol{\lambda}_k - \boldsymbol{\lambda}^*\|^2 + \frac{\beta}{2}\|\mathbf{B}\mathbf{y}_k - \mathbf{B}\mathbf{y}^*\|^2.$$

Letting $\frac{\sigma^2\beta t}{L} = \frac{\mu(1-t)}{\beta\|\mathbf{B}\|_2^2}$, we have $t = \frac{\mu L}{\mu L + \|\mathbf{B}\|_2^2\sigma^2\beta^2}$ and

$$\left(\frac{\mu\beta\sigma^2}{\mu L + \|\mathbf{B}\|_2^2\sigma^2\beta^2} + 1\right)\left(\frac{1}{2\beta}\|\boldsymbol{\lambda}_{k+1} - \boldsymbol{\lambda}^*\|^2 + \frac{\beta}{2}\|\mathbf{B}\mathbf{y}_{k+1} - \mathbf{B}\mathbf{y}^*\|^2\right)$$

$$\leq \frac{1}{2\beta}\|\boldsymbol{\lambda}_k - \boldsymbol{\lambda}^*\|^2 + \frac{\beta}{2}\|\mathbf{B}\mathbf{y}_k - \mathbf{B}\mathbf{y}^*\|^2.$$

Letting $\beta = \frac{\sqrt{\mu L}}{\sigma\|\mathbf{B}\|_2}$, which maximizes $\frac{\mu\beta\sigma^2}{\mu L + \|\mathbf{B}\|_2^2\sigma^2\beta^2} + 1$, we have $\frac{1}{\frac{\mu\beta\sigma^2}{\mu L + \|\mathbf{B}\|_2^2\sigma^2\beta^2} + 1} =$ $\frac{1}{1 + \frac{1}{2}\sqrt{\frac{\mu}{L}}\frac{\sigma}{\|\mathbf{B}\|_2}}.$ □

Theorem 3.9 shows that the convergence is linear.

3.4.5 Non-ergodic Convergence Rate

In Theorems 3.6–3.8, the convergence rates are all in the ergodic sense. Now, we discuss the non-ergodic convergence in this section. We only consider the scenario that f and g are both generally convex and nonsmooth.

3.4.5.1 Original ADMM

We first give the $O\left(\frac{1}{\sqrt{K}}\right)$ non-ergodic convergence rate of the original ADMM. The result was first proven in [13] and then extended in [6].

Theorem 3.10 *Let $\beta_k = \beta, \forall k$. For Algorithm 3.4, we have*

$$-\|\boldsymbol{\lambda}^*\|\sqrt{\frac{C}{\beta(K+1)}} \le f(\mathbf{x}_{K+1}) + g(\mathbf{y}_{K+1}) - f(\mathbf{x}^*) - g(\mathbf{y}^*)$$

$$\le \frac{C}{K+1} + \frac{2C}{\sqrt{K+1}} + \|\boldsymbol{\lambda}^*\|\sqrt{\frac{C}{\beta(K+1)}},$$

$$\|\mathbf{A}\mathbf{x}_{K+1} + \mathbf{B}\mathbf{y}_{K+1} - \mathbf{b}\| \le \sqrt{\frac{C}{\beta(K+1)}},$$

where $C = \frac{1}{\beta}\|\boldsymbol{\lambda}_0 - \boldsymbol{\lambda}^*\|^2 + \beta\|\mathbf{B}\mathbf{y}_0 - \mathbf{B}\mathbf{y}^*\|^2$.

Proof From Lemma 3.10, we know that $\frac{1}{2\beta}\|\boldsymbol{\lambda}_{k+1} - \boldsymbol{\lambda}_k\|^2 + \frac{\beta}{2}\|\mathbf{B}\mathbf{y}_{k+1} - \mathbf{B}\mathbf{y}_k\|^2$ is decreasing. From Lemmas 3.2 and 3.11 and the convexity of f and g, we have

$$\frac{1}{2\beta}\|\boldsymbol{\lambda}_{k+1} - \boldsymbol{\lambda}_k\|^2 + \frac{\beta}{2}\|\mathbf{B}\mathbf{y}_{k+1} - \mathbf{B}\mathbf{y}_k\|^2$$

$$\le \frac{1}{2\beta}\|\boldsymbol{\lambda}_k - \boldsymbol{\lambda}^*\|^2 - \frac{1}{2\beta}\|\boldsymbol{\lambda}_{k+1} - \boldsymbol{\lambda}^*\|^2$$

$$+ \frac{\beta}{2}\|\mathbf{B}\mathbf{y}_k - \mathbf{B}\mathbf{y}^*\|^2 - \frac{\beta}{2}\|\mathbf{B}\mathbf{y}_{k+1} - \mathbf{B}\mathbf{y}^*\|^2. \tag{3.31}$$

Summing over $k = 0, 1, \cdots, K - 1$, we have

$$\frac{1}{\beta}\|\boldsymbol{\lambda}_{K+1} - \boldsymbol{\lambda}_K\|^2 + \beta\|\mathbf{B}\mathbf{y}_{K+1} - \mathbf{B}\mathbf{y}_K\|^2$$

$$\le \frac{1}{K+1}\left(\frac{1}{\beta}\|\boldsymbol{\lambda}_0 - \boldsymbol{\lambda}^*\|^2 + \beta\|\mathbf{B}\mathbf{y}_0 - \mathbf{B}\mathbf{y}^*\|^2\right).$$

Then we have

$$\beta\|\mathbf{A}\mathbf{x}_{K+1} + \mathbf{B}\mathbf{y}_{K+1} - \mathbf{b}\| = \|\boldsymbol{\lambda}_{K+1} - \boldsymbol{\lambda}_K\| \leq \sqrt{\frac{\beta C}{K+1}},$$

$$\|\mathbf{B}\mathbf{y}_{K+1} - \mathbf{B}\mathbf{y}_K\| \leq \sqrt{\frac{C}{\beta(K+1)}}.$$

On the other hand, (3.31) gives

$$\frac{1}{2\beta}\|\boldsymbol{\lambda}_{k+1} - \boldsymbol{\lambda}^*\|^2 + \frac{\beta}{2}\|\mathbf{B}\mathbf{y}_{k+1} - \mathbf{B}\mathbf{y}^*\|^2 \leq \frac{1}{2\beta}\|\boldsymbol{\lambda}_k - \boldsymbol{\lambda}^*\|^2 + \frac{\beta}{2}\|\mathbf{B}\mathbf{y}_k - \mathbf{B}\mathbf{y}^*\|^2$$

$$\leq \frac{1}{2\beta}\|\boldsymbol{\lambda}_0 - \boldsymbol{\lambda}^*\|^2 + \frac{\beta}{2}\|\mathbf{B}\mathbf{y}_0 - \mathbf{B}\mathbf{y}^*\|^2$$

$$= \frac{1}{2}C.$$

So we have

$$\|\boldsymbol{\lambda}_{K+1} - \boldsymbol{\lambda}^*\| \leq \sqrt{\beta C},$$

$$\|\mathbf{B}\mathbf{y}_{K+1} - \mathbf{B}\mathbf{y}^*\| \leq \sqrt{\frac{C}{\beta}}.$$

Then from (3.24) and the convexity of f and g, we have

$$f(\mathbf{x}_{K+1}) - f(\mathbf{x}^*) + g(\mathbf{y}_{K+1}) - g(\mathbf{y}^*) + \langle \boldsymbol{\lambda}^*, \mathbf{A}\mathbf{x}_{K+1} + \mathbf{B}\mathbf{y}_{K+1} - \mathbf{b}\rangle$$

$$\leq \frac{1}{\beta}\|\boldsymbol{\lambda}_{K+1} - \boldsymbol{\lambda}^*\|\|\boldsymbol{\lambda}_{K+1} - \boldsymbol{\lambda}_K\| + \|\mathbf{B}\mathbf{y}_{K+1} - \mathbf{B}\mathbf{y}_K\|\|\boldsymbol{\lambda}_{K+1} - \boldsymbol{\lambda}_K\|$$

$$+ \beta\|\mathbf{B}\mathbf{y}_{K+1} - \mathbf{B}\mathbf{y}_K\|\|\mathbf{B}\mathbf{y}_{K+1} - \mathbf{B}\mathbf{y}^*\|$$

$$\leq \frac{C}{K+1} + \frac{2C}{\sqrt{K+1}}.$$

From Lemma 3.3, we can have the conclusion.   □

3.4.5.2 ADMM with Extrapolation and Increasing Penalty Parameter

Now we describe the results in [19], which gives an improved $O\left(\frac{1}{K}\right)$ non-ergodic convergence rate. Both the extrapolation and the increasing penalty parameters are used to build the method, which is described in Algorithm 3.5.

Algorithm 3.5 Accelerated alternating direction method of multiplier (Acc-ADMM)

Initialize $\theta_0 = 1$.

for $k = 1, 2, 3, \cdots$ **do**

    Solve θ_k via $\frac{1-\theta_k}{\theta_k} = \frac{1}{\theta_{k-1}} - \tau$,

    $\mathbf{v}_k = \mathbf{y}_k + \frac{\theta_k(1-\theta_{k-1})}{\theta_{k-1}}(\mathbf{y}_k - \mathbf{y}_{k-1})$,

    $\mathbf{x}_{k+1} = \mathrm{argmin}_{\mathbf{x}}\left(f(\mathbf{x}) + \langle \boldsymbol{\lambda}_k, \mathbf{Ax}\rangle + \frac{\beta}{2\theta_k}\|\mathbf{Ax} + \mathbf{Bv}_k - \mathbf{b}\|^2\right)$,

    $\mathbf{y}_{k+1} = \mathrm{argmin}_{\mathbf{y}}\left(g(\mathbf{y}) + \langle \boldsymbol{\lambda}_k, \mathbf{By}\rangle + \frac{\beta}{2\theta_k}\|\mathbf{Ax}_{k+1} + \mathbf{By} - \mathbf{b}\|^2\right)$,

    $\boldsymbol{\lambda}_{k+1} = \boldsymbol{\lambda}_k + \beta\tau(\mathbf{Ax}_{k+1} + \mathbf{By}_{k+1} - \mathbf{b})$.

end for

Define several auxiliary variables

$$\overline{\boldsymbol{\lambda}}_{k+1} = \boldsymbol{\lambda}_k + \frac{\beta}{\theta_k}\left(\mathbf{Ax}^{k+1} + \mathbf{Bv}^k - \mathbf{b}\right),$$

$$\hat{\boldsymbol{\lambda}}_k = \boldsymbol{\lambda}_k + \frac{\beta(1-\theta_k)}{\theta_k}\left(\mathbf{Ax}_k + \mathbf{By}_k - \mathbf{b}\right),$$

$$\mathbf{z}_{k+1} = \frac{1}{\theta_k}\mathbf{y}_{k+1} - \frac{1-\theta_k}{\theta_k}\mathbf{y}_k,$$

and let θ_k satisfy $\frac{1-\theta_{k+1}}{\theta_{k+1}} = \frac{1}{\theta_k} - \tau$, $\theta_0 = 1$, and $\theta_{-1} = 1/\tau$. Then we first give the following lemma.

Lemma 3.15 *For the definitions of* $\overline{\boldsymbol{\lambda}}_{k+1}$, $\hat{\boldsymbol{\lambda}}_k$, $\boldsymbol{\lambda}_k$, $\mathbf{z}_{k+1}$, $\mathbf{y}_{k+1}$, $\mathbf{v}_k$, *and* θ_k, *we have*

$$\hat{\boldsymbol{\lambda}}_{k+1} - \hat{\boldsymbol{\lambda}}_k = \frac{\beta}{\theta_k}[\mathbf{Ax}_{k+1} + \mathbf{By}_{k+1} - \mathbf{b} - (1-\theta_k)(\mathbf{Ax}_k + \mathbf{By}_k - \mathbf{b})],$$

$$\|\hat{\boldsymbol{\lambda}}_{k+1} - \overline{\boldsymbol{\lambda}}_{k+1}\| = \frac{\beta}{\theta_k}\|\mathbf{By}_{k+1} - \mathbf{Bv}_k\|,$$

$$\hat{\boldsymbol{\lambda}}_{K+1} - \hat{\boldsymbol{\lambda}}_0 = \frac{\beta}{\theta_K}(\mathbf{Ax}_{K+1} + \mathbf{By}_{K+1} - \mathbf{b}) + \beta\tau\sum_{k=1}^{K}(\mathbf{Ax}_k + \mathbf{By}_k - \mathbf{b}),$$

$$\mathbf{v}_k - (1-\theta_k)\mathbf{y}_k = \theta_k\mathbf{z}_k.$$

Proof From the definitions of $\hat{\boldsymbol{\lambda}}_k$ and $\boldsymbol{\lambda}_{k+1}$ and $\frac{1-\theta_{k+1}}{\theta_{k+1}} = \frac{1}{\theta_k} - \tau$, we have

$$\hat{\boldsymbol{\lambda}}_{k+1} = \boldsymbol{\lambda}_{k+1} + \beta\frac{1-\theta_{k+1}}{\theta_{k+1}}(\mathbf{Ax}_{k+1} + \mathbf{By}_{k+1} - \mathbf{b})$$

$$= \boldsymbol{\lambda}_{k+1} + \beta\left(\frac{1}{\theta_k} - \tau\right)(\mathbf{Ax}_{k+1} + \mathbf{By}_{k+1} - \mathbf{b})$$

$$= \boldsymbol{\lambda}_k + \beta \tau \left(\mathbf{A}\mathbf{x}_{k+1} + \mathbf{B}\mathbf{y}_{k+1} - \mathbf{b} \right) + \beta \left(\frac{1}{\theta_k} - \tau \right) (\mathbf{A}\mathbf{x}_{k+1} + \mathbf{B}\mathbf{y}_{k+1} - \mathbf{b})$$

$$= \boldsymbol{\lambda}_k + \frac{\beta}{\theta_k} (\mathbf{A}\mathbf{x}_{k+1} + \mathbf{B}\mathbf{y}_{k+1} - \mathbf{b}) \tag{3.32a}$$

$$= \hat{\boldsymbol{\lambda}}_k - \beta \frac{1 - \theta_k}{\theta_k} (\mathbf{A}\mathbf{x}_k + \mathbf{B}\mathbf{y}_k - \mathbf{b}) + \frac{\beta}{\theta_k} (\mathbf{A}\mathbf{x}_{k+1} + \mathbf{B}\mathbf{y}_{k+1} - \mathbf{b}) \tag{3.32b}$$

$$= \hat{\boldsymbol{\lambda}}_k + \frac{\beta}{\theta_k} [\mathbf{A}\mathbf{x}_{k+1} + \mathbf{B}\mathbf{y}_{k+1} - \mathbf{b} - (1 - \theta_k)(\mathbf{A}\mathbf{x}_k + \mathbf{B}\mathbf{y}_k - \mathbf{b})].$$

On the other hand, from (3.32a) and the definition of $\overline{\boldsymbol{\lambda}}_{k+1}$ we have

$$\| \hat{\boldsymbol{\lambda}}_{k+1} - \overline{\boldsymbol{\lambda}}_{k+1} \|_2 = \frac{\beta}{\theta_k} \| \mathbf{B}(\mathbf{y}_{k+1} - \mathbf{v}_k) \|_2 .$$

From (3.32b), $\frac{1-\theta_k}{\theta_k} = \frac{1}{\theta_{k-1}} - \tau$, and $\frac{1}{\theta_{-1}} = \tau$, we have

$$\hat{\boldsymbol{\lambda}}_{K+1} - \hat{\boldsymbol{\lambda}}_0 = \sum_{k=0}^{K} \left(\hat{\boldsymbol{\lambda}}_{k+1} - \hat{\boldsymbol{\lambda}}_k \right)$$

$$= \beta \sum_{k=0}^{K} \left[\frac{1}{\theta_k} (\mathbf{A}\mathbf{x}_{k+1} + \mathbf{B}\mathbf{y}_{k+1} - \mathbf{b}) - \frac{1 - \theta_k}{\theta_k} (\mathbf{A}\mathbf{x}_k + \mathbf{B}\mathbf{y}_k - \mathbf{b}) \right]$$

$$= \beta \sum_{k=0}^{K} \left[\frac{1}{\theta_k} (\mathbf{A}\mathbf{x}_{k+1} + \mathbf{B}\mathbf{y}_{k+1} - \mathbf{b}) - \frac{1}{\theta_{k-1}} (\mathbf{A}\mathbf{x}_k + \mathbf{B}\mathbf{y}_k - \mathbf{b}) \right.$$

$$\left. + \tau (\mathbf{A}\mathbf{x}_k + \mathbf{B}\mathbf{y}_k - \mathbf{b}) \right]$$

$$= \frac{\beta}{\theta_K} (\mathbf{A}\mathbf{x}_{K+1} + \mathbf{B}\mathbf{y}_{K+1} - \mathbf{b}) + \beta \tau \sum_{k=1}^{K} (\mathbf{A}\mathbf{x}_k + \mathbf{B}\mathbf{y}_k - \mathbf{b}) .$$

For the last identity, we have

$$(1 - \theta_k)\mathbf{y}_k + \theta_k \mathbf{z}_k = (1 - \theta_k)\mathbf{y}_k + \frac{\theta_k}{\theta_{k-1}} [\mathbf{y}_k - (1 - \theta_{k-1})\mathbf{y}_{k-1}]$$

$$= \mathbf{y}_k + \frac{\theta_k(1 - \theta_{k-1})}{\theta_{k-1}} (\mathbf{y}_k - \mathbf{y}_{k-1}).$$

The right-hand side is the definition of $\mathbf{v}_k$. $\square$

The following lemma plays the role of Lemma 2.4 in the unconstrained optimization.

Lemma 3.16 *Suppose that* $f(\mathbf{x})$ *and* $g(\mathbf{y})$ *are convex. Then for Algorithm 3.5, we have*

$$f(\mathbf{x}_{k+1}) + g(\mathbf{y}_{k+1}) - f(\mathbf{x}) - g(\mathbf{y})$$

$$\leq -\langle \overline{\lambda}_{k+1}, \mathbf{A}\mathbf{x}_{k+1} + \mathbf{B}\mathbf{y}_{k+1} - \mathbf{A}\mathbf{x} - \mathbf{B}\mathbf{y} \rangle - \frac{\beta}{\theta_k} \langle \mathbf{B}\mathbf{y}_{k+1} - \mathbf{B}\mathbf{v}_k, \mathbf{B}\mathbf{y}_{k+1} - \mathbf{B}\mathbf{y} \rangle .$$

$$(3.33)$$

Proof Let

$$\hat{\nabla} f(\mathbf{x}_{k+1}) \equiv -\mathbf{A}^T \lambda_k - \frac{\beta}{\theta_k} \mathbf{A}^T (\mathbf{A}\mathbf{x}_{k+1} + \mathbf{B}\mathbf{v}_k - \mathbf{b}) = -\mathbf{A}^T \overline{\lambda}_{k+1},$$

$$\hat{\nabla} g(\mathbf{y}_{k+1}) \equiv -\mathbf{B}^T \lambda_k - \frac{\beta}{\theta_k} \mathbf{B}^T (\mathbf{A}\mathbf{x}_{k+1} + \mathbf{B}\mathbf{y}_{k+1} - \mathbf{b})$$

$$= -\mathbf{B}^T \overline{\lambda}_{k+1} - \frac{\beta}{\theta_k} \mathbf{B}^T \mathbf{B}(\mathbf{y}_{k+1} - \mathbf{v}_k).$$

For Algorithm 3.5, we have $\hat{\nabla} f(\mathbf{x}_{k+1}) \in \partial f(\mathbf{x}_{k+1})$ and $\hat{\nabla} g(\mathbf{y}_{k+1}) \in \partial g(\mathbf{y}_{k+1})$. From the convexity of f and g, we have

$$f(\mathbf{x}_{k+1}) - f(\mathbf{x}) \leq \langle \hat{\nabla} f(\mathbf{x}_{k+1}), \mathbf{x}_{k+1} - \mathbf{x} \rangle = -\langle \overline{\lambda}_{k+1}, \mathbf{A}\mathbf{x}_{k+1} - \mathbf{A}\mathbf{x} \rangle$$

and

$$g(\mathbf{y}_{k+1}) - g(\mathbf{y}) \leq \langle \hat{\nabla} g(\mathbf{y}_{k+1}), \mathbf{y}_{k+1} - \mathbf{y} \rangle$$

$$= -\langle \overline{\lambda}_{k+1}, \mathbf{B}\mathbf{y}_{k+1} - \mathbf{B}\mathbf{y} \rangle - \frac{\beta}{\theta_k} \langle \mathbf{B}\mathbf{y}_{k+1} - \mathbf{B}\mathbf{v}_k, \mathbf{B}\mathbf{y}_{k+1} - \mathbf{B}\mathbf{y} \rangle .$$

Adding them together, we can have the conclusion. □

The following lemma plays a crucial role for the $O\left(\frac{1}{K}\right)$ non-ergodic convergence rate and it is close to the final conclusion except the constraint violation.

Lemma 3.17 *Suppose that* $f(\mathbf{x})$ *and* $g(\mathbf{y})$ *are convex. With the definitions in Lemma 3.15, for Algorithm 3.5 we have*

$$f(\mathbf{x}_{K+1}) + g(\mathbf{y}_{K+1}) - f(\mathbf{x}^*) - g(\mathbf{y}^*) + \langle \lambda^*, \mathbf{A}\mathbf{x}_{K+1} + \mathbf{B}\mathbf{y}_{K+1} - \mathbf{b} \rangle$$

$$\leq \theta_K \left(\frac{1}{2\beta} \| \hat{\lambda}_0 - \lambda^* \|^2 + \frac{\beta}{2} \| \mathbf{B}\mathbf{z}_0 - \mathbf{B}\mathbf{y}^* \|^2 \right)$$

$$(3.34)$$

and

$$\left\| \frac{1}{\theta_K}(\mathbf{Ax}_{K+1} + \mathbf{By}_{K+1} - \mathbf{b}) + \tau \sum_{k=1}^{K} (\mathbf{Ax}_k + \mathbf{By}_k - \mathbf{b}) \right\|$$

$$\leq \frac{2}{\beta} \|\hat{\boldsymbol{\lambda}}_0 - \boldsymbol{\lambda}^*\| + \|\mathbf{Bz}_0 - \mathbf{By}^*\|. \tag{3.35}$$

Proof Letting $\mathbf{x} = \mathbf{x}^*$ and $\mathbf{x} = \mathbf{x}_k$ in (3.33), respectively, we obtain two inequalities. Multiplying the first inequality by θ_k, multiplying the second by $1 - \theta_k$, adding them together, and using $\mathbf{Ax}^* + \mathbf{By}^* = \mathbf{b}$, we have

$$f(\mathbf{x}_{k+1}) + g(\mathbf{y}_{k+1}) - (1 - \theta_k)(f(\mathbf{x}_k) + g(\mathbf{y}_k)) - \theta_k(f(\mathbf{x}^*) + g(\mathbf{y}^*))$$

$$\leq -\langle \overline{\boldsymbol{\lambda}}_{k+1}, \mathbf{Ax}_{k+1} + \mathbf{By}_{k+1} - \mathbf{b} - (1 - \theta_k)(\mathbf{Ax}_k + \mathbf{By}_k - \mathbf{b})\rangle$$

$$- \frac{\beta}{\theta_k} \langle \mathbf{By}_{k+1} - \mathbf{Bv}_k, \mathbf{By}_{k+1} - (1 - \theta_k)\mathbf{By}_k - \theta_k\mathbf{By}^*\rangle.$$

Dividing both sides by θ_k, adding

$$\left\langle \boldsymbol{\lambda}^*, \frac{1}{\theta_k}(\mathbf{Ax}_{k+1} + \mathbf{By}_{k+1} - \mathbf{b}) - \frac{1 - \theta_k}{\theta_k}(\mathbf{Ax}_k + \mathbf{By}_k - \mathbf{b})\right\rangle$$

to both sides, and using $\mathbf{Ax} - \mathbf{Ax}^* = \mathbf{Ax} - \mathbf{b} + \mathbf{By}^*$ and Lemmas 3.15 and A.1, we have

$$\frac{f(\mathbf{x}_{k+1}) + g(\mathbf{y}_{k+1}) - f(\mathbf{x}^*) - g(\mathbf{y}^*) + \langle \boldsymbol{\lambda}^*, \mathbf{Ax}_{k+1} + \mathbf{By}_{k+1} - \mathbf{b}\rangle}{\theta_k}$$

$$- \frac{1 - \theta_k}{\theta_k}\left(f(\mathbf{x}_k) + g(\mathbf{y}_k) - f(\mathbf{x}^*) - g(\mathbf{y}^*) + \langle \boldsymbol{\lambda}^*, \mathbf{Ax}_k + \mathbf{By}_k - \mathbf{b}\rangle\right)$$

$$\leq -\frac{1}{\beta}\langle \overline{\boldsymbol{\lambda}}_{k+1} - \boldsymbol{\lambda}^*, \hat{\boldsymbol{\lambda}}_{k+1} - \hat{\boldsymbol{\lambda}}_k\rangle$$

$$- \frac{\beta}{\theta_k^2}\langle \mathbf{By}_{k+1} - \mathbf{Bv}_k, \mathbf{By}_{k+1} - (1 - \theta_k)\mathbf{By}_k - \theta_k\mathbf{By}^*\rangle$$

$$\overset{a}{=} \frac{1}{2\beta}\left(\|\hat{\boldsymbol{\lambda}}_k - \boldsymbol{\lambda}^*\|^2 - \|\hat{\boldsymbol{\lambda}}_{k+1} - \boldsymbol{\lambda}^*\|^2 - \|\hat{\boldsymbol{\lambda}}_k - \overline{\boldsymbol{\lambda}}_{k+1}\|^2 + \|\hat{\boldsymbol{\lambda}}_{k+1} - \overline{\boldsymbol{\lambda}}_{k+1}\|^2\right)$$

$$+ \frac{\beta}{2\theta_k^2}\left(\left\|\mathbf{Bv}_k - (1 - \theta_k)\mathbf{By}_k - \theta_k\mathbf{By}^*\right\|^2\right.$$

$$\left. - \left\|\mathbf{By}_{k+1} - (1 - \theta_k)\mathbf{By}_k - \theta_k\mathbf{By}^*\right\|^2 - \|\mathbf{By}_{k+1} - \mathbf{Bv}_k\|^2\right)$$

$$\leq \frac{1}{2\beta} \left(\|\hat{\boldsymbol{\lambda}}_k - \boldsymbol{\lambda}^*\|^2 - \|\hat{\boldsymbol{\lambda}}_{k+1} - \boldsymbol{\lambda}^*\|^2 \right)$$

$$+ \frac{\beta}{2} \left(\|\mathbf{Bz}_k - \mathbf{By}^*\|^2 - \|\mathbf{Bz}_{k+1} - \mathbf{By}^*\|^2 \right),$$

where $\overset{a}{=}$ uses (A.3) and (A.1). Using $\frac{1-\theta_k}{\theta_k} = \frac{1}{\theta_{k-1}} - \tau$ and $\theta_{-1} = 1/\tau$ and summing over $k = 0, 1, \cdots, K$, we have

$$\frac{f(\mathbf{x}_{K+1}) + g(\mathbf{y}_{K+1}) - f(\mathbf{x}^*) - g(\mathbf{y}^*) + \langle \boldsymbol{\lambda}^*, \mathbf{Ax}_{K+1} + \mathbf{By}_{K+1} - \mathbf{b} \rangle}{\theta_K}$$

$$+ \tau \sum_{k=1}^{K} \left(f(\mathbf{x}_k) + g(\mathbf{y}_k) - f(\mathbf{x}^*) - g(\mathbf{y}^*) + \langle \boldsymbol{\lambda}^*, \mathbf{Ax}_k + \mathbf{By}_k - \mathbf{b} \rangle \right)$$

$$\leq \frac{1}{2\beta} \left(\|\hat{\boldsymbol{\lambda}}_0 - \boldsymbol{\lambda}^*\|^2 - \|\hat{\boldsymbol{\lambda}}_{K+1} - \boldsymbol{\lambda}^*\|^2 \right) + \frac{\beta}{2} \|\mathbf{Bz}_0 - \mathbf{By}^*\|^2.$$

From Lemma 3.2, we have

$$\frac{f(\mathbf{x}_{K+1}) + g(\mathbf{y}_{K+1}) - f(\mathbf{x}^*) - g(\mathbf{y}^*) + \langle \boldsymbol{\lambda}^*, \mathbf{Ax}_{K+1} + \mathbf{By}_{K+1} - \mathbf{b} \rangle}{\theta_K}$$

$$\leq \frac{1}{2\beta} \left(\|\hat{\boldsymbol{\lambda}}_0 - \boldsymbol{\lambda}^*\|^2 - \|\hat{\boldsymbol{\lambda}}_{K+1} - \boldsymbol{\lambda}^*\|^2 \right) + \frac{\beta}{2} \|\mathbf{Bz}_0 - \mathbf{By}^*\|^2.$$

So we can have (3.34) and

$$\|\hat{\boldsymbol{\lambda}}_{K+1} - \boldsymbol{\lambda}^*\| \leq \sqrt{\|\hat{\boldsymbol{\lambda}}_0 - \boldsymbol{\lambda}^*\|^2 + \beta^2 \|\mathbf{Bz}_0 - \mathbf{By}^*\|^2}$$

$$\leq \|\hat{\boldsymbol{\lambda}}_0 - \boldsymbol{\lambda}^*\| + \beta \|\mathbf{Bz}_0 - \mathbf{By}^*\|,$$

which leads to

$$\|\hat{\boldsymbol{\lambda}}_{K+1} - \hat{\boldsymbol{\lambda}}_0\| \leq 2\|\hat{\boldsymbol{\lambda}}_0 - \boldsymbol{\lambda}^*\| + \beta \|\mathbf{Bz}_0 - \mathbf{By}^*\|.$$

From Lemma 3.15, we can have (3.35). $\square$

We need to bound the violation of constraint in the form of $\|\mathbf{Ax} + \mathbf{By} - \mathbf{b}\|$, rather than (3.35). The following lemma provides a useful tool for it.

Lemma 3.18 *Consider a sequence $\{\mathbf{a}_k\}_{k=1}^{\infty}$ of vectors, if $\{\mathbf{a}_k\}$ satisfies*

$$\left\| [1/\tau + K(1/\tau - 1)]\mathbf{a}_{K+1} + \sum_{k=1}^{K} \mathbf{a}_k \right\| \leq c, \quad \forall K = 0, 1, 2, \cdots,$$

where $0 < \tau < 1$. Then $\left\| \sum_{k=1}^{K} \mathbf{a}_k \right\| < c$ for all $K = 1, 2, \cdots$.

Proof Let $\mathbf{s}_K = \sum_{k=1}^{K} \mathbf{a}_k, \forall K \geq 1$, and $\mathbf{s}_0 = \mathbf{0}$. For each $K \geq 0$, there exists $\mathbf{c}_{K+1}$ with every entry $(\mathbf{c}_{K+1})_i \geq 0$ such that

$$-(\mathbf{c}_{K+1})_i \leq [1/\tau + K(1/\tau - 1)](\mathbf{a}_{K+1})_i + (\mathbf{s}_K)_i \leq (\mathbf{c}_{K+1})_i,$$

and $\|\mathbf{c}_{K+1}\| = c$. Then

$$\frac{-(\mathbf{c}_{K+1})_i - (\mathbf{s}_K)_i}{1/\tau + K(1/\tau - 1)} \leq (\mathbf{a}_{K+1})_i \leq \frac{(\mathbf{c}_{K+1})_i - (\mathbf{s}_K)_i}{1/\tau + K(1/\tau - 1)}, \forall K \geq 0,$$

where we use $1/\tau > 1$ and $1/\tau + K(1/\tau - 1) > 0$. Thus for all $K \geq 0$, we have

$$\begin{aligned}
(\mathbf{s}_{K+1})_i &= (\mathbf{a}_{K+1})_i + (\mathbf{s}_K)_i \\
&\leq \frac{(\mathbf{c}_{K+1})_i - (\mathbf{s}_K)_i}{1/\tau + K(1/\tau - 1)} + (\mathbf{s}_K)_i \\
&= \frac{(\mathbf{c}_{K+1})_i}{1/\tau + K(1/\tau - 1)} + \frac{(K+1)(1/\tau - 1)}{1/\tau + K(1/\tau - 1)}(\mathbf{s}_K)_i.
\end{aligned}$$

By recursion, we have

$$\begin{aligned}
(\mathbf{s}_{K+1})_i \\
\leq \frac{(\mathbf{c}_{K+1})_i}{1/\tau + K(1/\tau - 1)} \\
+ \frac{(K+1)(1/\tau - 1)}{1/\tau + K(1/\tau - 1)} \frac{(\mathbf{c}_K)_i}{1/\tau + (K-1)(1/\tau - 1)} \\
+ \frac{(K+1)(1/\tau - 1)}{1/\tau + K(1/\tau - 1)} \frac{K(1/\tau - 1)}{1/\tau + (K-1)(1/\tau - 1)} \frac{(\mathbf{c}_{K-1})_i}{1/\tau + (K-2)(1/\tau - 1)} \\
+ \cdots \\
+ \left[\prod_{j=2}^{K+1} \frac{j(1/\tau - 1)}{1/\tau + (j-1)(1/\tau - 1)}\right] \\
\times \left[\frac{(\mathbf{c}_1)_i}{1/\tau + 0(1/\tau - 1)} + \frac{1/\tau - 1}{1/\tau + 0(1/\tau - 1)}(\mathbf{s}_0)_i\right] \\
= \sum_{k=1}^{K+1} \frac{(\mathbf{c}_k)_i}{1/\tau + (k-1)(1/\tau - 1)} \prod_{j=k+1}^{K+1} \frac{j(1/\tau - 1)}{1/\tau + (j-1)(1/\tau - 1)},
\end{aligned}$$

where we set $\prod_{j=K+2}^{K+1} \frac{j(1/\tau-1)}{1/\tau+(j-1)(1/\tau-1)} = 1$. Define

$$r_k = \frac{1}{1/\tau + (k-1)(1/\tau-1)} \prod_{j=k+1}^{K+1} \frac{j(1/\tau-1)}{1/\tau+(j-1)(1/\tau-1)},$$

$$\forall k = 1, 2, \cdots, K+1.$$

Then we have $r_k > 0$ and $(\mathbf{s}_{K+1})_i \leq \sum_{k=1}^{K+1} r_k(\mathbf{c}_k)_i$. Similarly, we also have $(\mathbf{s}_{K+1})_i \geq -\sum_{k=1}^{K+1} r_k(\mathbf{c}_k)_i$. Thus

$$|(\mathbf{s}_{K+1})_i| \leq \sum_{k=1}^{K+1} r_k(\mathbf{c}_k)_i.$$

Further define

$$R_K = \sum_{k=1}^{K} \frac{1}{1/\tau + (k-1)(1/\tau-1)} \prod_{j=k+1}^{K} \frac{j(1/\tau-1)}{1/\tau+(j-1)(1/\tau-1)} = \sum_{k=1}^{K} r_k,$$

then

$$R_1 = \sum_{k=1}^{1} \frac{1}{1/\tau + (k-1)(1/\tau-1)} \prod_{j=k+1}^{1} \frac{j(1/\tau-1)}{1/\tau+(j-1)(1/\tau-1)} = \tau,$$

and

$$R_{K+1} = \frac{1}{1/\tau + K(1/\tau-1)} + \sum_{k=1}^{K} \frac{1}{1/\tau + (k-1)(1/\tau-1)}$$

$$\times \prod_{j=k+1}^{K+1} \frac{j(1/\tau-1)}{1/\tau+(j-1)(1/\tau-1)}$$

$$= \frac{1}{1/\tau + K(1/\tau-1)} + \frac{(K+1)(1/\tau-1)}{1/\tau + K(1/\tau-1)} \sum_{k=1}^{K} \frac{1}{1/\tau+(k-1)(1/\tau-1)}$$

$$\times \prod_{j=k+1}^{K} \frac{j(1/\tau-1)}{1/\tau+(j-1)(1/\tau-1)}$$

$$= \frac{1}{1/\tau + K(1/\tau-1)} + \frac{(K+1)(1/\tau-1)}{1/\tau + K(1/\tau-1)} R_K.$$

Next, we prove $R_K < 1, \forall K \geq 1$, by induction. It can be easily checked that $R_1 = \tau < 1$. Assume that $R_K < 1$ holds, then

$$R_{K+1} < \frac{1}{1/\tau + K(1/\tau - 1)} + \frac{(K+1)(1/\tau - 1)}{1/\tau + K(1/\tau - 1)} = 1.$$

So by induction we can have $R_K < 1, \forall K \geq 1$.

So for any $K \geq 0$, we have

$$[(s_{K+1})_i]^2 \leq \left(\sum_{k=1}^{K+1} r_k\right)^2 \left(\frac{\sum_{k=1}^{K+1} r_k(c_k)_i}{\sum_{k=1}^{K+1} r_k}\right)^2 \leq \left(\sum_{k=1}^{K+1} r_k\right)^2 \frac{\sum_{k=1}^{K+1} r_k((c_k)_i)^2}{\sum_{k=1}^{K+1} r_k}$$

$$< \sum_{k=1}^{K+1} r_k((c_k)_i)^2,$$

where we use the convexity of x^2 and $\sum_{k=1}^{K+1} r_k = R_{K+1} < 1$. So we have

$$\|s_{K+1}\|^2 = \sum_i ((s_{K+1})_i)^2 < \sum_{k=1}^{K+1} r_k \sum_i ((c_k)_i)^2 = \sum_{k=1}^{K+1} r_k c^2 < c^2,$$

where we use $\|c_k\| = c, \forall k \geq 1$, and $\sum_{k=1}^{K+1} r_k = R_{K+1} < 1$. So $\left\|\sum_{k=1}^{K+1} a_k\right\| = \|s_{K+1}\| < c, \forall K \geq 0$. □

Now, based on the previous results, we are ready to present the final conclusion.

Theorem 3.11 *Suppose that $f(x)$ and $g(y)$ are convex. For Algorithm 3.5, we have*

$$-\frac{2C_1\|\lambda^*\|}{1 + K(1 - \tau)} \leq f(x_{K+1}) + g(y_{K+1}) - f(x^*) - g(y^*)$$

$$\leq \frac{2C_1\|\lambda^*\|}{1 + K(1 - \tau)} + \frac{C_2}{1 + K(1 - \tau)},$$

and

$$\|Ax_{K+1} + By_{K+1} - b\| \leq \frac{2C_1}{1 + K(1 - \tau)},$$

where $C_1 = \frac{2}{\beta}\|\hat{\lambda}_0 - \lambda^*\| + \|Bz_0 - By^*\|$ and $C_2 = \frac{1}{2\beta}\|\hat{\lambda}_0 - \lambda^*\|^2 + \frac{\beta}{2}\|Bz_0 - By^*\|^2$.

Proof Since $\frac{1}{\theta_k} = \frac{1}{\theta_{k-1}}+1-\tau = \frac{1}{\theta_0}+k(1-\tau)$, we have $\theta_k = \frac{1}{\frac{1}{\theta_0}+k(1-\tau)} = \frac{1}{1+k(1-\tau)}$.
For simplicity, let $\mathbf{a}_k = \mathbf{Ax}_k + \mathbf{By}_k - \mathbf{b}$. Then from (3.35) we can have

$$\left\| [1/\tau + K(1/\tau - 1)]\mathbf{a}_{K+1} + \sum_{k=1}^{K} \mathbf{a}_k \right\|$$

$$\leq \frac{1}{\tau}\left(\frac{2}{\beta}\|\hat{\boldsymbol{\lambda}}_0 - \hat{\boldsymbol{\lambda}}^*\| + \|\mathbf{Bz}_0 - \mathbf{By}^*\|\right) \equiv \frac{1}{\tau}C_1, \forall K = 0, 1, \cdots. \quad (3.36)$$

From Lemma 3.18 we have $\left\|\sum_{k=1}^{K} \mathbf{a}_k\right\| \leq \frac{1}{\tau}C_1, \forall K = 1, 2, \cdots$. So $\|\mathbf{a}_{K+1}\| \leq \frac{2\frac{1}{\tau}C_1}{1/\tau+K(1/\tau-1)}, \forall K = 1, 2, \cdots$. Moreover, letting $K = 0$ in (3.36), we have $\|\mathbf{a}_1\| \leq C_1 \leq \frac{2\frac{1}{\tau}C_1}{1/\tau+0(1/\tau-1)}$. So

$$\|\mathbf{Ax}_{K+1} + \mathbf{By}_{K+1} - \mathbf{b}\| \leq \frac{2C_1}{1 + K(1 - \tau)}, \forall K = 0, 1, \cdots.$$

Then from (3.34) and Lemma 3.3, we can have the conclusion. $\qquad\square$

3.5 Primal-Dual Method

In this section, we consider the following convex-concave saddle point problem:

$$\min_{\mathbf{x}} \max_{\mathbf{y}} L(\mathbf{x}, \mathbf{y}) \equiv \langle \mathbf{Kx}, \mathbf{y}\rangle + g(\mathbf{x}) - h(\mathbf{y}),$$

where it includes problem (3.2) as a special case. Pock et al. [32] proposed a primal-dual method for minimizing a convex relaxation of the Mumford–Shah functional and [7] studied the same algorithm in a more general framework and established connections to other known algorithms. Chambolle and Pock [3] proved several convergence rates for the primal-dual method and [4] gave a more comprehensive study later. The primal-dual method we study here is described in Algorithm 3.6 and our main goal is to describe the convergence rates for several scenarios discussed in [3].

Let $(\mathbf{x}^*, \mathbf{y}^*)$ be a saddle point. Introduce the primal-dual gap:

$$G(\mathbf{x}, \mathbf{y}) = L(\mathbf{x}, \mathbf{y}^*) - L(\mathbf{x}^*, \mathbf{y}).$$

We prove the convergence rate in the sense of this primal-dual gap. We first give the following general result, which can be used in different scenarios.

Algorithm 3.6 Primal-dual method

Initialize $\bar{\mathbf{x}}_0 = \mathbf{x}_0 = \mathbf{x}_{-1}$.
for $k = 0, 1, 2, 3, \cdots$ **do**
    $\mathbf{y}_{k+1} = \arg\min_{\mathbf{y}} \left(h(\mathbf{y}) - \langle \mathbf{K}\bar{\mathbf{x}}_k, \mathbf{y} \rangle + \frac{1}{2\sigma_k} \|\mathbf{y} - \mathbf{y}_k\|^2 \right)$,
    $\mathbf{x}_{k+1} = \arg\min_{\mathbf{x}} \left(g(\mathbf{x}) + \langle \mathbf{K}\mathbf{x}, \mathbf{y}_{k+1} \rangle + \frac{1}{2\tau_k} \|\mathbf{x} - \mathbf{x}_k\|^2 \right)$,
    $\bar{\mathbf{x}}_{k+1} = \mathbf{x}_{k+1} + \theta_k (\mathbf{x}_{k+1} - \mathbf{x}_k)$.
end for

Lemma 3.19 *Suppose that g and h are strongly convex with moduli μ_g and μ_h, respectively. Then for Algorithm 3.6, we have*

$$L(\mathbf{x}_{k+1}, \mathbf{y}^*) - L(\mathbf{x}^*, \mathbf{y}_{k+1})$$

$$\leq \frac{1}{2\tau_k} \|\mathbf{x}^* - \mathbf{x}_k\|^2 - \left(\frac{1}{2\tau_k} + \frac{\mu_g}{2} \right) \|\mathbf{x}^* - \mathbf{x}_{k+1}\|^2 - \frac{1}{2\tau_k} \|\mathbf{x}_{k+1} - \mathbf{x}_k\|^2$$

$$+ \frac{1}{2\sigma_k} \|\mathbf{y}^* - \mathbf{y}_k\|^2 - \left(\frac{1}{2\sigma_k} + \frac{\mu_h}{2} \right) \|\mathbf{y}^* - \mathbf{y}_{k+1}\|^2 - \frac{1}{2\sigma_k} \|\mathbf{y}_{k+1} - \mathbf{y}_k\|^2$$

$$+ \langle \mathbf{K}\mathbf{x}_{k+1} - \mathbf{K}\mathbf{x}_k, \mathbf{y}^* - \mathbf{y}_{k+1} \rangle - \theta_{k-1} \langle \mathbf{K}\mathbf{x}_k - \mathbf{K}\mathbf{x}_{k-1}, \mathbf{y}^* - \mathbf{y}_k \rangle$$

$$+ \theta_{k-1} \langle \mathbf{K}\mathbf{x}_k - \mathbf{K}\mathbf{x}_{k-1}, \mathbf{y}_{k+1} - \mathbf{y}_k \rangle.$$

Proof From the optimality condition, we have

$$\mathbf{0} \in \partial h(\mathbf{y}_{k+1}) - \mathbf{K}\bar{\mathbf{x}}_k + \frac{1}{\sigma_k} (\mathbf{y}_{k+1} - \mathbf{y}_k),$$

$$\mathbf{0} \in \partial g(\mathbf{x}_{k+1}) + \mathbf{K}^T \mathbf{y}_{k+1} + \frac{1}{\tau_k} (\mathbf{x}_{k+1} - \mathbf{x}_k).$$

Then since h is μ_h-strongly convex and g is μ_g-strongly convex, we have

$$h(\mathbf{y}) \geq h(\mathbf{y}_{k+1}) - \frac{1}{\sigma_k} \langle \mathbf{y}_{k+1} - \mathbf{y}_k, \mathbf{y} - \mathbf{y}_{k+1} \rangle + \langle \mathbf{K}\bar{\mathbf{x}}_k, \mathbf{y} - \mathbf{y}_{k+1} \rangle$$

$$+ \frac{\mu_h}{2} \|\mathbf{y} - \mathbf{y}_{k+1}\|^2,$$

$$g(\mathbf{x}) \geq g(\mathbf{x}_{k+1}) - \frac{1}{\tau_k} \langle \mathbf{x}_{k+1} - \mathbf{x}_k, \mathbf{x} - \mathbf{x}_{k+1} \rangle - \langle \mathbf{K}^T \mathbf{y}_{k+1}, \mathbf{x} - \mathbf{x}_{k+1} \rangle$$

$$+ \frac{\mu_g}{2} \|\mathbf{x} - \mathbf{x}_{k+1}\|^2.$$

Adding them together and using (A.2), we have

$$g(\mathbf{x}) + h(\mathbf{y})$$

$$\geq g(\mathbf{x}_{k+1}) + h(\mathbf{y}_{k+1}) - \left\langle \mathbf{K}^T \mathbf{y}_{k+1}, \mathbf{x} - \mathbf{x}_{k+1} \right\rangle + \left\langle \mathbf{K}\bar{\mathbf{x}}_k, \mathbf{y} - \mathbf{y}_{k+1} \right\rangle$$

$$+ \frac{1}{2\sigma_k} \left(\|\mathbf{y}_{k+1} - \mathbf{y}_k\|^2 + \|\mathbf{y} - \mathbf{y}_{k+1}\|^2 - \|\mathbf{y} - \mathbf{y}_k\|^2 \right)$$

$$+ \frac{1}{2\tau_k} \left(\|\mathbf{x}_{k+1} - \mathbf{x}_k\|^2 + \|\mathbf{x} - \mathbf{x}_{k+1}\|^2 - \|\mathbf{x} - \mathbf{x}_k\|^2 \right)$$

$$+ \frac{\mu_g}{2} \|\mathbf{x} - \mathbf{x}_{k+1}\|^2 + \frac{\mu_h}{2} \|\mathbf{y} - \mathbf{y}_{k+1}\|^2.$$

Letting $\mathbf{x} = \mathbf{x}^*$ and $\mathbf{y} = \mathbf{y}^*$ and rearranging the terms, we have

$$L(\mathbf{x}_{k+1}, \mathbf{y}^*) - L(\mathbf{x}^*, \mathbf{y}_{k+1})$$

$$\leq \frac{1}{2\tau_k} \|\mathbf{x}^* - \mathbf{x}_k\|^2 - \left(\frac{1}{2\tau_k} + \frac{\mu_g}{2} \right) \|\mathbf{x}^* - \mathbf{x}_{k+1}\|^2 - \frac{1}{2\tau_k} \|\mathbf{x}_{k+1} - \mathbf{x}_k\|^2$$

$$+ \frac{1}{2\sigma_k} \|\mathbf{y}^* - \mathbf{y}_k\|^2 - \left(\frac{1}{2\sigma_k} + \frac{\mu_h}{2} \right) \|\mathbf{y}^* - \mathbf{y}_{k+1}\|^2 - \frac{1}{2\sigma_k} \|\mathbf{y}_{k+1} - \mathbf{y}_k\|^2$$

$$+ \left\langle \mathbf{K}\mathbf{x}_{k+1}, \mathbf{y}^* \right\rangle - \left\langle \mathbf{K}\mathbf{x}^*, \mathbf{y}_{k+1} \right\rangle + \left\langle \mathbf{y}_{k+1}, \mathbf{K}\mathbf{x}^* - \mathbf{K}\mathbf{x}_{k+1} \right\rangle - \left\langle \mathbf{K}\bar{\mathbf{x}}_k, \mathbf{y}^* - \mathbf{y}_{k+1} \right\rangle.$$

Since

$$\left\langle \mathbf{K}\mathbf{x}_{k+1}, \mathbf{y}^* \right\rangle - \left\langle \mathbf{K}\mathbf{x}^*, \mathbf{y}_{k+1} \right\rangle + \left\langle \mathbf{y}_{k+1}, \mathbf{K}\mathbf{x}^* - \mathbf{K}\mathbf{x}_{k+1} \right\rangle - \left\langle \mathbf{K}\bar{\mathbf{x}}_k, \mathbf{y}^* - \mathbf{y}_{k+1} \right\rangle$$

$$= \left\langle \mathbf{K}\mathbf{x}_{k+1} - \mathbf{K}\bar{\mathbf{x}}_k, \mathbf{y}^* - \mathbf{y}_{k+1} \right\rangle$$

$$= \left\langle \mathbf{K}\mathbf{x}_{k+1} - \mathbf{K}\mathbf{x}_k, \mathbf{y}^* - \mathbf{y}_{k+1} \right\rangle - \theta_{k-1} \left\langle \mathbf{K}\mathbf{x}_k - \mathbf{K}\mathbf{x}_{k-1}, \mathbf{y}^* - \mathbf{y}_{k+1} \right\rangle$$

$$= \left\langle \mathbf{K}\mathbf{x}_{k+1} - \mathbf{K}\mathbf{x}_k, \mathbf{y}^* - \mathbf{y}_{k+1} \right\rangle - \theta_{k-1} \left\langle \mathbf{K}\mathbf{x}_k - \mathbf{K}\mathbf{x}_{k-1}, \mathbf{y}^* - \mathbf{y}_k \right\rangle$$

$$+ \theta_{k-1} \left\langle \mathbf{K}\mathbf{x}_k - \mathbf{K}\mathbf{x}_{k-1}, \mathbf{y}_{k+1} - \mathbf{y}_k \right\rangle,$$

we can have the conclusion. $\square$

3.5.1 Case 1: $\mu_g = \mu_h = 0$

We first consider the scenario that both g and h are generally convex. The $O\left(\frac{1}{K}\right)$ convergence rate is given in the following theorem.

Theorem 3.12 *Let* $\hat{\mathbf{x}}_K = \frac{1}{K+1} \sum_{k=0}^{K} \mathbf{x}_{k+1}$, $\hat{\mathbf{y}}_K = \frac{1}{K+1} \sum_{k=0}^{K} \mathbf{y}_{k+1}$, $\theta_k = 1, \forall k$, *and* $\sigma_k = \tau_k = \sigma \leq \frac{1}{\|\mathbf{K}\|_2}$. *Then for Algorithm 3.6, we have*

$$L(\hat{\mathbf{x}}_K, \mathbf{y}^*) - L(\mathbf{x}^*, \hat{\mathbf{y}}_K) \leq \frac{1}{2\sigma(K+1)} \left(\|\mathbf{x}_0 - \mathbf{x}^*\|^2 + \|\mathbf{y}_0 - \mathbf{y}^*\|^2 \right).$$

Proof From Lemma 3.19, we have

$$L(\mathbf{x}_{k+1}, \mathbf{y}^*) - L(\mathbf{x}^*, \mathbf{y}_{k+1})$$

$$\leq \frac{1}{2\sigma} \left(\|\mathbf{x}^* - \mathbf{x}_k\|^2 - \|\mathbf{x}^* - \mathbf{x}_{k+1}\|^2 + \|\mathbf{y}^* - \mathbf{y}_k\|^2 - \|\mathbf{y}^* - \mathbf{y}_{k+1}\|^2 \right)$$

$$- \frac{1}{2\sigma} \|\mathbf{x}_{k+1} - \mathbf{x}_k\|^2 - \frac{1}{2\sigma} \|\mathbf{y}_{k+1} - \mathbf{y}_k\|^2$$

$$+ \left\langle \mathbf{K}\mathbf{x}_{k+1} - \mathbf{K}\mathbf{x}_k, \mathbf{y}^* - \mathbf{y}_{k+1} \right\rangle - \left\langle \mathbf{K}\mathbf{x}_k - \mathbf{K}\mathbf{x}_{k-1}, \mathbf{y}^* - \mathbf{y}_k \right\rangle$$

$$+ \left\langle \mathbf{K}\mathbf{x}_k - \mathbf{K}\mathbf{x}_{k-1}, \mathbf{y}_{k+1} - \mathbf{y}_k \right\rangle.$$

Since

$$\left\langle \mathbf{K}\mathbf{x}_k - \mathbf{K}\mathbf{x}_{k-1}, \mathbf{y}_{k+1} - \mathbf{y}_k \right\rangle \leq \|\mathbf{K}\|_2 \|\mathbf{x}_k - \mathbf{x}_{k-1}\| \|\mathbf{y}_{k+1} - \mathbf{y}_k\|$$

$$\leq \frac{\|\mathbf{K}\|_2}{2} \|\mathbf{x}_k - \mathbf{x}_{k-1}\|^2 + \frac{\|\mathbf{K}\|_2}{2} \|\mathbf{y}_{k+1} - \mathbf{y}_k\|^2,$$

we can have

$$L(\mathbf{x}_{k+1}, \mathbf{y}^*) - L(\mathbf{x}^*, \mathbf{y}_{k+1})$$

$$\leq \frac{1}{2\sigma} \left(\|\mathbf{x}_k - \mathbf{x}^*\|^2 + \|\mathbf{y}_k - \mathbf{y}^*\|^2 - \|\mathbf{x}_{k+1} - \mathbf{x}^*\|^2 - \|\mathbf{y}_{k+1} - \mathbf{y}^*\|^2 \right)$$

$$+ \left\langle \mathbf{K}\mathbf{x}_{k+1} - \mathbf{K}\mathbf{x}_k, \mathbf{y}^* - \mathbf{y}_{k+1} \right\rangle - \left\langle \mathbf{K}\mathbf{x}_k - \mathbf{K}\mathbf{x}_{k-1}, \mathbf{y}^* - \mathbf{y}_k \right\rangle$$

$$+ \frac{\|\mathbf{K}\|_2}{2} \left(\|\mathbf{x}_k - \mathbf{x}_{k-1}\|^2 - \|\mathbf{x}_{k+1} - \mathbf{x}_k\|^2 \right).$$

Summing over $k = 0, 1, \cdots, K$, using $\mathbf{x}_0 = \mathbf{x}_{-1}$, dividing both sides with $K + 1$, and using the convexity of g and h, we have

$$L(\hat{\mathbf{x}}_K, \mathbf{y}^*) - L(\mathbf{x}^*, \hat{\mathbf{y}}_K)$$

$$\leq \frac{1}{K+1} \left(\frac{1}{2\sigma} \|\mathbf{x}_0 - \mathbf{x}^*\|^2 + \frac{1}{2\sigma} \|\mathbf{y}_0 - \mathbf{y}^*\|^2 \right.$$

$$\left. - \frac{1}{2\sigma} \|\mathbf{y}_{K+1} - \mathbf{y}^*\|^2 + \|\mathbf{K}\|_2 \|\mathbf{x}_{K+1} - \mathbf{x}_K\| \|\mathbf{y}^* - \mathbf{y}_{K+1}\| \right.$$

$$
\left. - \frac{\|\mathbf{K}\|_2}{2} \|\mathbf{x}_{K+1} - \mathbf{x}_K\|^2 \right)
$$

$$
\leq \frac{1}{2\sigma(K+1)} \left(\|\mathbf{x}_0 - \mathbf{x}^*\|^2 + \|\mathbf{y}_0 - \mathbf{y}^*\|^2 \right),
$$

finishing the proof. □

Chen et al. [5] proposed an accelerated primal-dual method when one of the objectives is L-smooth. Similar to the accelerated ADMM in Sect. 3.4.1, [5] can also partially accelerate the primal-dual method with a better dependence on L and the entire complexity remains $O\left(\frac{1}{K}\right)$.

3.5.2 Case 2: $\mu_g > 0$, $\mu_h = 0$

Then we consider the scenario that g is strongly convex. The convergence rate can be improved to $O\left(\frac{1}{K^2}\right)$.[3]

Theorem 3.13 *Assume that g is μ_g-strongly convex. Let $\hat{\mathbf{x}}_K = \left(\sum_{k=0}^{K} \frac{1}{\tau_k}\right)^{-1} \sum_{k=0}^{K} \frac{1}{\tau_k}$*
$\mathbf{x}_{k+1}$, $\hat{\mathbf{y}}_K = \left(\sum_{k=0}^{K} \frac{1}{\tau_k}\right)^{-1} \sum_{k=0}^{K} \frac{1}{\tau_k} \mathbf{y}_{k+1}$, $\tau_k = \frac{1}{(2k+1)\mu_g}$, $\theta_k = \frac{\tau_{k+1}}{\tau_k}$, and $\sigma_k = \frac{1}{\tau_k \|\mathbf{K}\|_2^2}$.
Then for Algorithm 3.6, we have

$$
L(\hat{\mathbf{x}}_K, \mathbf{y}^*) - L(\mathbf{x}^*, \hat{\mathbf{y}}_K)
$$

$$
\leq \frac{1}{2\mu_g(K+1)^2} \left(\mu_g^2 \|\mathbf{x}_0 - \mathbf{x}^*\|^2 + \|\mathbf{K}\|_2^2 \|\mathbf{y}_0 - \mathbf{y}^*\|^2 \right). \tag{3.37}
$$

Proof From Lemma 3.19 and using

$$
\theta_{k-1} \langle \mathbf{K}\mathbf{x}_k - \mathbf{K}\mathbf{x}_{k-1}, \mathbf{y}_{k+1} - \mathbf{y}_k \rangle \leq \frac{\|\mathbf{K}\|_2^2 \theta_{k-1}^2 \sigma_k}{2} \|\mathbf{x}_k - \mathbf{x}_{k-1}\|^2
$$

$$
+ \frac{1}{2\sigma_k} \|\mathbf{y}_{k+1} - \mathbf{y}_k\|^2, \tag{3.38}
$$

---

[3] Similar to Sect. 3.4.2, we have the faster rate due to the stronger assumption, rather than the acceleration technique.

we have

$$L(\mathbf{x}_{k+1}, \mathbf{y}^*) - L(\mathbf{x}^*, \mathbf{y}_{k+1})$$

$$\leq \frac{1}{2\tau_k}\|\mathbf{x}^* - \mathbf{x}_k\|^2 - \left(\frac{1}{2\tau_k} + \frac{\mu_g}{2}\right)\|\mathbf{x}^* - \mathbf{x}_{k+1}\|^2 + \frac{1}{2\sigma_k}\|\mathbf{y}^* - \mathbf{y}_k\|^2$$

$$- \frac{1}{2\sigma_k}\|\mathbf{y}^* - \mathbf{y}_{k+1}\|^2 + \frac{\|\mathbf{K}\|_2^2\theta_{k-1}^2\sigma_k}{2}\|\mathbf{x}_k - \mathbf{x}_{k-1}\|^2 - \frac{1}{2\tau_k}\|\mathbf{x}_{k+1} - \mathbf{x}_k\|^2$$

$$+ \langle \mathbf{K}\mathbf{x}_{k+1} - \mathbf{K}\mathbf{x}_k, \mathbf{y}^* - \mathbf{y}_{k+1}\rangle - \theta_{k-1}\langle \mathbf{K}\mathbf{x}_k - \mathbf{K}\mathbf{x}_{k-1}, \mathbf{y}^* - \mathbf{y}_k\rangle. \qquad (3.39)$$

Letting

$$\frac{1}{\tau_k} + \mu_g \geq \frac{\tau_k}{\tau_{k+1}^2}, \quad \tau_{k+1}\sigma_{k+1} = \tau_k\sigma_k = \cdots = \tau_0\sigma_0 = \frac{1}{\|\mathbf{K}\|_2^2}, \quad \theta_{k-1} = \frac{\tau_k}{\tau_{k-1}}, \quad (3.40)$$

and dividing both sides of (3.39) by τ_k, we have

$$\frac{1}{\tau_k}L(\mathbf{x}_{k+1}, \mathbf{y}^*) - \frac{1}{\tau_k}L(\mathbf{x}^*, \mathbf{y}_{k+1})$$

$$\leq \frac{1}{2\tau_k^2}\|\mathbf{x}^* - \mathbf{x}_k\|^2 - \frac{1}{2\tau_{k+1}^2}\|\mathbf{x}^* - \mathbf{x}_{k+1}\|^2 + \frac{1}{2\sigma_k\tau_k}\|\mathbf{y}^* - \mathbf{y}_k\|^2$$

$$- \frac{1}{2\sigma_{k+1}\tau_{k+1}}\|\mathbf{y}^* - \mathbf{y}_{k+1}\|^2 + \frac{1}{2\tau_{k-1}^2}\|\mathbf{x}_k - \mathbf{x}_{k-1}\|^2 - \frac{1}{2\tau_k^2}\|\mathbf{x}_{k+1} - \mathbf{x}_k\|^2$$

$$+ \frac{1}{\tau_k}\langle \mathbf{K}\mathbf{x}_{k+1} - \mathbf{K}\mathbf{x}_k, \mathbf{y}^* - \mathbf{y}_{k+1}\rangle - \frac{1}{\tau_{k-1}}\langle \mathbf{K}\mathbf{x}_k - \mathbf{K}\mathbf{x}_{k-1}, \mathbf{y}^* - \mathbf{y}_k\rangle, \quad (3.41)$$

where we use

$$\frac{\|\mathbf{K}\|_2^2\theta_{k-1}^2\sigma_k}{2\tau_k} = \frac{\|\mathbf{K}\|_2^2\tau_k\sigma_k}{2\tau_{k-1}^2} = \frac{1}{2\tau_{k-1}^2}.$$

Summing (3.41) over $k = 0, 1, \cdots, K$, observing $\mathbf{x}_0 = \mathbf{x}_{-1}$, and using the convexity of f, g, and h and the definitions of $\hat{\mathbf{x}}_K$ and $\hat{\mathbf{y}}_K$, we have

$$\left(\sum_{k=0}^{K}\frac{1}{\tau_k}\right)\left[L(\hat{\mathbf{x}}_K, \mathbf{y}^*) - L(\mathbf{x}^*, \hat{\mathbf{y}}_K)\right]$$

$$\leq \frac{1}{2\tau_0^2}\|\mathbf{x}_0 - \mathbf{x}^*\|^2 + \frac{1}{2\sigma_0\tau_0}\|\mathbf{y}_0 - \mathbf{y}^*\|^2 - \frac{\|\mathbf{K}\|_2^2}{2}\|\mathbf{y}_{K+1} - \mathbf{y}^*\|^2$$

$$+ \frac{\|\mathbf{K}\|_2}{\tau_K} \|\mathbf{x}_{K+1} - \mathbf{x}_K\| \|\mathbf{y}^* - \mathbf{y}_{K+1}\| - \frac{1}{2\tau_K^2} \|\mathbf{x}_{K+1} - \mathbf{x}_K\|^2$$

$$\leq \frac{1}{2\sigma_0 \tau_0} \|\mathbf{y}_0 - \mathbf{y}^*\|^2 + \frac{1}{2\tau_0^2} \|\mathbf{x}_0 - \mathbf{x}^*\|^2.$$

The choices of τ_k, θ_k, and σ_k satisfy the requirements in (3.40) and make the above reduce to (3.37). $\square$

3.5.3   Case 3: $\mu_g = 0$, $\mu_h > 0$

The results for the scenario that h is strongly convex can be obtained by studying problem of $-\min_{\mathbf{x}} \max_{\mathbf{y}} L(\mathbf{x}, \mathbf{y})$ and using the conclusions in the previous section. We omit the details.

3.5.4   Case 4: $\mu_g > 0$, $\mu_h > 0$

At last, we consider the scenario that both g and h are strongly convex and prove the faster convergence rate by setting the parameters carefully.

Theorem 3.14 *Assume that g is μ_g-strongly convex and h is μ_h-strongly convex. Let* $\theta_k = \theta = \frac{1}{1 + \frac{\sqrt{\mu_g \mu_h}}{\|\mathbf{K}\|_2}}$, $\forall k$, $\hat{\mathbf{x}}_K = \left(\sum_{k=0}^{K} \frac{1}{\theta^k} \right)^{-1} \sum_{k=0}^{K} \frac{1}{\theta^k} \mathbf{x}_{k+1}$, $\hat{\mathbf{y}}_K = \left(\sum_{k=0}^{K} \frac{1}{\theta^k} \right)^{-1} \sum_{k=0}^{K} \frac{1}{\theta^k} \mathbf{y}_{k+1}$, $\sigma_k = \sigma = \frac{1}{\|\mathbf{K}\|_2} \sqrt{\frac{\mu_g}{\mu_h}}$, *and* $\tau_k = \tau = \frac{1}{\|\mathbf{K}\|_2} \sqrt{\frac{\mu_h}{\mu_g}}$, *then we have*

$$L(\hat{\mathbf{x}}_K, \mathbf{y}^*) - L(\mathbf{x}^*, \hat{\mathbf{y}}_K) \leq \theta^K \left(\frac{1}{2\tau} \|\mathbf{x}^* - \mathbf{x}_0\|^2 + \frac{1}{2\sigma} \|\mathbf{y}^* - \mathbf{y}_0\|^2 \right).$$

Proof From Lemma 3.19 and using (3.38) with $\theta_{k-1} = \theta$, we have

$$L(\mathbf{x}_{k+1}, \mathbf{y}^*) - L(\mathbf{x}^*, \mathbf{y}_{k+1})$$

$$\leq \frac{1}{2\tau} \|\mathbf{x}^* - \mathbf{x}_k\|^2 - \left(\frac{1}{2\tau} + \frac{\mu_g}{2} \right) \|\mathbf{x}^* - \mathbf{x}_{k+1}\|^2 + \frac{1}{2\sigma} \|\mathbf{y}^* - \mathbf{y}_k\|^2$$

$$- \left(\frac{1}{2\sigma} + \frac{\mu_h}{2} \right) \|\mathbf{y}^* - \mathbf{y}_{k+1}\|^2 + \frac{\|\mathbf{K}\|_2^2 \theta^2 \sigma}{2} \|\mathbf{x}_k - \mathbf{x}_{k-1}\|^2 - \frac{1}{2\tau} \|\mathbf{x}_{k+1} - \mathbf{x}_k\|^2$$

$$+ \langle \mathbf{K}\mathbf{x}_{k+1} - \mathbf{K}\mathbf{x}_k, \mathbf{y}^* - \mathbf{y}_{k+1} \rangle - \theta \langle \mathbf{K}\mathbf{x}_k - \mathbf{K}\mathbf{x}_{k-1}, \mathbf{y}^* - \mathbf{y}_k \rangle.$$

Letting

$$\frac{1}{2\theta\tau} \le \frac{1}{2\tau} + \frac{\mu_g}{2}, \qquad \frac{1}{2\theta\sigma} \le \frac{1}{2\sigma} + \frac{\mu_h}{2}, \qquad \|\mathbf{K}\|_2^2\theta\sigma \le \frac{1}{\tau},$$

which are satisfied by the definitions of τ, σ, and θ, we have

$$L(\mathbf{x}_{k+1}, \mathbf{y}^*) - L(\mathbf{x}^*, \mathbf{y}_{k+1})$$

$$\le \frac{1}{2\tau}\|\mathbf{x}^* - \mathbf{x}_k\|^2 + \frac{1}{2\sigma}\|\mathbf{y}^* - \mathbf{y}_k\|^2$$

$$- \frac{1}{\theta}\left(\frac{1}{2\tau}\|\mathbf{x}^* - \mathbf{x}_{k+1}\|^2 + \frac{1}{2\sigma}\|\mathbf{y}^* - \mathbf{y}_{k+1}\|^2\right)$$

$$+ \frac{\theta}{2\tau}\|\mathbf{x}_k - \mathbf{x}_{k-1}\|^2 - \frac{1}{2\tau}\|\mathbf{x}_{k+1} - \mathbf{x}_k\|^2$$

$$+ \langle \mathbf{K}\mathbf{x}_{k+1} - \mathbf{K}\mathbf{x}_k, \mathbf{y}^* - \mathbf{y}_{k+1}\rangle - \theta\langle \mathbf{K}\mathbf{x}_k - \mathbf{K}\mathbf{x}_{k-1}, \mathbf{y}^* - \mathbf{y}_k\rangle.$$

Dividing both sides by θ^k, we have

$$\frac{1}{\theta^k}L(\mathbf{x}_{k+1}, \mathbf{y}^*) - \frac{1}{\theta^k}L(\mathbf{x}^*, \mathbf{y}_{k+1})$$

$$\le \frac{1}{\theta^k}\left(\frac{1}{2\tau}\|\mathbf{x}^* - \mathbf{x}_k\|^2 + \frac{1}{2\sigma}\|\mathbf{y}^* - \mathbf{y}_k\|^2\right)$$

$$- \frac{1}{\theta^{k+1}}\left(\frac{1}{2\tau}\|\mathbf{x}^* - \mathbf{x}_{k+1}\|^2 + \frac{1}{2\sigma}\|\mathbf{y}^* - \mathbf{y}_{k+1}\|^2\right)$$

$$+ \frac{1}{\theta^{k-1}}\frac{1}{2\tau}\|\mathbf{x}_k - \mathbf{x}_{k-1}\|^2 - \frac{1}{\theta^k}\frac{1}{2\tau}\|\mathbf{x}_{k+1} - \mathbf{x}_k\|^2$$

$$+ \frac{1}{\theta^k}\langle \mathbf{K}\mathbf{x}_{k+1} - \mathbf{K}\mathbf{x}_k, \mathbf{y}^* - \mathbf{y}_{k+1}\rangle - \frac{1}{\theta^{k-1}}\langle \mathbf{K}\mathbf{x}_k - \mathbf{K}\mathbf{x}_{k-1}, \mathbf{y}^* - \mathbf{y}_k\rangle.$$

Summing over $k = 0, 1, \cdots, K$, and using $\mathbf{x}_0 = \mathbf{x}_{-1}$ and the convexity of f and g, we have

$$\left(\sum_{k=0}^{K}\frac{1}{\theta^k}\right)\left[L(\hat{\mathbf{x}}_K, \mathbf{y}^*) - L(\mathbf{x}^*, \hat{\mathbf{y}}_K)\right]$$

$$\le \frac{1}{2\tau}\|\mathbf{x}^* - \mathbf{x}_0\|^2 + \frac{1}{2\sigma}\|\mathbf{y}^* - \mathbf{y}_0\|^2$$

$$- \frac{1}{\theta^K}\frac{1}{2\tau}\|\mathbf{x}_{K+1} - \mathbf{x}_K\|^2 - \frac{1}{\theta^{K+1}}\frac{1}{2\sigma}\|\mathbf{y}^* - \mathbf{y}_{K+1}\|^2$$

$$+ \frac{1}{\theta^K}\langle \mathbf{K}\mathbf{x}_{K+1} - \mathbf{K}\mathbf{x}_K, \mathbf{y}^* - \mathbf{y}_{K+1}\rangle.$$

Using $\|\mathbf{K}\|_2^2\theta\sigma \le \frac{1}{\tau}$ and $\sum_{k=0}^{K}\frac{1}{\theta^k} > \frac{1}{\theta^K}$, we can have the conclusion. $\qquad\square$

3.6 Faster Frank–Wolfe Algorithm

The Frank–Wolfe method [8], also called the conditional gradient method, is a first-order projection-free method for minimizing a convex function over a convex set. It works by iteratively solving a linear optimization problem and remaining inside the feasible set. It avoids the projection step and has the advantage of keeping the "sparsity" of its solution, in the sense that the solution is a linear combination of finite extreme points (Definition A.6) of the feasible set. So it is particularly suitable for solving sparse and low-rank problems. For these reasons, the Frank–Wolfe method has drawn growing interest in recent years, especially in matrix completion, structural SVM, object tracking, sparse PCA, metric learning, and many other settings (e.g., [16]).

The convergence rate of the Frank–Wolfe method is $O(1/k)$ for generally convex optimization [15]. In this section, we follow [9] to show that under stronger assumptions, the convergence rate of the Frank–Wolfe algorithm can be improved to $O(1/k^2)$ (since this rate is obtained with stronger conditions, we do not called it "accelerated."). More specifically, we need to assume that the objective function is smooth and satisfies the quadratic functional growth condition and the set is a strongly convex set.

Definition 3.1 (Quadratic Functional Growth) A function satisfies the quadratic functional growth condition over set $\mathcal{K}$ if

$$f(\mathbf{x}) - f(\mathbf{x}^*) \geq \frac{\mu}{2}\|\mathbf{x} - \mathbf{x}^*\|^2, \quad \forall \mathbf{x} \in \mathcal{K},$$

where $\mathbf{x}^*$ is the minimizer of $f(\mathbf{x})$ over $\mathcal{K}$.

If f is a strongly convex function over $\mathcal{K}$ (which implies that $\mathcal{K}$ is convex), it naturally satisfies the quadratic functional growth condition. Thus the quadratic functional growth condition is weaker than the condition of strong convexity. Another concept is the strongly convex set. Let $\| \cdot \|$ and $\| \cdot \|^*$ be a pair of dual norms (Definition A.3) over $\mathbb{R}^n$.

Definition 3.2 (Strongly Convex Set) We say that a convex set $\mathcal{K} \subset \mathbb{R}^n$ is α-strongly convex with respect to $\| \cdot \|$ if for any $\mathbf{x}$ and $\mathbf{y} \in \mathcal{K}$, any $\gamma \in [0, 1]$, and any vector $\mathbf{z} \in \mathbb{R}^n$ such that $\|\mathbf{z}\| = 1$, it holds that

$$\gamma\mathbf{x} + (1 - \gamma)\mathbf{y} + \frac{\alpha\gamma(1 - \gamma)}{2}\|\mathbf{x} - \mathbf{y}\|^2\mathbf{z} \in \mathcal{K}.$$

Many convex sets in sparse and low-rank optimization are strongly convex sets [9]. Examples include the ℓ_p ball for $p \in (1, 2]$ and the Schatten-p ball for $p \in (1, 2]$.

We consider the following general constrained convex problem:

$$\min_{\mathbf{x}} f(\mathbf{x}), \quad s.t. \quad \mathbf{x} \in \mathcal{K}$$

under the following assumptions:

Assumption 3.1

1. $f(\mathbf{x})$ *satisfies the quadratic functional growth condition.*
2. $f(\mathbf{x})$ *is L-smooth.*
3. $\mathcal{K}$ *is an $\alpha_{\mathcal{K}}$-strongly convex set.*

We describe the Frank–Wolfe method in Algorithm 3.7. We can see that $\mathbf{x}_{k+1}$ is a convex combination of $\mathbf{x}_k$ and $\mathbf{p}_k$, with $\mathbf{p}_k \in \mathcal{K}$. Thus we can simply prove $\mathbf{x}_k \in \mathcal{K}, \forall k \geq 0$ by induction. In the traditional Frank–Wolfe algorithm, η_k is set as $\frac{2}{k+2}$ [15]. Here we give a complex setting to fit the proof.

Algorithm 3.7 Faster Frank–Wolfe method

Initialize $\mathbf{x}_0 \in \mathcal{K}$.
for $k = 0, 1, 2, 3, \cdots$ **do**
   $\mathbf{p}_k = \text{argmin}_{\mathbf{p} \in \mathcal{K}} \langle \mathbf{p}, \nabla f(\mathbf{x}_k) \rangle$,
   $\eta_k = \min\left\{1, \frac{\alpha_{\mathcal{K}} \|\nabla f(\mathbf{x}_k)\|^*}{4L}\right\}$,
   $\mathbf{x}_{k+1} = \mathbf{x}_k + \eta_k(\mathbf{p}_k - \mathbf{x}_k)$.
end for

We first give the following lemma to describe the decreasing property of $h_k = f(\mathbf{x}_k) - f(\mathbf{x}^*)$ at each iteration.

Lemma 3.20 *Suppose that Assumption 3.1 holds. For Algorithm 3.7, we have*

$$h_{k+1} \leq h_k \max\left\{\frac{1}{2}, 1 - \frac{\alpha_{\mathcal{K}} \|\nabla f(\mathbf{x}_k)\|^*}{8L}\right\}.$$

Proof From the optimality of $\mathbf{p}_k$, we have

$$\langle \mathbf{p}_k - \mathbf{x}_k, \nabla f(\mathbf{x}_k) \rangle \leq \langle \mathbf{x}^* - \mathbf{x}_k, \nabla f(\mathbf{x}_k) \rangle \leq f(\mathbf{x}^*) - f(\mathbf{x}_k) = -h_k.$$

Denote $\mathbf{w}_k = \text{argmin}_{\|\mathbf{w}\|=1} \langle \mathbf{w}, \nabla f(\mathbf{x}_k) \rangle$, then we have $\langle \mathbf{w}, \nabla f(\mathbf{x}_k) \rangle = -\|\nabla f(\mathbf{x}_k)\|^*$. Using the strong convexity of the set $\mathcal{K}$, we have $\widetilde{\mathbf{p}}_k = \frac{1}{2}(\mathbf{p}_k + \mathbf{x}_k) + \frac{\alpha_{\mathcal{K}}}{8}\|\mathbf{x}_k - \mathbf{p}_k\|^2 \mathbf{w}_k \in \mathcal{K}$. So we have

$$\langle \mathbf{p}_k - \mathbf{x}_k, \nabla f(\mathbf{x}_k) \rangle \leq \langle \widetilde{\mathbf{p}}_k - \mathbf{x}_k, \nabla f(\mathbf{x}_k) \rangle$$

$$= \frac{1}{2} \langle \mathbf{p}_k - \mathbf{x}_k, \nabla f(\mathbf{x}_k) \rangle + \frac{\alpha_{\mathcal{K}}\|\mathbf{x}_k - \mathbf{p}_k\|^2}{8} \langle \mathbf{w}_k, \nabla f(\mathbf{x}_k) \rangle$$

$$\leq -\frac{1}{2}h_k - \frac{\alpha_{\mathcal{K}}\|\mathbf{x}_k - \mathbf{p}_k\|^2}{8}\|\nabla f(\mathbf{x}_k)\|^*. \qquad (3.42)$$

On the other hand, from the smoothness of f we have

$$f(\mathbf{x}_{k+1}) \leq f(\mathbf{x}_k) + \eta_k \langle \mathbf{p}_k - \mathbf{x}_k, \nabla f(\mathbf{x}_k) \rangle + \frac{L\eta_k^2}{2} \|\mathbf{p}_k - \mathbf{x}_k\|^2.$$

Subtracting $f(\mathbf{x}^*)$ from both sides, we have

$$h_{k+1} \leq h_k + \eta_k \langle \mathbf{p}_k - \mathbf{x}_k, \nabla f(\mathbf{x}_k) \rangle + \frac{L\eta_k^2}{2} \|\mathbf{p}_k - \mathbf{x}_k\|^2.$$

Plugging (3.42), we have

$$h_{k+1} \leq h_k \left(1 - \frac{\eta_k}{2}\right) + \frac{\|\mathbf{p}_k - \mathbf{x}_k\|^2}{2} \left(L\eta_k^2 - \frac{\eta_k \alpha_{\mathcal{K}} \|\nabla f(\mathbf{x}_k)\|^*}{4}\right).$$

If $\frac{\alpha_{\mathcal{K}} \|\nabla f(\mathbf{x}_k)\|^*}{4} \geq L$, then $\eta_k = 1$. So we have

$$h_{k+1} \leq \frac{h_k}{2}.$$

Otherwise, $\eta_k = \frac{\alpha_{\mathcal{K}} \|\nabla f(\mathbf{x}_k)\|^*}{4L}$ and we have

$$h_{k+1} \leq h_k \left(1 - \frac{\alpha_{\mathcal{K}} \|\nabla f(\mathbf{x}_k)\|^*}{8L}\right),$$

which completes the proof.   □

Based on Lemma 3.20, we can prove the convergence rate of the Frank–Wolfe algorithm in the following theorem.

Theorem 3.15 *Suppose that Assumption 3.1 holds. Let $M = \frac{\alpha_{\mathcal{K}} \sqrt{\mu}}{8\sqrt{2L}}$ and denote $D = \max_{\mathbf{x}, \mathbf{y} \in \mathcal{K}} \|\mathbf{x} - \mathbf{y}\|$ as the diameter of $\mathcal{K}$. For Algorithm 3.7, we have*

$$f(\mathbf{x}_k) - f(\mathbf{x}^*) \leq \frac{\max\left\{\frac{9}{2} L D^2, 4(f(\mathbf{x}_0) - f(\mathbf{x}^*)), 18M^{-2}\right\}}{(k+2)^2}.$$

Proof From the quadratic functional growth condition, we have

$$f(\mathbf{x}) - f(\mathbf{x}^*) \geq \frac{\mu}{2} \|\mathbf{x} - \mathbf{x}^*\|^2.$$

So we have

$$
\begin{aligned}
f(\mathbf{x}) - f(\mathbf{x}^*) &\leq \langle \mathbf{x} - \mathbf{x}^*, \nabla f(\mathbf{x}) \rangle \\
&\leq \|\mathbf{x} - \mathbf{x}^*\| \|\nabla f(\mathbf{x})\|^* \\
&\leq \sqrt{\frac{2}{\mu}(f(\mathbf{x}) - f(\mathbf{x}^*))} \|\nabla f(\mathbf{x})\|^*,
\end{aligned}
$$

which leads to

$$
\sqrt{\frac{\mu}{2}(f(\mathbf{x}) - f(\mathbf{x}^*))} \leq \|\nabla f(\mathbf{x})\|^*.
$$

Using Lemma 3.20, we have

$$
h_{k+1} \leq h_k \max\left\{ \frac{1}{2}, 1 - M\sqrt{h_k} \right\}. \tag{3.43}
$$

Now we use induction to prove $h_k \leq \frac{C}{(k+2)^2}$, where $C = \max\left\{ \frac{9}{2}LD^2, 4(f(\mathbf{x}_0) - f(\mathbf{x}^*)), 18M^{-2} \right\}$. It holds for $k = 0$ trivially. Assume that $h_k \leq \frac{C}{(k+2)^2}$ for $k = t$. We then consider $k = t + 1$.

If $\frac{1}{2} \geq 1 - M\sqrt{h_t}$, we have

$$
h_{t+1} \leq \frac{h_t}{2} \leq \frac{C}{2(t+2)^2} \leq \frac{C}{(t+3)^2}.
$$

If $\frac{1}{2} < 1 - M\sqrt{h_t}$ and $h_t \leq \frac{C}{2(t+2)^2}$, similar to the above analysis, it holds that

$$
h_{t+1} \leq h_t \leq \frac{C}{2(t+2)^2} \leq \frac{C}{(t+3)^2}.
$$

If $\frac{1}{2} < 1 - M\sqrt{h_t}$ and $h_t > \frac{C}{2(t+2)^2}$. From (3.43), we have

$$
\begin{aligned}
h_{t+1} &\leq h_t \left(1 - M\sqrt{h_t}\right) \\
&\leq \frac{C}{(t+2)^2}\left(1 - M\sqrt{\frac{C}{2}}\frac{1}{t+2}\right) \\
&\leq \frac{C}{(t+2)^2}\left(1 - \frac{3}{t+2}\right) \leq \frac{C}{(t+3)^2}.
\end{aligned}
$$

The proof is completed.   □

References

1. D.P. Bertsekas, *Nonlinear Programming*, 2nd edn. (Athena Scientific, Belmont, MA, 1999)
2. S. Boyd, N. Parikh, E. Chu, B. Peleato, J. Eckstein, Distributed optimization and statistical learning via the alternating direction method of multipliers. Found. Trends Mach. Learn. **3**(1), 1–122 (2011)
3. A. Chambolle, T. Pock, A first-order primal-dual algorithm for convex problems with applications to imaging. J. Math. Imag. Vis. **40**(1), 120–145 (2011)
4. A. Chambolle, T. Pock, On the ergodic convergence rates of a first-order primal-dual algorithm. Math. Program. **159**(1–2), 253–287 (2016)
5. Y. Chen, G. Lan, Y. Ouyang, Optimal primal-dual methods for a class of saddle point problems. SIAM J. Optim. **24**(4), 1779–1814 (2014)
6. D. Davis, W. Yin, Convergence rate analysis of several splitting schemes, in *Splitting Methods in Communication, Imaging, Science, and Engineering* (Springer, New York, 2016), pp. 115–163
7. E. Esser, X. Zhang, T.F. Chan, A general framework for a class of first order primal-dual algorithms for convex optimization in imaging science. SIAM J. Imag. Sci. **3**(4), 1015–1046 (2010)
8. M. Frank, P. Wolfe, An algorithm for quadratic programming. Nav. Res. Logist. Q. **3**(1–2), 95–110 (1956)
9. D. Garber, E. Hazan, Faster rates for the Frank-Wolfe method over strongly-convex sets, in *Proceedings of the 32nd International Conference on Machine Learning*, Lille, (2015), pp. 541–549
10. P. Giselsson, S. Boyd, Linear convergence and metric selection in Douglas Rachford splitting and ADMM. IEEE Trans. Automat. Contr. **62**(2), 532–544 (2017)
11. B. He, X. Yuan, On the acceleration of augmented Lagrangian method for linearly constrained optimization (2010). Preprint. http://www.optimization-online.org/DB_FILE/2010/10/2760.pdf
12. B. He, X. Yuan, On the $O(1/t)$ convergence rate of the Douglas-Rachford alternating direction method. SIAM J. Numer. Anal. **50**(2), 700–709 (2012)
13. B. He, X. Yuan, On non-ergodic convergence rate of Douglas-Rachford alternating directions method of multipliers. Numer. Math. **130**(3), 567–577 (2015)
14. B. He, L.-Z. Liao, D. Han, H. Yang, A new inexact alternating directions method for monotone variational inequalities. Math. Program. **92**(1), 103–118 (2002)
15. M. Jaggi, Revisiting Frank-Wolfe: projection free sparse convex optimization, in *Proceedings of the 31th International Conference on Machine Learning*, Atlanta, (2013), pp. 427–435
16. M. Jaggi, M. Sulovsk, A simple algorithm for nuclear norm regularized problems, in *Proceedings of the 27th International Conference on Machine Learning*, Haifa, (2010), pp. 471–478
17. G. Lan, R.D. Monteiro, Iteration-complexity of first-order penalty methods for convex programming. Math. Program. **138**(1–2), 115–139 (2013)
18. H. Li, Z. Lin, On the complexity analysis of the primal solutions for the accelerated randomized dual coordinate ascent. J. Mach. Learn. Res. (2020). http://jmlr.org/papers/v21/18-425.html
19. H. Li, Z. Lin, Accelerated alternating direction method of multipliers: an optimal $O(1/K)$ nonergodic analysis. J. Sci. Comput. **79**(2), 671–699 (2019)
20. H. Li, C. Fang, Z. Lin, Convergence rates analysis of the quadratic penalty method and its applications to decentralized distributed optimization (2017). Preprint. arXiv:1711.10802
21. Z. Lin, M. Chen, Y. Ma, The augmented Lagrange multiplier method for exact recovery of corrupted low-rank matrices (2010). Preprint. arXiv:1009.5055
22. Z. Lin, R. Liu, H. Li, Linearized alternating direction method with parallel splitting and adaptive penalty for separable convex programs in machine learning. Mach. Learn. **99**(2), 287–325 (2015)

23. J. Lu, M. Johansson, Convergence analysis of approximate primal solutions in dual first-order methods. SIAM J. Optim. **26**(4), 2430–2467 (2016)
24. C. Lu, H. Li, Z. Lin, S. Yan, Fast proximal linearized alternating direction method of multiplier with parallel splitting, in *Proceedings of the 30th AAAI Conference on Artificial Intelligence*, Phoenix, (2016), pp. 739–745
25. D.G. Luenberger, Convergence rate of a penalty-function scheme. J. Optim. Theory Appl. **7**(1), 39–51 (1971)
26. I. Necoara, V. Nedelcu, Rate analysis of inexact dual first-order methods application to dual decomposition. IEEE Trans. Automat. Contr. **59**(5), 1232–1243 (2014)
27. I. Necoara, A. Patrascu, Iteration complexity analysis of dual first-order methods for conic convex programming. Optim. Methods Softw. **31**(3), 645–678 (2016)
28. I. Necoara, A. Patrascu, F. Glineur, Complexity of first-order inexact Lagrangian and penalty methods for conic convex programming. Optim. Methods Softw. **34**(2), 305–335 (2019)
29. V.H. Nguyen, J.-J. Strodiot, Convergence rate results for a penalty function method, in *Optimization Techniques* (Springer, New York, 1978), pp. 101–106
30. Y. Ouyang, Y. Chen, G. Lan, E. Pasiliao Jr., An accelerated linearized alternating direction method of multipliers. SIAM J. Imag. Sci. **8**(1), 644–681 (2015)
31. P. Patrinos, A. Bemporad, An accelerated dual gradient projection algorithm for embedded linear model predictive control. IEEE Trans. Automat. Contr. **59**(1), 18–33 (2013)
32. T. Pock, D. Cremers, H. Bischof, A. Chambolle, An algorithm for minimizing the Mumford-Shah functional, in *Proceedings of the 12th International Conference on Computer Vision*, Kyoto, (2009), pp. 1133–1140
33. B.T. Polyak, The convergence rate of the penalty function method. USSR Comput. Math. Math. Phys. **11**(1), 1–12 (1971)
34. R. Shefi, M. Teboulle, Rate of convergence analysis of decomposition methods based on the proximal method of multipliers for convex minimization. SIAM J. Optim. **24**(1), 269–297 (2014)
35. W. Tian, X. Yuan, An alternating direction method of multipliers with a worst-case $o(1/n^2)$ convergence rate. Math. Comput. **88**(318), 1685–1713 (2019)
36. P. Tseng, On accelerated proximal gradient methods for convex-concave optimization. Technical report, University of Washington, Seattle (2008)
37. X. Wang, X. Yuan, The linearized alternating direction method of multipliers for Dantzig selector. SIAM J. Sci. Comput. **34**(5), A2792–A2811 (2012)
38. Y. Xu, Accelerated first-order primal-dual proximal methods for linearly constrained composite convex programming. SIAM J. Optim. **27**(3), 1459–1484 (2017)

Chapter 4
Accelerated Algorithms for Nonconvex Optimization

Nonconvex optimization has gained extensive attention recently in the machine learning community. In this section, we present the accelerated gradient methods for nonconvex optimization. The topics studied include the general convergence results under the Kurdyka–Łojasiewicz (KŁ) condition (Definition A.36), how to achieve the critical point quickly, and how to escape the saddle point quickly.

4.1 Proximal Gradient with Momentum

Consider the following composite minimization problem:

$$\min_{\mathbf{x}} F(\mathbf{x}) \equiv f(\mathbf{x}) + g(\mathbf{x}), \tag{4.1}$$

where f is differentiable (it can be nonconvex) and g can be both nonconvex and nonsmooth. Examples of problem (4.1) include sparse and low-rank learning with nonconvex regularizers, e.g., ℓ_p-norm [10], Capped-ℓ_1 penalty [27], Log-Sum Penalty [6], Minimax Concave Penalty [26], Geman Penalty [13], Smoothly Clipped Absolute Deviation [9], and Schatten-p norm [22].

Popular methods for problem (4.1) include the general iterative shrinkage and thresholding (GIST) [15], inertial forward-backward [5], iPiano [25], and the proximal gradient with momentum [21], where GIST is the general proximal gradient (PG) method and the later three are extensions of PG with momentum. The accelerated PG method was studied in [14, 20, 24], where the convergence to the critical point is guaranteed for nonconvex programs and acceleration is proven for convex programs. As a comparison, acceleration is not ensured in [5, 21, 25].

© Springer Nature Singapore Pte Ltd. 2020
Z. Lin et al., *Accelerated Optimization for Machine Learning*,
https://doi.org/10.1007/978-981-15-2910-8_4

The KŁ inequality [2–4] is a powerful tool proposed recently for the analysis of nonconvex programming, where the convergence of the whole sequence can be proven, rather than the subsequence only. Moreover, the convergence rate can also be proven when the desingularizing function (Definition A.35) φ used in the KŁ property has some special form [11].

In this section, we use the method discussed in [21] as an example to give the general convergence results for nonconvex programs and stronger results under the KŁ condition. The method is described in Algorithm 4.1.

Algorithm 4.1 Proximal gradient (PG) with momentum

Initialize $\mathbf{y}_0 = \mathbf{x}_0$, $\beta \in (0, 1)$, and $\eta < \frac{1}{L}$.
for $k = 0, 1, 2, 3, \cdots$ **do**
$\quad \mathbf{x}_k = \text{Prox}_{\eta g}(\mathbf{y}_k - \eta \nabla f(\mathbf{y}_k))$,
$\quad \mathbf{v}_k = \mathbf{x}_k + \beta(\mathbf{x}_k - \mathbf{x}_{k-1})$,
$\quad$ **if** $F(\mathbf{x}_k) \leq F(\mathbf{v}_k)$ **then**
$\quad\quad \mathbf{y}_{k+1} = \mathbf{x}_k$,
$\quad$ **else**
$\quad\quad \mathbf{y}_{k+1} = \mathbf{v}_k$.
$\quad$ **end if**
end for

In this section, we make the following assumptions.

Assumption 4.1

1) $f(\mathbf{x})$ is a proper (Definition A.30) and L-smooth function and $g(\mathbf{x})$ is proper and lower semicontinuous (Definition A.31).
2) $F(\mathbf{x})$ is coercive (Definition A.32).
3) $F(\mathbf{x})$ has the KŁ property.

4.1.1 Convergence Theorem

We first give the general convergence result of Algorithm 4.1 in the following theorem. Generally speaking, we prove that every accumulation point is a critical point.

Theorem 4.1 *Assume that 1) and 2) of Assumption 4.1 hold, then with $\eta < \frac{1}{L}$ the sequence $\{\mathbf{x}_k\}$ generated by Algorithm 4.1 satisfies*

1. $\{\mathbf{x}_k\}$ is a bounded sequence.
2. The set of accumulation points Ω of $\{\mathbf{x}_k\}$ forms a compact set (Definition A.27), on which the objective function F is constant.
3. All elements of Ω are critical points of $F(\mathbf{x})$.

Proof From the L-smoothness of f we have

$$F(\mathbf{x}_k) \leq g(\mathbf{x}_k) + f(\mathbf{y}_k) + \langle \nabla f(\mathbf{y}_k), \mathbf{x}_k - \mathbf{y}_k \rangle + \frac{L}{2} \|\mathbf{x}_k - \mathbf{y}_k\|^2$$

$$\overset{a}{\leq} g(\mathbf{y}_k) - \langle \nabla f(\mathbf{y}_k), \mathbf{x}_k - \mathbf{y}_k \rangle - \frac{1}{2\eta} \|\mathbf{x}_k - \mathbf{y}_k\|^2$$

$$+ f(\mathbf{y}_k) + \langle \nabla f(\mathbf{y}_k), \mathbf{x}_k - \mathbf{y}_k \rangle + \frac{L}{2} \|\mathbf{x}_k - \mathbf{y}_k\|^2$$

$$= F(\mathbf{y}_k) - \left(\frac{1}{2\eta} - \frac{L}{2} \right) \|\mathbf{x}_k - \mathbf{y}_k\|^2$$

$$= F(\mathbf{y}_k) - \alpha \|\mathbf{x}_k - \mathbf{y}_k\|^2,$$

where $\alpha = \frac{1}{2\eta} - \frac{L}{2}$ and $\overset{a}{\leq}$ uses the definition of $\mathbf{x}_k$, thus

$$g(\mathbf{x}_k) + \frac{1}{2\eta} \|\mathbf{x}_k - (\mathbf{y}_k - \eta \nabla f(\mathbf{y}_k))\|^2 \leq g(\mathbf{y}_k) + \frac{1}{2\eta} \|\mathbf{y}_k - (\mathbf{y}_k - \eta \nabla f(\mathbf{y}_k))\|^2$$

$$= g(\mathbf{y}_k) + \frac{1}{2\eta} \|\eta \nabla f(\mathbf{y}_k)\|^2.$$

So we have

$$F(\mathbf{y}_{k+1}) \leq F(\mathbf{x}_k) \leq F(\mathbf{y}_k) - \alpha \|\mathbf{x}_k - \mathbf{y}_k\|^2 \leq F(\mathbf{x}_{k-1}) - \alpha \|\mathbf{x}_k - \mathbf{y}_k\|^2. \quad (4.2)$$

Since $F(\mathbf{x}) > -\infty$, we conclude that $F(\mathbf{x}_k)$ and $F(\mathbf{y}_k)$ converge to the same limit F^*, i.e.,

$$\lim_{k \to \infty} F(\mathbf{x}_k) = \lim_{k \to \infty} F(\mathbf{y}_k) = F^*. \quad (4.3)$$

On the other hand, we also have $F(\mathbf{x}_k) \leq F(\mathbf{x}_0)$ and $F(\mathbf{y}_k) \leq F(\mathbf{x}_0)$ for all k. Thus $\{\mathbf{x}_k\}$ and $\{\mathbf{y}_k\}$ are bounded and have bounded accumulation points. From (4.2), we also have

$$\alpha \|\mathbf{x}_k - \mathbf{y}_k\|^2 \leq F(\mathbf{y}_k) - F(\mathbf{y}_{k+1}).$$

Summing over $k = 0, 1, \cdots, k$ and letting $k \to \infty$, we have

$$\alpha \sum_{k=0}^{\infty} \|\mathbf{x}_k - \mathbf{y}_k\|^2 \leq F(\mathbf{y}_k) - \inf F < \infty.$$

It further implies that $\|\mathbf{x}_k - \mathbf{y}_k\| \to 0$. Thus $\{\mathbf{x}_k\}$ and $\{\mathbf{y}_k\}$ share the same accumulation points Ω. Since Ω is closed and bounded, we conclude that Ω is compact.

Define

$$\mathbf{u}_k = \nabla f(\mathbf{x}_k) - \nabla f(\mathbf{y}_k) - \frac{1}{\eta}(\mathbf{x}_k - \mathbf{y}_k).$$

From the optimality condition, we have

$$-\nabla f(\mathbf{y}_k) - \frac{1}{\eta}(\mathbf{x}_k - \mathbf{y}_k) \in \partial g(\mathbf{x}_k).$$

Then

$$\mathbf{u}_k = \nabla f(\mathbf{x}_k) - \nabla f(\mathbf{y}_k) - \frac{1}{\eta}(\mathbf{x}_k - \mathbf{y}_k) \in \partial F(\mathbf{x}_k).$$

We further have

$$\|\mathbf{u}_k\| = \left\| \nabla f(\mathbf{x}_k) - \nabla f(\mathbf{y}_k) - \frac{1}{\eta}(\mathbf{x}_k - \mathbf{y}_k) \right\|$$

$$\leq \left(L + \frac{1}{\eta} \right) \|\mathbf{y}_k - \mathbf{x}_k\| \to 0. \tag{4.4}$$

Consider any $\mathbf{z} \in \Omega$ and we write $\mathbf{x}_k \to \mathbf{z}$ and $\mathbf{y}_k \to \mathbf{z}$ by restricting to a subsequence. By the definition of the proximal mapping, we have

$$\langle \nabla f(\mathbf{y}_k), \mathbf{x}_k - \mathbf{y}_k \rangle + \frac{1}{2\eta}\|\mathbf{x}_k - \mathbf{y}_k\|^2 + g(\mathbf{x}_k)$$

$$\leq \langle \nabla f(\mathbf{y}_k), \mathbf{z} - \mathbf{y}_k \rangle + \frac{1}{2\eta}\|\mathbf{z} - \mathbf{y}_k\|^2 + g(\mathbf{z}).$$

Taking limsup on both sides and noting that $\mathbf{x}_k - \mathbf{y}_k \to \mathbf{0}$ and $\mathbf{y}_k \to \mathbf{z}$, we obtain that $\limsup_{k\to\infty} g(\mathbf{x}_k) \leq g(\mathbf{z})$. Since g is lower semicontinuous and $\mathbf{x}_k \to \mathbf{z}$, it follows that $\limsup_{k\to\infty} g(\mathbf{x}_k) \geq g(\mathbf{z})$. Combining both inequalities, we conclude that $\lim_{k\to\infty} g(\mathbf{x}_k) = g(\mathbf{z})$. Note that the continuity of f yields $\lim_{k\to\infty} f(\mathbf{x}_k) = f(\mathbf{z})$. We then conclude that $\lim_{k\to\infty} F(\mathbf{x}_k) = F(\mathbf{z})$. Since $\lim_{k\to\infty} F(\mathbf{x}_k) = F^*$

by (4.3), we conclude that

$$F(\mathbf{z}) = F^*, \forall \mathbf{z} \in \Omega.$$

Hence, F is constant on the compact set Ω.

Now we have established $\mathbf{x}_k \rightarrow \mathbf{z}$, $F(\mathbf{x}_k) \rightarrow F(\mathbf{z})$, and $\partial F(\mathbf{x}_k) \ni \mathbf{u}_k \rightarrow \mathbf{0}$. Recalling the definition of limiting subdifferential (Definition A.33), we conclude that $\mathbf{0} \in \partial F(\mathbf{z})$ for all $\mathbf{z} \in \Omega$. □

With the KŁ property, we can have a stronger convergence result, that is to say, the whole sequence generated by the algorithm converges to a critical point, rather than only the subsequence proven in Theorem 4.1.

Theorem 4.2 *Assume that 1)–3) of Assumption 4.1 hold, then with $\eta < \frac{1}{L}$ the sequence $\{\mathbf{x}_k\}$ generated by Algorithm 4.1 satisfies*

1. *$\{\mathbf{x}_k\}$ converges;*
2. *$\{\mathbf{x}_k\}$ converges to a critical point $\mathbf{x}^* \in \Omega$.*

Proof From (4.2), we have

$$F(\mathbf{x}_k) \leq F(\mathbf{x}_{k-1}) - \alpha \|\mathbf{x}_k - \mathbf{y}_k\|^2.$$

Moreover, from (4.4) we have

$$\text{dist}(\mathbf{0}, \partial F(\mathbf{x}_k)) \leq \left(\frac{1}{\eta} + L\right) \|\mathbf{x}_k - \mathbf{y}_k\|.$$

We have shown in the proof of Theorem 4.1 that $F(\mathbf{x}_{k+1}) \leq F(\mathbf{x}_k)$, $F(\mathbf{x}_k) \rightarrow F^*$, and $\text{dist}(\mathbf{x}_k, \Omega) \rightarrow 0$. Thus for any $\epsilon > 0$ and $\delta > 0$, there exists K_0 such that

$$\mathbf{x}_k \in \{\mathbf{x}, \text{dist}(\mathbf{x}, \Omega) \leq \epsilon\} \cap [F^* < F(\mathbf{x}) < F^* + \delta], \forall k \geq K_0.$$

From the uniform KŁ property (Lemma A.3), there exists a desingularizing function φ such that

$$\varphi'(F(\mathbf{x}_k) - F^*)\text{dist}(\mathbf{0}, \partial F(\mathbf{x}_k)) \geq 1. \tag{4.5}$$

So

$$\varphi'(F(\mathbf{x}_k) - F^*) \geq \frac{1}{\text{dist}(\mathbf{0}, \partial F(\mathbf{x}_k))} \geq \frac{1}{\left(\frac{1}{\eta} + L\right) \|\mathbf{x}_k - \mathbf{y}_k\|}.$$

On the other hand, since φ is concave, we have

$$\varphi(F(\mathbf{x}_{k+1}) - F^*) \leq \varphi(F(\mathbf{x}_k) - F^*) + \varphi'(F(\mathbf{x}_k) - F^*)(F(\mathbf{x}_{k+1}) - F(\mathbf{x}_k))$$

$$\leq \varphi(F(\mathbf{x}_k) - F^*) + \frac{F(\mathbf{x}_{k+1}) - F(\mathbf{x}_k)}{\left(\frac{1}{\eta} + L\right)\|\mathbf{x}_k - \mathbf{y}_k\|}$$

$$\leq \varphi(F(\mathbf{x}_k) - F^*) - \frac{\alpha\|\mathbf{x}_{k+1} - \mathbf{y}_{k+1}\|^2}{\left(\frac{1}{\eta} + L\right)\|\mathbf{x}_k - \mathbf{y}_k\|}.$$

So

$$\|\mathbf{x}_{k+1} - \mathbf{y}_{k+1}\| \leq \sqrt{c\|\mathbf{x}_k - \mathbf{y}_k\|(\Psi_k - \Psi_{k+1})} \leq \frac{1}{2}\|\mathbf{x}_k - \mathbf{y}_k\| + \frac{c}{2}(\Psi_k - \Psi_{k+1}),$$

where $c = \frac{\frac{1}{\eta} + L}{\alpha}$ and $\Psi_k = \varphi(F(\mathbf{x}_k) - F^*)$. Summing over $k = 1, 2, \cdots, \infty$, we have

$$\sum_{k=2}^{\infty} \|\mathbf{x}_k - \mathbf{y}_k\| \leq \|\mathbf{x}_1 - \mathbf{y}_1\| + c\Psi_1.$$

From the definition of $\mathbf{y}_k$ in Algorithm 4.1, we have $\|\mathbf{x}_k - \mathbf{y}_k\| = \|\mathbf{x}_k - \mathbf{x}_{k-1}\|$ or $\|\mathbf{x}_k - \mathbf{y}_k\| = \|\mathbf{x}_k - \mathbf{x}_{k-1} - \beta(\mathbf{x}_{k-1} - \mathbf{x}_{k-2})\|$. So $\|\mathbf{x}_k - \mathbf{x}_{k-1}\| - \beta\|\mathbf{x}_{k-1} - \mathbf{x}_{k-2}\| \leq \|\mathbf{x}_k - \mathbf{y}_k\|$. Thus,

$$\sum_{k=2}^{\infty} \|\mathbf{x}_k - \mathbf{x}_{k-1}\| - \sum_{k=1}^{\infty} \beta\|\mathbf{x}_k - \mathbf{x}_{k-1}\| \leq \|\mathbf{x}_1 - \mathbf{y}_1\| + c\Psi_1.$$

So

$$(1 - \beta)\sum_{k=2}^{\infty} \|\mathbf{x}_k - \mathbf{x}_{k-1}\| \leq \beta\|\mathbf{x}_1 - \mathbf{x}_0\| + \|\mathbf{x}_1 - \mathbf{y}_1\| + c\Psi_1.$$

So $\{\mathbf{x}_k\}$ is a Cauchy sequence and hence is a convergent sequence. The second result follows immediately from Theorem 4.1. □

Moreover, with the KŁ property we can give the convergence rate of Algorithm 4.1. As a comparison, the convergence rate of gradient descent for the general nonconvex problem is in the form of $\min_{0 \leq k \leq K} \|\text{dist}(\mathbf{0}, \partial F(\mathbf{x}_k))\|^2 \leq O(1/K)$ [23].

Theorem 4.3 *Assume that 1)–3) of Assumption 4.1 hold and the desingularizing function φ has the form of $\varphi(t) = \frac{C}{\theta}t^\theta$ for some $C > 0$, $\theta \in (0, 1]$. Let $F^* = F(\mathbf{x})$ for all $\mathbf{x} \in \Omega$ and $r_k = F(\mathbf{x}_k) - F^*$, then with $\eta < \frac{1}{L}$ the sequence $\{\mathbf{x}_k\}$ generated by Algorithm 4.1 satisfies*

1. If $\theta = 1$, then there exists k_1 such that $F(\mathbf{x}_k) = F^*$ for all $k > k_1$ and the algorithm terminates in finite steps;
2. If $\theta \in [\frac{1}{2}, 1)$, then there exists k_2 such that for all $k > k_2$,

$$F(\mathbf{x}_k) - F^* \leq \left(\frac{d_1 C^2}{1 + d_1 C^2}\right)^{k-k_2} r_{k_2};$$

3. If $\theta \in (0, \frac{1}{2})$, then there exists k_3 such that for all $k > k_3$,

$$F(\mathbf{x}_k) - F^* \leq \left[\frac{C}{(k - k_3)d_2(1 - 2\theta)}\right]^{\frac{1}{1-2\theta}},$$

where $d_1 = \left(\frac{1}{\eta} + L\right)^2 / \left(\frac{1}{2\eta} - \frac{L}{2}\right)$ and $d_2 = \min\left\{\frac{1}{2d_1 C}, \frac{C}{1-2\theta}\left(2^{\frac{2\theta-1}{2\theta-2}} - 1\right)r_0^{2\theta-1}\right\}$.

Proof Throughout the proof we assume that $F(\mathbf{x}_k) \neq F^*$ for all k. From (4.5) we have

$$1 \leq \left[\varphi'(F(\mathbf{x}_k) - F^*)\text{dist}(\mathbf{0}, \partial F(\mathbf{x}_k))\right]^2$$

$$\leq [\varphi'(r_k)]^2 \left(\frac{1}{\eta} + L\right)^2 \|\mathbf{x}_k - \mathbf{y}_k\|^2$$

$$\leq [\varphi'(r_k)]^2 \left(\frac{1}{\eta} + L\right)^2 \frac{F(\mathbf{x}_{k-1}) - F(\mathbf{x}_k)}{\alpha}$$

$$= d_1 [\varphi'(r_k)]^2 (r_{k-1} - r_k),$$

for all $k > k_0$. Because φ has the form of $\varphi(t) = \frac{C}{\theta} t^\theta$, we have $\varphi'(t) = C t^{\theta-1}$. So we have

$$1 \leq d_1 C^2 r_k^{2\theta-2}(r_{k-1} - r_k). \tag{4.6}$$

1. Case $\theta = 1$.
   In this case, (4.6) becomes

   $$1 \leq d_1 C^2 (r_k - r_{k+1}).$$

   Because $r_k \to 0$, $d_1 > 0$, and $C > 0$, this is a contradiction. So there exists k_1 such that $r_k = 0$ for all $k > k_1$. Thus the algorithm terminates in finite steps.
2. Case $\theta \in [\frac{1}{2}, 1)$.
   In this case, $0 < 2 - 2\theta \leq 1$. As $r_k \to 0$, there exists $\hat{k}_3$ such that $r_k^{2-2\theta} \geq r_k$ for all $k > \hat{k}_3$. Then (4.6) becomes

   $$r_k \leq d_1 C^2 (r_{k-1} - r_k).$$

So we have

$$r_k \leq \frac{d_1 C^2}{1 + d_1 C^2} r_{k-1},$$

for all $k_2 > \max\{k_0, \hat{k}_3\}$ and

$$r_k \leq \left(\frac{d_1 C^2}{1 + d_1 C^2}\right)^{k-k_2} r_{k_2}.$$

3. Case $\theta \in (0, \frac{1}{2})$.

   In this case, $2\theta - 2 \in (-2, -1)$ and $2\theta - 1 \in (-1, 0)$. As $r_{k-1} > r_k$, we have $r_{k-1}^{2\theta-2} < r_k^{2\theta-2}$ and $r_0^{2\theta-1} < \cdots < r_{k-1}^{2\theta-1} < r_k^{2\theta-1}$.

   Define $\phi(t) = \frac{C}{1-2\theta} t^{2\theta-1}$, then $\phi'(t) = -Ct^{2\theta-2}$.

   If $r_k^{2\theta-2} \leq 2r_{k-1}^{2\theta-2}$, then

$$\phi(r_k) - \phi(r_{k-1}) = \int_{r_{k-1}}^{r_k} \phi'(t)dt = C \int_{r_k}^{r_{k-1}} t^{2\theta-2}dt$$

$$\geq C(r_{k-1} - r_k)r_{k-1}^{2\theta-2}$$

$$\geq \frac{C}{2}(r_{k-1} - r_k)r_k^{2\theta-2} \stackrel{a}{\geq} \frac{1}{2d_1 C},$$

for all $k > k_0$, where $\stackrel{a}{\geq}$ uses (4.6).

If $r_k^{2\theta-2} \geq 2r_{k-1}^{2\theta-2}$, then $r_k^{2\theta-1} \geq 2^{\frac{2\theta-1}{2\theta-2}} r_{k-1}^{2\theta-1}$ and

$$\phi(r_k) - \phi(r_{k-1}) = \frac{C}{1-2\theta}\left(r_k^{2\theta-1} - r_{k-1}^{2\theta-1}\right)$$

$$\geq \frac{C}{1-2\theta}\left(2^{\frac{2\theta-1}{2\theta-2}} - 1\right)r_{k-1}^{2\theta-1}$$

$$= qr_{k-1}^{2\theta-1} \geq qr_0^{2\theta-1},$$

where $q = \frac{C}{1-2\theta}\left(2^{\frac{2\theta-1}{2\theta-2}} - 1\right)$. Let $d_2 = \min\left\{\frac{1}{2d_1 C}, qr_0^{2\theta-1}\right\}$, we have

$$\phi(r_k) - \phi(r_{k-1}) \geq d_2,$$

for all $k > k_0$ and

$$\phi(r_k) \geq \phi(r_k) - \phi(r_{k_0}) = \sum_{i=k_0+1}^{k} [\phi(r_i) - \phi(r_{i-1})] \geq (k - k_0)d_2.$$

So we have

$$r_k^{2\theta-1} \geq \frac{(k-k_0)d_2(1-2\theta)}{C}$$

and thus

$$r_k \leq \left[\frac{C}{(k-k_0)d_2(1-2\theta)} \right]^{\frac{1}{1-2\theta}}.$$

Letting $k_3 = k_0$, we have

$$F(\mathbf{x}_k) - F^* = r_k \leq \left[\frac{C}{(k-k_3)d_2(1-2\theta)} \right]^{\frac{1}{1-2\theta}}, \forall k \geq k_3. \qquad \square$$

4.1.2 Another Method: Monotone APG

Algorithm 4.2 Monotone APG

Initialize $\mathbf{y}_0 = \mathbf{x}_0$, $\beta \in (0, 1)$, and $\eta < \frac{1}{L}$.

for $k = 0, 1, 2, 3, \cdots$ **do**

$\quad \mathbf{y}_k = \mathbf{x}_k + \frac{t_{k-1}}{t_k}(\mathbf{z}_k - \mathbf{x}_k) + \frac{t_{k-1}-1}{t_k}(\mathbf{x}_k - \mathbf{x}_{k-1})$,

$\quad \mathbf{z}_{k+1} = \text{Prox}_{\alpha_y g}(\mathbf{y}_k - \alpha_y \nabla f(\mathbf{y}_k))$,

$\quad \mathbf{v}_{k+1} = \text{Prox}_{\alpha_x g}(\mathbf{x}_k - \alpha_x \nabla f(\mathbf{x}_k))$,

$\quad t_{k+1} = \frac{\sqrt{4t_k^2+1}+1}{2}$,

$\quad \mathbf{x}_{k+1} = \begin{cases} \mathbf{z}_{k+1}, & \text{if } F(\mathbf{z}_{k+1}) \leq F(\mathbf{v}_{k+1}), \\ \mathbf{v}_{k+1}, & \text{otherwise.} \end{cases}$

end for

In this section, we introduce another accelerated PG method for nonconvex optimization, namely the monotone APG [20]. We describe it in Algorithm 4.2. Algorithm 4.2 compares the accelerated PG step and the non-accelerated PG step and chooses the one which has a smaller objective function value. Compared with Algorithm 4.1, Algorithm 4.2 has the beauty that the provable acceleration with $O\left(\frac{1}{K^2}\right)$ rate is guaranteed for convex programs, while Algorithm 4.1 has no convergence rate analysis for convex problems if the KŁ property is not assumed. However, for Algorithm 4.2 with the KŁ property we cannot prove that the whole sequence converges to a critical point, i.e., Theorem 4.2, when the desingularizing function is general. But when the desingularizing function is also of the special form $\varphi(t) = \frac{C}{\theta}t^\theta$, the same convergence rate as Theorem 4.3 describes can also be guaranteed for Algorithm 4.2 [20].

4.2   AGD Achieves Critical Points Quickly

Although Algorithm 4.1 uses momentum, there is no provable advantage over the gradient descent. Recently, there is a trend to analyze the accelerated gradient descent for nonconvex programs with provable guarantees, e.g., the problems of how to achieve critical points quickly and how to escape saddle points quickly. In this section, we describe the result in [7] that the accelerated gradient descent achieves critical points quickly for nonconvex problems.

In the following sections, we present two building blocks of the result: a monitored variation of AGD and a negative curvature descent step. Then we combine these components to obtain the desired conclusion.

4.2.1   AGD as a Convexity Monitor

We describe the first component in Algorithm 4.3, where f is an L-smooth function and is *conjectured* to be σ-strongly convex (so the σ-strong convexity needs to be verified in Find-Witness-Pair()), and Nesterov's accelerated gradient descent (AGD, Algorithm 2.1) for strongly convex functions is employed. At every iteration, the method invokes Certify-Progress() to test whether the optimization is progressing as it should be for strongly convex functions. In particular, it tests whether the norm of gradient decreases exponentially quickly. If the test fails, Find-Witness-Pair() produces points $\mathbf{u}$ and $\mathbf{v}$ such that f violates the σ-strong convexity. Otherwise, we proceed until we find a point $\mathbf{x}$ such that $\|\nabla f(\mathbf{x})\| \leq \epsilon$.

The effectiveness of Algorithm 4.3 is based on the following guarantee on the performance of AGD, which can be obtained from the proof of Theorem 2.2, where $\mathbf{x}^*$ is simply replaced by $\mathbf{w}$.

Lemma 4.1 *Let f be L-smooth. If the sequences $\{\mathbf{x}_j\}$ and $\{\mathbf{y}_j\}$ and $\mathbf{w}$ satisfy*

$$f(\mathbf{x}_j) \geq f(\mathbf{y}_j) + \langle \nabla f(\mathbf{y}_j), \mathbf{x}_i - \mathbf{y}_j \rangle + \frac{\sigma}{2}\|\mathbf{x}_i - \mathbf{y}_i\|^2,$$
$$f(\mathbf{w}) \geq f(\mathbf{y}_j) + \langle \nabla f(\mathbf{y}_j), \mathbf{w} - \mathbf{y}_j \rangle + \frac{\sigma}{2}\|\mathbf{w} - \mathbf{y}_i\|^2, \qquad j = 0, 1, \cdots, t-1, \quad (4.7)$$

then

$$f(\mathbf{x}_t) - f(\mathbf{w}) \leq \left(1 - \frac{1}{\sqrt{\kappa}}\right)^t \psi(\mathbf{w}), \qquad (4.8)$$

where $\psi(\mathbf{w}) = f(\mathbf{x}_0) - f(\mathbf{w}) + \frac{\sigma}{2}\|\mathbf{x}_0 - \mathbf{w}\|^2$.

Algorithm 4.3 AGD-Until-Guilty($f, \mathbf{x}_0, \epsilon, L, \sigma$)

$\kappa = L/\sigma$, $\omega = \frac{\sqrt{\kappa}-1}{\sqrt{\kappa}+1}$, and $\mathbf{y}_0 = \mathbf{x}_0$,

for $t = 1, 2, 3, \cdots$ **do**

$\quad \mathbf{x}_t = \mathbf{y}_{t-1} - \frac{1}{L}\nabla f(\mathbf{y}_{t-1})$,

$\quad \mathbf{y}_t = \mathbf{x}_t + \omega(\mathbf{x}_t - \mathbf{x}_{t-1})$,

$\quad \mathbf{w}_t \leftarrow$Certify-Progress($f, \mathbf{x}_0, \mathbf{x}_t, L, \sigma, \kappa, t$).

$\quad$**if** $\mathbf{w}_t \neq$ null **then**

$\quad\quad$($\mathbf{u}, \mathbf{v}$) $\leftarrow$Find-Witness-Pair($f, \mathbf{x}_{0:t}, \mathbf{y}_{0:t}, \mathbf{w}_t, \sigma$),

$\quad\quad$return ($\mathbf{x}_{0:t}, \mathbf{y}_{0:t}, \mathbf{u}, \mathbf{v}$).

$\quad$**end if**

$\quad$**if** $\|\nabla f(\mathbf{x}_t)\| \leq \epsilon$ **then**

$\quad\quad$return ($\mathbf{x}_{0:t}, \mathbf{y}_{0:t}$, null).

$\quad$**end if**

end for

****

function Certify-Progress($f, \mathbf{x}_0, \mathbf{x}_t, L, \sigma, \kappa, t$)

if $f(\mathbf{x}_t) > f(\mathbf{x}_0)$ **then**

$\quad$return $\mathbf{x}_0$.

end if

$\mathbf{w}_t = \mathbf{x}_t - \frac{1}{L}\nabla f(\mathbf{x}_t)$,

if $\|\nabla f(\mathbf{x}_t)\|^2 > 2L\left(1 - \frac{1}{\sqrt{\kappa}}\right)^t \psi(\mathbf{w}_t)$ **then**

$\quad$return $\mathbf{w}_t$.

else

$\quad$return null.

end if

****

function Find-Witness-Pair($f, \mathbf{x}_{0:t}, \mathbf{y}_{0:t}, \mathbf{w}_t, \sigma$)

for $j = 0, 1, \cdots, t - 1$ **do**

$\quad$**for** $\mathbf{u} = \mathbf{x}_j, \mathbf{w}_t$ **do**

$\quad\quad$**if** $f(\mathbf{u}) < f(\mathbf{y}_j) + \langle\nabla f(\mathbf{y}_j), \mathbf{u} - \mathbf{y}_j\rangle + \frac{\sigma}{2}\|\mathbf{u} - \mathbf{y}_j\|^2$ **then**

$\quad\quad\quad$return ($\mathbf{u}, \mathbf{y}_j$).

$\quad\quad$**end if**

$\quad$**end for**

end for

Specifically, letting $\mathbf{w} = \mathbf{w}_t = \mathbf{x}_t - \frac{1}{L}\nabla f(\mathbf{x}_t)$, we can have $\frac{1}{2L}\|\nabla f(\mathbf{x}_t)\|^2 \leq f(\mathbf{x}_t) - f(\mathbf{w}_t)$ ((A.6) of Proposition A.6). After combining with (4.8), it leads to

$$\frac{1}{2L}\|\nabla f(\mathbf{x}_t)\|^2 \leq \left(1 - \frac{1}{\sqrt{\kappa}}\right)^t \psi(\mathbf{w}_t).$$

If Certify-Progress() always tells that the optimization progresses as we have expected, then $f(\mathbf{w}_t) \leq f(\mathbf{x}_t) \leq f(\mathbf{x}_0)$. From the coerciveness assumption we know that $\|\mathbf{w}_t - \mathbf{x}_0\|$ is bounded by some constant C and $\psi(\mathbf{w}_t)$ is bounded by $f(\mathbf{x}_0) - f(\mathbf{x}^*) + \sigma C^2/2$. Thus $\|\nabla f(\mathbf{x}_t)\|$ decreases exponentially quickly and Algorithm 4.3 can always terminate.

With the above results in hand, we summarize the guarantees of Algorithm 4.3 as follows.

Lemma 4.2 *Let f be L-smooth and t be the number of iterations that Algorithm 4.3 terminates. Then t satisfies*

$$t \leq 1 + \max\left\{0, \sqrt{\frac{L}{\sigma}}\log\left(\frac{2L\psi(\mathbf{w}_{t-1})}{\epsilon^2}\right)\right\}. \tag{4.9}$$

If $\mathbf{w}_t \neq$ null, then $(\mathbf{u}, \mathbf{v}) \neq$ null and

$$f(\mathbf{u}) < f(\mathbf{v}) + \langle\nabla f(\mathbf{v}), \mathbf{u} - \mathbf{v}\rangle + \frac{\sigma}{2}\|\mathbf{u} - \mathbf{v}\|^2 \tag{4.10}$$

for some $\mathbf{v} = \mathbf{y}_j$ and $\mathbf{u} = \mathbf{x}_j$ or $\mathbf{u} = \mathbf{w}_t$, $0 \leq j < t$. Moreover,

$$\max\{f(\mathbf{x}_1), \cdots, f(\mathbf{x}_{t-1}), f(\mathbf{u})\} \leq f(\mathbf{x}_0). \tag{4.11}$$

Proof The algorithm does not terminate at iteration $t - 1$. So we have

$$\epsilon^2 \overset{a}{<} \|\nabla f(\mathbf{x}_{t-1})\|^2 \overset{b}{\leq} 2L\left(1 - \frac{1}{\sqrt{\kappa}}\right)^{t-1}\psi(\mathbf{w}_{t-1}) \leq 2Le^{-(t-1)/\sqrt{\kappa}}\psi(\mathbf{w}_{t-1}),$$

where $\overset{a}{<}$ uses the fact that the second "if condition" in AGD-Until-Guilty() fails when the algorithm does not terminate and $\overset{b}{\leq}$ uses that the second "if condition" in Certify-Progress() fails. So we have (4.9).

When $\mathbf{w}_t \neq$ null, suppose

$$f(\mathbf{u}) \geq f(\mathbf{v}) + \langle\nabla f(\mathbf{v}), \mathbf{u} - \mathbf{v}\rangle + \frac{\sigma}{2}\|\mathbf{u} - \mathbf{v}\|^2$$

holds for all $\mathbf{v} = \mathbf{y}_j$ and $\mathbf{u} = \mathbf{x}_j$ or $\mathbf{u} = \mathbf{w}_t$, $0 \leq j < t$. Namely, (4.7) holds for $\mathbf{w} = \mathbf{w}_t$. Suppose $\mathbf{w}_t = \mathbf{x}_0$, then we have

$$f(\mathbf{x}_t) - f(\mathbf{w}_t) \overset{a}{>} 0 \overset{b}{=} \left(1 - \frac{1}{\sqrt{\kappa}}\right)^t\psi(\mathbf{w}_t),$$

where $\overset{a}{>}$ uses the fact that the first "if condition" in Certify-Progress() proceeds and $\overset{b}{=}$ uses $\psi(\mathbf{w}_t) = \psi(\mathbf{x}_0) = 0$. This contradicts (4.8). Similarly, suppose $\mathbf{w}_t = \mathbf{x}_t - \frac{1}{L}\nabla f(\mathbf{x}_t)$, then we have

$$f(\mathbf{x}_t) - f(\mathbf{w}_t) \geq \frac{1}{2L}\|\nabla f(\mathbf{x}_t)\|^2 \overset{a}{>} \left(1 - \frac{1}{\sqrt{\kappa}}\right)^t\psi(\mathbf{w}_t),$$

again contradicting (4.8), where $\overset{a}{>}$ uses the fact that the second "if condition" in Certify-Progress() proceeds. Thus there must exist some $\mathbf{y}_j$ and $\mathbf{x}_j$ or $\mathbf{w}_t$ such that (4.7) is violated. Thus Find-Witness-Pair() always produces points $(\mathbf{u}, \mathbf{v}) \neq$ null.

From the first "if condition" in Certify-Progress(), it is clear that $f(\mathbf{x}_s) \leq$ $f(\mathbf{x}_0), s = 0, 1, \cdots, t - 1$. If $\mathbf{u} = \mathbf{x}_s$ for some $0 \leq s \leq t - 1$, then $f(\mathbf{u}) \leq f(\mathbf{x}_0)$ holds trivially. If $\mathbf{u} = \mathbf{w}_t$, then the first "if condition" in Certify-Progress() proceeds, i.e., $f(\mathbf{x}_t) \leq f(\mathbf{x}_0)$. So we have $f(\mathbf{w}_t) \leq f(\mathbf{x}_t) \leq f(\mathbf{x}_0)$. So (4.11) holds. □

4.2.2 Negative Curvature Descent

Algorithm 4.4 Exploit-NC-Pair($f, \mathbf{u}, \mathbf{v}, \eta$)

$\delta = \frac{\mathbf{u} - \mathbf{v}}{\|\mathbf{u} - \mathbf{v}\|}$,
$\mathbf{u}_+ = \mathbf{u} + \eta\delta$,
$\mathbf{u}_- = \mathbf{u} - \eta\delta$,
return $\mathbf{z} = \text{argmin}_{\mathbf{u}_+, \mathbf{u}_-} f(\mathbf{z})$.

The second component is the exploitation of negative curvature to decrease function values, which is described in Algorithm 4.4. The property of Algorithm 4.4 is described in the following lemma.

Lemma 4.3 *Let f be L_1-smooth and have L_2-Lipschitz continuous Hessians (Definition A.14) and $\mathbf{u}$ and $\mathbf{v}$ satisfy*

$$f(\mathbf{u}) < f(\mathbf{v}) + \langle \nabla f(\mathbf{v}), \mathbf{u} - \mathbf{v} \rangle - \frac{\sigma}{2} \|\mathbf{u} - \mathbf{v}\|^2. \tag{4.12}$$

If $\|\mathbf{u} - \mathbf{v}\| \leq \frac{\sigma}{2L_2}$ and $\eta \leq \frac{\sigma}{L_2}$, then for Algorithm 4.4 we have

$$f(\mathbf{z}) \leq f(\mathbf{u}) - \frac{\sigma\eta^2}{12}.$$

Proof From (4.12) and the basic calculus, we have

$$-\frac{\sigma}{2} \|\mathbf{u} - \mathbf{v}\|^2 \geq f(\mathbf{u}) - f(\mathbf{v}) - \langle \nabla f(\mathbf{v}), \mathbf{u} - \mathbf{v} \rangle$$

$$= \int_0^1 \langle \nabla f(\mathbf{v} + t(\mathbf{u} - \mathbf{v})), \mathbf{u} - \mathbf{v} \rangle \, dt - \langle \nabla f(\mathbf{v}), \mathbf{u} - \mathbf{v} \rangle$$

$$= \int_0^{\|\mathbf{u} - \mathbf{v}\|} \langle \nabla f(\mathbf{v} + \tau\delta), \delta \rangle \, d\tau - \|\mathbf{u} - \mathbf{v}\| \langle \nabla f(\mathbf{v}), \delta \rangle$$

$$= \int_0^{\|\mathbf{u} - \mathbf{v}\|} \langle \nabla f(\mathbf{v} + \tau\delta) - \nabla f(\mathbf{v}), \delta \rangle \, d\tau$$

$$= \int_0^{\|\mathbf{u} - \mathbf{v}\|} \left(\int_0^\tau \delta^T \nabla^2 f(\mathbf{v} + \theta\delta)\delta d\theta \right) d\tau$$

$$\geq \frac{\|\mathbf{u} - \mathbf{v}\|^2}{2} c,$$

where $\boldsymbol{\delta} = \frac{\mathbf{u} - \mathbf{v}}{\|\mathbf{u} - \mathbf{v}\|}$ and $c = \min_{0 \le \tau \le \|\mathbf{u} - \mathbf{v}\|} \{\boldsymbol{\delta}^T \nabla f^2 (\mathbf{v} + \tau \boldsymbol{\delta}) \boldsymbol{\delta}\}$. Then $c \le -\sigma$. On the one hand,

$$\boldsymbol{\delta}^T \nabla^2 f(\mathbf{u}) \boldsymbol{\delta} - c = \boldsymbol{\delta}^T \left(\nabla^2 f(\mathbf{u}) - \nabla^2 f(\mathbf{v} + \tau^* (\mathbf{u} - \mathbf{v})) \right) \boldsymbol{\delta}$$

$$\le \|\nabla^2 f(\mathbf{u}) - \nabla^2 f(\mathbf{v} + \tau^* (\mathbf{u} - \mathbf{v}))\|$$

$$\le L_2 \|\mathbf{u} - [\mathbf{v} + \tau^* (\mathbf{u} - \mathbf{v})]\| \le L_2 \|\mathbf{u} - \mathbf{v}\|.$$

So we have $\boldsymbol{\delta}^T \nabla^2 f(\mathbf{u}) \boldsymbol{\delta} \le -\frac{\sigma}{2}$. On the other hand, from (A.8) we have

$$f(\mathbf{u}_\pm) \le f(\mathbf{u}) + \langle \nabla f(\mathbf{u}), \mathbf{u}_\pm - \mathbf{u} \rangle + \frac{1}{2} (\mathbf{u}_\pm - \mathbf{u})^T \nabla f^2(\mathbf{u})(\mathbf{u}_\pm - \mathbf{u})$$

$$+ \frac{L_2}{6} \|\mathbf{u}_\pm - \mathbf{u}\|^3$$

$$= f(\mathbf{u}) \pm \langle \nabla f(\mathbf{u}), \boldsymbol{\delta} \rangle + \frac{\eta^2}{2} \boldsymbol{\delta}^T \nabla^2 f(\mathbf{u}) \boldsymbol{\delta} + \frac{L_2}{6} |\eta|^3$$

$$\le f(\mathbf{u}) \pm \langle \nabla f(\mathbf{u}), \boldsymbol{\delta} \rangle - \frac{\sigma \eta^2}{12},$$

where we use $\eta \le \frac{\sigma}{L_2}$. Since $\pm \langle \nabla f(\mathbf{u}), \boldsymbol{\delta} \rangle$ must be negative for either $\mathbf{u}_+$ or $\mathbf{u}_-$, we have $f(\mathbf{z}) \le f(\mathbf{u}) - \frac{\sigma \eta^2}{12}$. $\qquad \square$

4.2.3 Accelerating Nonconvex Optimization

Algorithm 4.5 Nonconvex AGD for achieving critical points, NC-AGD-CP$(f, \mathbf{p}_0, \epsilon, L_1, \sigma, \eta)$

---

for $k = 1, 2, 3, \cdots, K$ **do**
  $\hat{f}(\mathbf{x}) = f(\mathbf{x}) + \sigma \|\mathbf{x} - \mathbf{p}_{k-1}\|^2$,
  $(\mathbf{x}_{0:t}, \mathbf{y}_{0:t}, \mathbf{u}, \mathbf{v}) \leftarrow$ AGD-Until-Guilty$(\hat{f}, \mathbf{p}_{k-1}, \frac{\epsilon}{10}, L_1 + 2\sigma, \sigma)$.
  if $(\mathbf{u}, \mathbf{v}) =$ null **then**
    $\mathbf{p}_k = \mathbf{x}_t$.
  else
    $\mathbf{b}_1 = \text{argmin}_{\mathbf{u}, \mathbf{x}_0, \cdots, \mathbf{x}_t} f(\mathbf{x})$,
    $\mathbf{b}_2 = \text{Expoint-NC-Pair}(f, \mathbf{u}, \mathbf{v}, \eta)$,
    $\mathbf{p}_k = \text{argmin}_{\mathbf{b}_1, \mathbf{b}_2} f(\mathbf{x})$.
  end if
  if $\|\nabla f(\mathbf{p}_k)\| \le \epsilon$ **then**
    return $\mathbf{p}_k$.
  end if
end for

We now combine the above building blocks and give the accelerated method in Algorithm 4.5. When f is not very nonconvex, in other words, almost convex [7], i.e.,

$$f(\mathbf{u}) \geq f(\mathbf{v}) + \langle \nabla f(\mathbf{v}), \mathbf{u} - \mathbf{v} \rangle - \frac{\sigma}{2} \|\mathbf{u} - \mathbf{v}\|^2, \qquad (4.13)$$

then function $\hat{f}$ is $\frac{\sigma}{2}$-strongly convex and AGD-Until-Guilty produces an $\mathbf{x}_t$ with $\|\nabla \hat{f}(\mathbf{x}_t)\| \leq \epsilon$ and $(\mathbf{u}, \mathbf{v}) \neq null$. Otherwise, when f is very nonconvex, we can use Exploit-NC-Pair to find a new descent direction.

The following lemma verifies the condition of Exploit-NC-Pair.

Lemma 4.4 *Let f be L_1-smooth and $\tau > 0$. In Algorithm 4.5, if $(\mathbf{u}, \mathbf{v}) \neq null$ and $f(\mathbf{b}_1) \geq f(\mathbf{x}_0) - \sigma\tau^2$, then $\|\mathbf{u} - \mathbf{v}\| \leq 4\tau$.*

Proof Since $\mathbf{p}_{k-1}$ is the initial point of $\hat{f}$ in AGD-Until-Guilty(), $\mathbf{p}_{k-1} = \mathbf{x}_0$. Then from (4.11), we have $\hat{f}(\mathbf{x}_i) \leq \hat{f}(\mathbf{x}_0) = f(\mathbf{x}_0), i = 1, \cdots, t - 1$. From $f(\mathbf{x}_i) \geq f(\mathbf{b}_1) \geq f(\mathbf{x}_0) - \sigma\tau^2$, we have

$$\sigma \|\mathbf{x}_i - \mathbf{x}_0\|^2 = \hat{f}(\mathbf{x}_i) - f(\mathbf{x}_i) \leq f(\mathbf{x}_0) - f(\mathbf{x}_i) \leq \sigma\tau^2,$$

which implies $\|\mathbf{x}_i - \mathbf{x}_0\| \leq \tau$. Since $\hat{f}(\mathbf{u}) \leq \hat{f}(\mathbf{x}_0)$ and $f(\mathbf{u}) \geq f(\mathbf{b}_1)$, we also have $\|\mathbf{u} - \mathbf{x}_0\| \leq \tau$. From $\mathbf{y}_i = \mathbf{x}_i + \omega(\mathbf{x}_i - \mathbf{x}_{i-1})$, we have

$$\|\mathbf{y}_i - \mathbf{x}_0\| \leq (1 + \omega)\|\mathbf{x}_i - \mathbf{x}_0\| + \omega\|\mathbf{x}_{i-1} - \mathbf{x}_0\| \leq 3\tau.$$

Since $\mathbf{v} = \mathbf{y}_i$ for some i (see Lemma 4.2), we have $\|\mathbf{u} - \mathbf{v}\| \leq \|\mathbf{u} - \mathbf{x}_0\| + \|\mathbf{y}_i - \mathbf{x}_0\| \leq 4\tau$. □

The following central lemma provides a progress guarantee for each iteration of Algorithm 4.5.

Lemma 4.5 *Let f be L_1-smooth and have L_2-Lipschitz continuous Hessians and $\eta = \frac{\sigma}{L_2}$. Then for Algorithm 4.5 with $k \leq K - 1$, we have*

$$f(\mathbf{p}_k) \leq f(\mathbf{p}_{k-1}) - \min\left\{\frac{\epsilon^2}{5\sigma}, \frac{\sigma^3}{64L_2^2}\right\}. \qquad (4.14)$$

Proof Case 1: $(\mathbf{u}, \mathbf{v}) = null$. In this case, $\mathbf{p}_k = \mathbf{x}_k$ and $\|\nabla \hat{f}(\mathbf{p}_k)\| \leq \epsilon/10$. On the other hand, $\|\nabla f(\mathbf{p}_k)\| > \epsilon$, thus we have

$$9\epsilon/10 \leq \|\nabla f(\mathbf{p}_k)\| - \|\nabla \hat{f}(\mathbf{p}_k)\| \leq \|\nabla f(\mathbf{p}_k) - \nabla \hat{f}(\mathbf{p}_k)\| = 2\sigma\|\mathbf{p}_k - \mathbf{p}_{k-1}\|.$$

We also have $\hat{f}(\mathbf{p}_k) = \hat{f}(\mathbf{x}_k) \leq \hat{f}(\mathbf{x}_0) = \hat{f}(\mathbf{p}_{k-1}) = f(\mathbf{p}_{k-1})$. So

$$f(\mathbf{p}_k) = \hat{f}(\mathbf{p}_k) - \sigma\|\mathbf{p}_k - \mathbf{p}_{k-1}\|^2 \leq f(\mathbf{p}_{k-1}) - \sigma\left(\frac{9\epsilon}{20\sigma}\right)^2 \leq f(\mathbf{p}_{k-1}) - \frac{\epsilon^2}{5\sigma}.$$

Case 2: $(\mathbf{u}, \mathbf{v}) \neq$ null. From (4.10), we have

$$\hat{f}(\mathbf{u}) < \hat{f}(\mathbf{v}) + \left\langle \nabla \hat{f}(\mathbf{v}), \mathbf{u} - \mathbf{v} \right\rangle + \frac{\sigma}{2}\|\mathbf{u} - \mathbf{v}\|^2,$$

which leads to

$$f(\mathbf{v}) + \langle \nabla f(\mathbf{v}), \mathbf{u} - \mathbf{v} \rangle - \frac{\sigma}{2}\|\mathbf{u} - \mathbf{v}\|^2 - f(\mathbf{u})$$
$$= \hat{f}(\mathbf{v}) + \left\langle \nabla \hat{f}(\mathbf{v}), \mathbf{u} - \mathbf{v} \right\rangle + \frac{\sigma}{2}\|\mathbf{u} - \mathbf{v}\|^2 - \hat{f}(\mathbf{u}) > 0.$$

So $\mathbf{u}$ and $\mathbf{v}$ satisfy the condition in Lemma 4.3. If $f(\mathbf{b}_1) \leq f(\mathbf{x}_0) - \frac{\sigma^3}{64L_2^2}$, then we are done. If $f(\mathbf{b}_1) > f(\mathbf{x}_0) - \frac{\sigma^3}{64L_2^2}$, then we have $\|\mathbf{u} - \mathbf{v}\| \leq \frac{\sigma}{2L_2}$ from Lemma 4.4. So from Lemma 4.3, we have

$$f(\mathbf{b}_2) \leq f(\mathbf{u}) - \frac{\sigma^3}{12L_2^2} \leq f(\mathbf{p}_{k-1}) - \frac{\sigma^3}{12L_2^2},$$

where we use $f(\mathbf{u}) \leq \hat{f}(\mathbf{u}) \leq \hat{f}(\mathbf{x}_0) = f(\mathbf{p}_{k-1})$. $\square$

Using the above lemma, we can give the final theorem with guarantee of acceleration.

Theorem 4.4 *Let f be L_1-smooth and have L_2-Lipschitz continuous Hessians, $\Delta = f(\mathbf{p}_0) - \inf_{\mathbf{z}} f(\mathbf{z})$, and $\sigma = 2\sqrt{L_2\epsilon}$. Then Algorithm 4.5 finds a point $\mathbf{p}_K$ such that $\|\nabla f(\mathbf{p}_K)\| \leq \epsilon$ with at most*

$$O\left(\frac{L_1^{1/2}L_2^{1/4}\Delta}{\epsilon^{7/4}} \log \frac{(L_1 + \sqrt{L_2\epsilon})\Delta}{\epsilon^2}\right)$$

gradient computations.

Proof From (4.14), we have

$$\Delta \geq f(\mathbf{p}_0) - f(\mathbf{p}_{K-1}) = \sum_{k=1}^{K-1}[f(\mathbf{p}_{k-1}) - f(\mathbf{p}_k)] \geq (K-1)\min\left\{\frac{\epsilon^2}{5\sigma}, \frac{\sigma^3}{64L_2^2}\right\}$$

$$\geq (K-1)\frac{\epsilon^{3/2}}{10L_2^{1/2}}.$$

So

$$K \leq 1 + 10\frac{L_2^{1/2}\Delta}{\epsilon^{3/2}}.$$

From (4.9) with the Lipschitz constant of $\hat{f}$ being $L = L_1 + 2\sigma = L_2 + 4\sqrt{L_2\epsilon}$, we have

$$
T \le 1 + \max\left\{0, \sqrt{\frac{L_1 + 4\sqrt{L_2\epsilon}}{2\sqrt{L_2\epsilon}}} \log\left(\frac{2(L_1 + 4\sqrt{L_2\epsilon})\psi(\mathbf{w}_{T-1})}{\epsilon^2}\right)\right\}
$$

$$
= O\left(\frac{L_1^{1/2}}{L_2^{1/4}\epsilon^{1/4}} \log \frac{(L_1 + \sqrt{L_2\epsilon})\Delta}{\epsilon^2}\right),
$$

where we use

$$
\psi(\mathbf{z}) = \hat{f}(\mathbf{x}_0) - \hat{f}(\mathbf{z}) + \frac{\sigma}{2}\|\mathbf{z} - \mathbf{x}_0\|^2 \overset{a}{=} f(\mathbf{x}_0) - f(\mathbf{z}) - \frac{\sigma}{2}\|\mathbf{z} - \mathbf{x}_0\|^2 \overset{b}{\le} \Delta,
$$

in which $\overset{a}{=}$ is because $\mathbf{x}_0 = \mathbf{p}_{k-1}$, $\hat{f}(\mathbf{x}_0) = f(\mathbf{x}_0)$, and $\hat{f}(\mathbf{z}) = f(\mathbf{z}) + \sigma\|\mathbf{z} - \mathbf{x}_0\|^2$ and $\overset{b}{\le}$ is because $f(\mathbf{x}_0) = f(\mathbf{p}_{k-1}) \le f(\mathbf{p}_0)$ by (4.14). So the total number of gradient computations is

$$
KT = O\left(\frac{L_1^{1/2}L_2^{1/4}\Delta}{\epsilon^{7/4}} \log \frac{(L_1 + \sqrt{L_2\epsilon})\Delta}{\epsilon^2}\right). \qquad\qquad \square
$$

4.3 AGD Escapes Saddle Points Quickly

In optimization and machine learning literature, recently there has been substantial work on the convergence of gradient-like methods to local optima for nonconvex problems. Typical ones include [12] which showed that stochastic gradient descent converges to a second-order local optimum, [19] which showed that gradient descent generically converges to a second-order local optimum, and [16] which showed that perturbed gradient descent can escape saddle points almost for free. [8] employed a combination of (regularized) accelerated gradient descent and the Lanczos method to obtain better rates for first-order methods, [17] proposed a single-loop accelerated method, and [1] proposed a careful implementation of the Nesterov–Polyak method, using accelerated methods for fast approximate matrix inversion.

In this section, we describe the method in [8]. The algorithm alternates between finding directions of negative curvature of f and solving structured subproblems that are nearly convex. Similar to Sect. 4.2, the analysis in this section also depends on two building blocks, which uses the accelerated gradient descent for the almost convex case and finds the negative curvature for the very nonconvex case, respectively.

4.3.1 Almost Convex Case

We first give the following lemma, which is a direct consequence of (A.6) and (A.11).

Lemma 4.6 *Assume that f is γ-strongly convex and L-smooth. Then*

$$2\gamma(f(\mathbf{x}) - f(\mathbf{x}^*)) \leq \|\nabla f(\mathbf{x})\|^2 \leq 2L(f(\mathbf{x}) - f(\mathbf{x}^*)).$$

We consider the case that f is σ-almost convex, defined in (4.13). Similar to Algorithm 4.5, we can add a regularizing term $\sigma\|\mathbf{x} - \mathbf{y}\|^2$ to make f σ-strongly convex and solve it quickly using accelerated gradient descent (AGD, Algorithm 2.1). We describe the method in Algorithm 4.6, which finds a $\mathbf{z}_j$ such that $\|\nabla f(\mathbf{z}_j)\| \leq \epsilon$. In Algorithm 4.6, we call subroutine AGD to minimize a σ-strongly convex and $(L + 2\sigma)$-smooth function g_j, initialized at $\mathbf{z}_j$, such that $g_j(\mathbf{z}_{j+1}) - \min_{\mathbf{z}} g_j(\mathbf{z}) \leq \frac{(\epsilon')^2}{2L+4\sigma}$.

Algorithm 4.6 Almost convex AGD, AC-AGD($f, \mathbf{z}_1, \epsilon, \sigma, L$)

---
for $j = 1, 2, 3, \cdots, J$ **do**
    If $\|\nabla f(\mathbf{z}_j)\| \leq \epsilon$, then return $\mathbf{z}_j$,
    Define $g_j(\mathbf{z}) = f(\mathbf{z}) + \sigma\|\mathbf{z} - \mathbf{z}_j\|^2$ and $\epsilon' = \epsilon\sqrt{\sigma/[50(L + 2\sigma)]}$,
    $\mathbf{z}_{j+1} = \text{AGD}(g_j, \mathbf{z}_j, (\epsilon')^2/(2L + 4\sigma), L + 2\sigma, \sigma)$.
end for

Lemma 4.7 *Assume that f is σ-almost convex and L-smooth. Then Algorithm 4.6 terminates in the total running time of*

$$\left[\sqrt{\frac{L}{\sigma}} + \frac{5\sqrt{L\sigma}}{\epsilon^2}(f(\mathbf{z}_1) - f(\mathbf{z}_J))\right]\log 1/\epsilon.$$

Proof g_j is σ-strongly convex and $(L + 2\sigma)$-smooth. Then from Theorem 2.1, we know that after $O\left(\sqrt{\kappa}\log\frac{2L+4\sigma}{(\epsilon')^2}\right) = O(\sqrt{\kappa}\log 1/\epsilon')$ iterations, we have $g_j(\mathbf{z}_{j+1}) - \min_{\mathbf{z}} g_j(\mathbf{z}) \leq \frac{(\epsilon')^2}{2L+4\sigma}$ and

$$\|\nabla g_j(\mathbf{z}_{j+1})\| \overset{a}{\leq} \sqrt{2(L + 2\sigma)\frac{(\epsilon')^2}{2L + 4\sigma}} = \epsilon' = \epsilon\sqrt{\frac{\sigma}{50(L + 2\sigma)}} \leq \frac{\epsilon}{10},$$

where we use Lemma 4.6 in $\overset{a}{\leq}$.

On the other hand, $\|\nabla g_j(\mathbf{z}_j)\| = \|\nabla f(\mathbf{z}_j)\| \geq \epsilon$ for $j < J$. So $\|\nabla g_j(\mathbf{z}_{j+1})\|^2 < \frac{\sigma\epsilon^2}{L+2\sigma} \leq \frac{\sigma}{L+2\sigma}\|\nabla g_j(\mathbf{z}_j)\|^2$ and

$$g_j(\mathbf{z}_{j+1}) - g_j(\mathbf{z}_j^*) \leq \frac{1}{2\sigma}\|\nabla g_j(\mathbf{z}_{j+1})\|^2 \leq \frac{1}{2(L+2\sigma)}\|\nabla g_j(\mathbf{z}_j)\|^2 \leq g_j(\mathbf{z}_j) - g_j(\mathbf{z}_j^*),$$

where $\mathbf{z}_j^* = \operatorname{argmin}_{\mathbf{z}} g_j(\mathbf{z})$ and we use Lemma 4.6. So $g_j(\mathbf{z}_{j+1}) \leq g_j(\mathbf{z}_j)$ and

$$\begin{aligned} f(\mathbf{z}_{j+1}) = g_j(\mathbf{z}_{j+1}) - \sigma\|\mathbf{z}_{j+1} - \mathbf{z}_j\|^2 &\leq g_j(\mathbf{z}_j) - \sigma\|\mathbf{z}_{j+1} - \mathbf{z}_j\|^2 \\ &= f(\mathbf{z}_j) - \sigma\|\mathbf{z}_{j+1} - \mathbf{z}_j\|^2. \quad (4.15) \end{aligned}$$

Summing over $j = 1, 2, \cdots, J - 1$, we have

$$\sigma \sum_{j=1}^{J-1} \|\mathbf{z}_{j+1} - \mathbf{z}_j\|^2 \leq f(\mathbf{z}_1) - f(\mathbf{z}_J).$$

On the other hand, since

$$2\sigma\|\mathbf{z}_{j+1} - \mathbf{z}_j\| = \|\nabla f(\mathbf{z}_{j+1}) - \nabla g_j(\mathbf{z}_{j+1})\| \geq \|\nabla f(\mathbf{z}_{j+1})\| - \|\nabla g_j(\mathbf{z}_{j+1})\| \geq \epsilon - \frac{\epsilon}{10},$$

we can have

$$f(\mathbf{z}_1) - f(\mathbf{z}_J) \geq \sigma(J-1)\frac{0.81\epsilon^2}{4\sigma^2} \geq (J-1)\frac{\epsilon^2}{5\sigma}. \quad (4.16)$$

Thus $J \leq 1 + \frac{5\sigma}{\epsilon^2}(f(\mathbf{z}_1) - f(\mathbf{z}_J))$ and the total running time is

$$\begin{aligned} \left[1 + \frac{5\sigma}{\epsilon^2}(f(\mathbf{z}_1) - f(\mathbf{z}_J))\right]&\sqrt{\frac{L}{\sigma}}\log 1/\epsilon \\ &= \left[\sqrt{\frac{L}{\sigma}} + \frac{5\sqrt{L\sigma}}{\epsilon^2}(f(\mathbf{z}_1) - f(\mathbf{z}_J))\right]\log 1/\epsilon. \quad (4.17) \end{aligned}$$

The proof is complete. $\square$

4.3.2 Very Nonconvex Case

When f is very nonconvex, i.e., satisfying (4.12), similar to Algorithm 4.5, we find the negative curvature and use it as the descent direction. However, since our purpose is different from that of Sect. 4.2, the negative curvature procedure is different from Algorithm 4.4. We need a tool with the following guarantee:

If f is L_1-smooth and $\lambda_{\min}(\nabla^2 f(\mathbf{x})) \leq -\alpha$, where $\alpha > 0$ is a small constant, then we can find a $\mathbf{v}$ such that $\|\mathbf{v}\| = 1$ and $\mathbf{v}^T \nabla^2 f(\mathbf{x})\mathbf{v} \leq -O(\alpha)$ in $O\left(\sqrt{\frac{L_1}{\alpha}} \log 1/\delta'\right)$ time with probability at least $1 - \delta'$.

A number of methods computing approximate leading eigenvectors can be used as the tool, e.g., the Lanczos method [18] and noisy accelerated gradient descent [17].

Using the tool to find the leading eigenvector, we can describe the negative curvature descent in Algorithm 4.7.

Algorithm 4.7 Negative curvature descent, NCD($f, \mathbf{z}_1, L_2, \alpha, \delta$)

Let $\delta' = \dfrac{\delta}{1 + \frac{3L_2^2}{\alpha^3}[f(\mathbf{z}_1) - \min_{\mathbf{z}} f(\mathbf{z})]}$.

for $j = 1, 2, 3, \cdots, J$ **do**

   If $\lambda_{\min}(\nabla^2 f(\mathbf{z}_j)) \geq -\alpha$, then return $\mathbf{z}_j$,

   Find $\mathbf{v}_j$ such that $\|\mathbf{v}_j\| = 1$ and $\mathbf{v}_j^T \nabla^2 f(\mathbf{z}_j)\mathbf{v}_j \leq -O(\alpha)$ with probability at least $1 - \delta'$,

   $\mathbf{z}_{j+1} = \mathbf{z}_j - \eta_j \mathbf{v}_j$, where $\eta_j = \dfrac{\left|\mathbf{v}_j^T \nabla^2 f(\mathbf{z}_j)\mathbf{v}_j\right|}{L_2} \text{sign}\left(\mathbf{v}_j^T \nabla f(\mathbf{z}_j)\right)$.

end for

We provide a formal guarantee for Algorithm 4.7 in the following lemma.

Lemma 4.8 *Assume that f has L_1-Lipschitz continuous gradients and L_2-Lipschitz continuous Hessians. Then with probability of at least $1 - \delta$, Algorithm 4.7 terminates in the total running time of*

$$\left[1 + \frac{3L_2^2}{\alpha^3}(f(\mathbf{z}_1) - f(\mathbf{z}_J))\right] O\left(\sqrt{\frac{L_1}{\alpha}} \log 1/\delta'\right).$$

Proof Since f has L_2-Lipschitz continuous Hessians, by Proposition A.8 we have

$$f(\mathbf{z}_j - \eta_j \mathbf{v}_j) - f(\mathbf{z}_j) + \eta_j \mathbf{v}_j^T \nabla f(\mathbf{z}_j) - \frac{\eta_j^2}{2} \mathbf{v}_j^T \nabla^2 f(\mathbf{z}_j)\mathbf{v}_j \leq \frac{L_2 \eta_j^3}{6}\|\mathbf{v}_j\|^3.$$

From the definition of η_j, we have $\eta_j \mathbf{v}_j^T \nabla f(\mathbf{z}_j) \geq 0$. Then

$$\begin{aligned}
f(\mathbf{z}_{j+1}) - f(\mathbf{z}_j) &\leq \frac{\eta_j^2}{2} \mathbf{v}_j^T \nabla^2 f(\mathbf{z}_j)\mathbf{v}_j + \frac{L_2|\eta_j|^3}{6}\|\mathbf{v}_j\|^3 \\
&= -\frac{\left|\mathbf{v}_j^T \nabla^2 f(\mathbf{z}_j)\mathbf{v}_j\right|^3}{3L_2^2} \leq -\frac{\alpha^3}{3L_2^2}.
\end{aligned} \tag{4.18}$$

Summing over $j = 1, \cdots, J - 1$, we have

$$(J - 1)\frac{\alpha^3}{3L_2^2} \leq f(\mathbf{z}_1) - f(\mathbf{z}_J). \tag{4.19}$$

So

$$J \leq 1 + \frac{3L_2^2}{\alpha^3}[f(\mathbf{z}_1) - f(\mathbf{z}_J)].$$

The probability of failure is $J\delta' \leq \delta$ and the total running time is

$$\left[1 + \frac{3L_2^2}{\alpha^3}(f(\mathbf{z}_1) - f(\mathbf{z}_J))\right] O\left(\sqrt{\frac{L_1}{\alpha}}\log 1/\delta'\right),$$

as claimed.   □

4.3.3   AGD for Nonconvex Problems

Based on the above two building blocks, we can give the accelerated gradient descent for nonconvex optimization in Algorithm 4.8 and the complexity guarantee in Theorem 4.5. Similar to Algorithm 4.5, Algorithm 4.8 alternates between procedures NCD and AC-AGD. However, since NCD terminates at a point where f is locally almost convex, we define a new function $f_k(\mathbf{x})$ that is globally almost convex. The details can be found in Sect. 4.3.3.1.

Algorithm 4.8 Nonconvex AGD for escaping saddle points, NC-AGD-SP($f,L_1,L_2,\epsilon,\delta,\Delta_f$)

$\alpha = \sqrt{L_2\epsilon}$, $K = \frac{3L_2^2\Delta_f}{\alpha^3}$, and $\delta'' = \frac{\delta}{K}$.
for $k = 1, 2, 3, \cdots, K$ **do**
  $\hat{\mathbf{x}}_k = \text{NCD}(f, \mathbf{x}_k, L_2, \alpha, \delta'')$,
  If $\|\nabla f(\hat{\mathbf{x}}_k)\| \leq \epsilon$, return $\hat{\mathbf{x}}_k$,
  Set $f_k(\mathbf{x}) = f(\mathbf{x}) + L_1([\|\mathbf{x} - \hat{\mathbf{x}}_k\| - \alpha/L_2]_+)^2$,
  $\mathbf{x}_{k+1} = \text{AC-AGD}(f_k, \hat{\mathbf{x}}_k, \epsilon/2, 3\alpha, 5L_1)$.
end for

Theorem 4.5 *Assume that f has L_1-Lipschitz continuous gradients and L_2-Lipschitz continuous Hessians. Let $\alpha = \sqrt{L_2\epsilon}$, then with probability of at least $1 - \delta$, Algorithm 4.8 terminates*

1. *in $1 + \frac{15\sqrt{L_2}\Delta_f}{\epsilon^{3/2}}$ outer iterations,*
2. *in $\left(\frac{L_2^{1/4}L_1^{1/2}\Delta_f}{\epsilon^{7/4}}\right)\log\frac{1}{\delta\epsilon}$ running time,*

such that $\|\nabla f(\mathbf{x}_k)\| \leq \epsilon$ *and* $\lambda_{\min}(\nabla^2 f(\mathbf{x}_k)) \geq -\sqrt{\epsilon}$, *where* $\Delta_f = f(\hat{\mathbf{x}}_0) - \min_{\mathbf{x}} f(\mathbf{x})$.

We will prove Theorem 4.5 in three steps in the following sections, respectively.

4.3.3.1   Locally Almost Convex → Globally Almost Convex

The following lemma illustrates how to transform a locally almost convex function into a globally almost convex function.

Lemma 4.9 *Let* $f_k(\mathbf{x}) = f(\mathbf{x}) + L_1 \left(\left[\|\mathbf{x} - \hat{\mathbf{x}}_k\| - \frac{\alpha}{L_2} \right]_+ \right)^2$, *where* $[x]_+ = \max(x, 0)$. *If* $\lambda_{\min} \nabla^2 f(\hat{\mathbf{x}}_k) \geq -\alpha$, *then* f_k *is* 3α-*almost convex and* $5L_1$-*smooth*.

Proof Let $\rho(\mathbf{x}) = L_1 \left(\left[\|\mathbf{x}\| - \frac{\alpha}{L_2} \right]_+ \right)^2$, then

$$\nabla \rho(\mathbf{x}) = 2L_1 \frac{\mathbf{x}}{\|\mathbf{x}\|} \left[\|\mathbf{x}\| - \frac{\alpha}{L_2} \right]_+.$$

We can have that $\nabla \rho(\mathbf{x})$ is continuous, $\nabla \rho(\mathbf{x})$ is differentiable except at $\|\mathbf{x}\| = \frac{\alpha}{L_2}$, and $\nabla^2 \rho(\mathbf{x}) = 0$ for $\|\mathbf{x}\| < \frac{\alpha}{L_2}$ and $\nabla^2 \rho(\mathbf{x}) = 2L_1 \left[\mathbf{I} + \frac{\alpha}{L_2} \left(\frac{\mathbf{x}\mathbf{x}^T}{\|\mathbf{x}\|^3} - \frac{\mathbf{I}}{\|\mathbf{x}\|} \right) \right]$ for $\|\mathbf{x}\| > \frac{\alpha}{L_2}$. So when $\|\mathbf{x}\| > \frac{\alpha}{L_2}$, we have

$$\nabla^2 \rho(\mathbf{x}) \succeq 2L_1 \left(\mathbf{I} - \frac{\alpha}{L_2} \frac{\mathbf{I}}{\|\mathbf{x}\|} \right) \succeq \mathbf{0},$$

$$\nabla^2 \rho(\mathbf{x}) \preceq 2L_1 \left(\mathbf{I} + \frac{\alpha}{L_2} \frac{\mathbf{x}\mathbf{x}^T}{\|\mathbf{x}\|^3} \right) \preceq 2L_1 \left(\mathbf{I} + \frac{\alpha}{L_2} \frac{\mathbf{I}}{\|\mathbf{x}\|} \right) \preceq 4L_1 \mathbf{I}.$$

Thus $f_k(\mathbf{x})$ is $5L_1$-smooth.

When $\|\mathbf{x} - \hat{\mathbf{x}}_k\| > \frac{2\alpha}{L_2}$, we have

$$\nabla^2 f_k(\mathbf{x}) \succeq \nabla^2 f(\mathbf{x}) + 2L_1 \left(\mathbf{I} - \frac{\alpha}{L_2} \frac{\mathbf{I}}{\|\mathbf{x} - \hat{\mathbf{x}}_k\|} \right) \succ \mathbf{0}.$$

When $\|\mathbf{x} - \hat{\mathbf{x}}_k\| \leq \frac{2\alpha}{L_2}$, we have

$$\nabla^2 f(\mathbf{x}) \overset{a}{\succeq} \nabla^2 f(\hat{\mathbf{x}}_k) - L_2 \|\mathbf{x} - \hat{\mathbf{x}}_k\| \mathbf{I} \succeq -3\alpha \mathbf{I},$$

where $\overset{a}{\succeq}$ uses the Lipschitz continuity of $\nabla^2 f$. So

$$\nabla^2 f_k(\mathbf{x}) \succeq \nabla^2 f(\mathbf{x}) \succeq -3\alpha \mathbf{I}.$$

Namely, $f_k(\mathbf{x})$ is 3α-almost convex.   □

4.3.3.2 Outer Iterations

We present the complexity of the outer iterations of Algorithm 4.8 in the following lemma.

Lemma 4.10 *Assume that f has L_1-Lipschitz continuous gradients and L_2-Lipschitz continuous Hessians. Let $\alpha = \sqrt{L_2\epsilon}$, then Algorithm 4.8 terminates in $\left(1 + \frac{15\sqrt{L_2}\Delta_f}{\epsilon^{3/2}}\right)$ outer iterations.*

Proof For NCD, we have

$$\frac{\alpha^3}{3L_2^2} \leq f(\mathbf{z}_1) - f(\mathbf{z}_2) \leq f(\mathbf{z}_1) - f(\mathbf{z}_J),$$

where we use (4.18). Then

$$\frac{\alpha^3}{3L_2^2} \leq f(\mathbf{x}_k) - f(\hat{\mathbf{x}}_k). \tag{4.20}$$

For AC-AGD, we have $f_k(\mathbf{x}_{k+1}) \leq f_k(\hat{\mathbf{x}}_k)$, $f_k(\hat{\mathbf{x}}_k) = f(\hat{\mathbf{x}}_k)$, and $f_k(\mathbf{x}_{k+1}) \geq f(\mathbf{x}_{k+1})$, where we use (4.15) and the definition of f_k. So $f(\mathbf{x}_{k+1}) \leq f(\hat{\mathbf{x}}_k)$ and

$$\frac{\alpha^3}{3L_2^2} \leq f(\hat{\mathbf{x}}_{k-1}) - f(\hat{\mathbf{x}}_k),$$

where we use (4.20). Summing over $k = 1, \cdots, K$, we have

$$K\frac{\alpha^3}{3L_2^2} \leq f(\hat{\mathbf{x}}_0) - f(\hat{\mathbf{x}}_K).$$

So

$$K \leq \frac{3L_2^2}{\alpha^3}(f(\hat{\mathbf{x}}_0) - f(\hat{\mathbf{x}}_K)) \leq \frac{3L_2^2}{\alpha^3}(f(\hat{\mathbf{x}}_0) - \min_{\mathbf{x}} f(\mathbf{x})) = \frac{3L_2^2\Delta_f}{\alpha^3}.$$

On the other hand, for AC-AGD we have

$$\frac{\epsilon^2}{15\alpha} \leq f_k(\mathbf{z}_1) - f_k(\mathbf{z}_2) \leq f_k(\mathbf{z}_1) - f_k(\mathbf{z}_J),$$

where we use (4.16) and $\sigma = 3\alpha$. So

$$\frac{\epsilon^2}{15\alpha} \leq f_k(\hat{\mathbf{x}}_k) - f_k(\mathbf{x}_{k+1}) = f(\hat{\mathbf{x}}_k) - f_k(\mathbf{x}_{k+1})$$

$$\leq f(\hat{\mathbf{x}}_k) - f(\mathbf{x}_{k+1}) \overset{a}{\leq} f(\hat{\mathbf{x}}_k) - f(\hat{\mathbf{x}}_{k+1}),$$

where $\overset{a}{\leq}$ uses (4.20). Summing over $k = 1, \cdots, K - 1$, we have

$$(K - 1)\frac{\epsilon^2}{15\alpha} \leq f(\hat{\mathbf{x}}_1) - f(\hat{\mathbf{x}}_K) \leq \Delta_f.$$

So

$$K \leq 1 + \frac{15\alpha\Delta_f}{\epsilon^2} = 1 + \frac{15\sqrt{L_2}\Delta_f}{\epsilon^{3/2}}. \tag{4.21}$$

$\square$

4.3.3.3 Inner Iterations

Now, we consider the total inner iterations, which directly leads to the second conclusion of Theorem 4.5.

Lemma 4.11 *Assume that f has L_1-Lipschitz continuous gradients and L_2-Lipschitz continuous Hessians. Let $\alpha = \sqrt{L_2\epsilon}$, then Algorithm 4.8 terminates in the running time of*

$$O\left(\frac{L_1^{1/2}L_2^{1/4}\Delta_f}{\epsilon^{7/4}}\log\frac{1}{\delta\epsilon}\right).$$

Proof Consider the inner iterations of NCD. Let j_k be the inner iterations at the k-th outer iteration. From (4.19), we have

$$\sum_{k=1}^{K}(j_k - 1) \leq \sum_{k=1}^{K}\frac{3L_2^2}{\alpha^3}(f(\mathbf{x}_k) - f(\hat{\mathbf{x}}_k))$$

$$\leq \sum_{k=1}^{K}\frac{3L_2^2}{\alpha^3}(f(\hat{\mathbf{x}}_{k-1}) - f(\hat{\mathbf{x}}_k))$$

$$= \frac{3L_2^2}{\alpha^3}(f(\hat{\mathbf{x}}_0) - f(\hat{\mathbf{x}}_K)) \leq \frac{3L_2^2\Delta_f}{\alpha^3}.$$

So

$$\sum_{k=1}^{K}j_k \leq \frac{3L_2^2\Delta_f}{\alpha^3} + K \leq \frac{18\sqrt{L_2}\Delta_f}{\epsilon^{3/2}}$$

and the total running time of NCD is

$$\frac{18\sqrt{L_2}\Delta_f}{\epsilon^{3/2}} O\left(\sqrt{\frac{L_1}{\alpha}} \log 1/\delta'\right) = O\left(\frac{L_1^{1/2}L_2^{1/4}\Delta_f}{\epsilon^{7/4}} \log \frac{1}{\delta\alpha}\right), \qquad (4.22)$$

where we use the fact that δ' defined in Algorithm 4.7 is at the same order of $\delta\alpha^3$.

Next, we consider the inner iterations of AC-AGD. From (4.17), we have that the running time is at the order of

$$\sum_{k=1}^{K-1}\left[\sqrt{\frac{L_1}{\alpha}} + \frac{\sqrt{L_1\alpha}}{\epsilon^2}(f(\hat{\mathbf{x}}_k) - f(\mathbf{x}_{k+1}))\right]\log 1/\epsilon$$

$$\leq \sum_{k=1}^{K-1}\left[\sqrt{\frac{L_1}{\alpha}} + \frac{\sqrt{L_1\alpha}}{\epsilon^2}(f(\hat{\mathbf{x}}_k) - f(\hat{\mathbf{x}}_{k+1}))\right]\log 1/\epsilon$$

$$\leq \left(K\sqrt{\frac{L_1}{\alpha}} + \frac{\Delta_f\sqrt{L_1\alpha}}{\epsilon^2}\right)\log 1/\epsilon$$

$$\overset{a}{\leq} \frac{16L_1^{1/2}L_2^{1/4}\Delta_f}{\epsilon^{7/4}} \log 1/\epsilon, \qquad (4.23)$$

where we have dropped some constant coefficients and $\overset{a}{\leq}$ uses (4.21) and $\alpha = \sqrt{L_2\epsilon}$. Adding (4.22) and (4.23) and using $\alpha = \sqrt{L_2\epsilon}$, the total running time of Algorithm 4.8 is

$$O\left(\frac{L_1^{1/2}L_2^{1/4}\Delta_f}{\epsilon^{7/4}} \log \frac{1}{\delta\epsilon}\right),$$

which completes the proof. $\qquad\square$

References

1. N. Agarwal, Z. Allen-Zhu, B. Bullins, E. Hazan, T. Ma, Finding approximate local minima for nonconvex optimization in linear time, in *Proceedings of the 49th Annual ACM SIGACT Symposium on Theory of Computing*, Montreal, (2017), pp. 1195–1200
2. H. Attouch, J. Bolte, P. Redont, A. Soubeyran, Proximal alternating minimization and projection methods for nonconvex problems: an approach based on the Kurdyka-Łojasiewicz inequality. Math. Oper. Res. **35**(2), 438–457 (2010)
3. H. Attouch, J. Bolte, B.F. Svaiter, Convergence of descent methods for semi-algebraic and tame problems: proximal algorithms, forward-backward splitting, and regularized Gauss-Seidel methods. Math. Program. **137**(1–2), 91–129 (2013)
4. J. Bolte, S. Sabach, M. Teboulle, Proximal alternating linearized minimization for nonconvex and nonsmooth problems. Math. Program. **146**(1–2), 459–494 (2014)

5. R.I. Boț, E.R. Csetnek, S.C. László, An inertial forward-backward algorithm for the minimization of the sum of two nonconvex functions. EURO J. Comput. Optim. **4**(1), 3–25 (2016)
6. E.J. Candès, M.B. Wakin, S.P. Boyd, Enhancing sparsity by reweighted l_1 minimization. J. Fourier Anal. Appl. **14**(5), 877–905 (2008)
7. Y. Carmon, J.C. Duchi, O. Hinder, A. Sidford, Convex until proven guilty: dimension-free acceleration of gradient descent on non-convex functions, in *Proceedings of the 34th International Conference on Machine Learning*, Sydney, (2017), pp. 654–663
8. Y. Carmon, J.C. Duchi, O. Hinder, A. Sidford, Accelerated methods for nonconvex optimization. SIAM J. Optim. **28**(2), 1751–1772 (2018)
9. J. Fan, R. Li, Variable selection via nonconcave penalized likelihood and its oracle properties. J. Am. Stat. Assoc. **96**(456), 1348–1360 (2001)
10. S. Foucart, M.-J. Lai, Sparsest solutions of underdetermined linear systems via l_q minimization for $0 < q \leq 1$. Appl. Comput. Harmon. Anal. **26**(3), 395–407 (2009)
11. P. Frankel, G. Garrigos, J. Peypouquet, Splitting methods with variable metric for Kurdyka-Łojasiewicz functions and general convergence rates. J. Optim. Theory Appl. **165**(3), 874–900 (2015)
12. R. Ge, F. Huang, C. Jin, Y. Yuan, Escaping from saddle points – online stochastic gradient for tensor decomposition, in *Proceedings of the 28th Conference on Learning Theory*, Lille, (2015), pp. 797–842
13. D. Geman, C. Yang, Nonlinear image recovery with half-quadratic regularization. IEEE Trans. Image Process. **4**(7), 932–946 (1995)
14. S. Ghadimi, G. Lan, Accelerated gradient methods for nonconvex nonlinear and stochastic programming. Math. Program. **156**(1–2), 59–99 (2016)
15. P. Gong, C. Zhang, Z. Lu, J. Huang, J. Ye, A general iterative shrinkage and thresholding algorithm for non-convex regularized optimization problems, in *Proceedings of the 30th International Conference on Machine Learning*, Atlanta, (2013), pp. 37–45
16. C. Jin, R. Ge, P. Netrapalli, S.M. Kakade, M.I. Jordan, How to escape saddle points efficiently, in *Proceedings of the 34th International Conference on Machine Learning*, Sydney, (2017), pp. 1724–1732
17. C. Jin, P. Netrapalli, M.I. Jordan, Accelerated gradient descent escapes saddle points faster than gradient descent, in *Proceedings of the 31th Conference On Learning Theory*, Stockholm, (2018), pp. 1042–1085
18. J. Kuczyński, H. Woźniakowski, Estimating the largest eigenvalue by the power and Lanczos algorithms with a random start. SIAM J. Matrix Anal. Appl. **13**(4), 1094–1122 (1992)
19. J.D. Lee, M. Simchowitz, M.I. Jordan, B. Recht, Gradient descent only converges to minimizers, in *Proceedings of the 29th Conference on Learning Theory*, New York, (2016), pp. 1246–1257
20. H. Li, Z. Lin, Accelerated proximal gradient methods for nonconvex programming, in *Advances in Neural Information Processing Systems*, Montreal, vol. 28 (2015), pp. 379–387
21. Q. Li, Y. Zhou, Y. Liang, P.K. Varshney, Convergence analysis of proximal gradient with momentum for nonconvex optimization, in *Proceedings of the 34th International Conference on Machine Learning*, Sydney, (2017), pp. 2111–2119
22. K. Mohan, M. Fazel, Iterative reweighted algorithms for matrix rank minimization. J. Mach. Learn. Res. **13**(1), 3441–3473 (2012)
23. Y. Nesterov, *Introductory Lectures on Convex Optimization: A Basic Course* (Springer, New York, 2004)
24. Y. Nesterov, A. Gasnikov, S. Guminov, P. Dvurechensky, Primal-dual accelerated gradient descent with line search for convex and nonconvex optimization problems (2018). Preprint. arXiv:1809.05895

25. P. Ochs, Y. Chen, T. Brox, T. Pock, iPiano: inertial proximal algorithm for nonconvex optimization. SIAM J. Imag. Sci. **7**(2), 1388–1419 (2014)
26. C.-H. Zhang, Nearly unbiased variable selection under minimax concave penalty. Ann. Stat. **38**(2), 894–942 (2010)
27. T. Zhang, Analysis of multi-stage convex relaxation for sparse regularization. J. Mach. Learn. Res. **11**, 1081–1107 (2010)

Chapter 5
Accelerated Stochastic Algorithms

In machine learning and statistics communities, lots of large-scale problems can be formulated as the following optimization problem:

$$\min_{\mathbf{x}} f(\mathbf{x}) \equiv \mathbb{E}[F(\mathbf{x}; \xi)], \tag{5.1}$$

where $F(\mathbf{x}; \xi)$ is a stochastic component indexed by a random number ξ. A special case that is of central interest is that $f(\mathbf{x})$ can be written as a sum of functions. If we denote each component function as $f_i(\mathbf{x})$, then (5.1) can be restated as

$$\min_{\mathbf{x}} f(\mathbf{x}) \equiv \frac{1}{n} \sum_{i=1}^{n} f_i(\mathbf{x}), \tag{5.2}$$

where n is the number of individual functions. When n is finite, (5.2) is an offline problem, with typical examples including empirical risk minimization (ERM). n can also go to infinity and we refer to this case as an online (streaming) problem.

Obtaining the full gradient of (5.2) might be expensive when n is large and even inaccessible when $n = \infty$. Instead, a standard manner is to estimate the full gradient via one or several randomly sampled counterparts from individual functions. The obtained algorithms to solve problem (5.2) are referred to as stochastic algorithms, involving the following characteristics:

1. in theory, most convergence properties are studied in the forms of expectation (concentration);
2. in real experiments, the algorithms are often much faster than the batch (deterministic) ones.

© Springer Nature Singapore Pte Ltd. 2020
Z. Lin et al., *Accelerated Optimization for Machine Learning*,
https://doi.org/10.1007/978-981-15-2910-8_5

Because the updates access the gradient individually, the time complexity to reach a tolerable accuracy can be evaluated by the total number of calls for individual functions. Formally, we refer to it as the Incremental First-order Oracle (IFO) calls, with definition as follows:

Definition 5.1 For problem (5.2), an IFO takes an index $i \in [n]$ and a point $\mathbf{x} \in \mathbb{R}^d$, and returns the pair $(f_i(\mathbf{x}), \nabla f_i(\mathbf{x}))$.

As has been discussed in Chap. 2, the momentum (acceleration) technique ensures a theoretically faster convergence rate for deterministic algorithms. We might ask whether the momentum technique can accelerate stochastic algorithms. Before we answer the question, we first put our attention on the stochastic algorithms itself. What is the main challenge in analyzing the stochastic algorithms? Definitely, it is the noise of the gradients. Specifically, the variance of the noisy gradient will not go to zero through the updates, which fundamentally slows down the convergence rate. So a more involved question is to ask whether the momentum technique can reduce the negative effect of noise? Unfortunately, the existing results answer the question with "No." The momentum technique cannot reduce the variance but instead can accumulate the noise. What really reduces the negative effect of noise is the technique called variance reduction (VR) [10]. When applying the VR technique, the algorithms are transformed to act like a deterministic algorithm and then can be further fused with momentum. Now we are able to answer our first question. The answer is towards positivity: in some cases, however, not all (e.g., when n is large), the momentum technique fused with VR ensures a provably faster rate! Besides, another effect of the momentum technique is that it can accumulate the noise. Thus one can reduce the variance together after it is aggregated by the momentum. By doing this, the mini-batch sampling size is increased, which is very helpful in distributed optimization (see Chap. 6).

From a high level view, we summarize the way to accelerate stochastic methods into two steps:

1. transforming the algorithm into a "near deterministic" one by VR and
2. fusing the momentum trick to achieve a faster rate of convergence.

We conclude the advantages of momentum technique for stochastic algorithms below:

- Ensure faster convergence rates (by order) when n is sufficiently small.
- Ensure larger mini-batch sizes for distributed optimization.

In the following sections, we will concretely introduce how to use the momentum technique to accelerate algorithms according to different properties of $f(\mathbf{x})$. In particular, we roughly split them into three cases:

Algorithm 5.1 Accelerated stochastic coordinate descent (ASCD) [8]

Input θ_k, step size $\gamma = \frac{1}{L_c}$, $\mathbf{x}^0 = \mathbf{0}$, and $\mathbf{z}^0 = \mathbf{0}$.
for $k = 0$ **to** K **do**
1    $\mathbf{y}^k = (1 - \theta_k)\mathbf{x}^k + \theta_k\mathbf{z}^k$,
2   Randomly choose an index i_k from $[n]$,
3    $\delta = \operatorname{argmin}_\delta \left(h_{i_k}(\mathbf{z}^k_{i_k} + \delta) + \langle \nabla_{i_k} f(\mathbf{y}^k), \delta \rangle + \frac{n\theta_k}{2\gamma}\delta^2 \right)$,
4    $\mathbf{z}^{k+1}_{i_k} = \mathbf{z}^k_{i_k} + \delta$, with other coordinates unchanged,
5    $\mathbf{x}^{k+1} = (1 - \theta_k)\mathbf{x}^k + n\theta_k\mathbf{z}^{k+1} - (n - 1)\theta_k\mathbf{z}^k$.
end for
Output $\mathbf{x}^{K+1}$.

- each $f_i(\mathbf{x})$ is convex, which we call the individually convex (IC) case.
- each $f_i(\mathbf{x})$ can be nonconvex but $f(\mathbf{x})$ is convex, which we call the individually nonconvex (INC) case.
- $f(\mathbf{x})$ is nonconvex (NC).

At last, we extend some results to linearly constrained problems.

5.1 The Individually Convex Case

We consider a more general form of (5.2):

$$\min_{\mathbf{x}} F(\mathbf{x}) \equiv h(\mathbf{x}) + \frac{1}{n}\sum_{i=1}^{n} f_i(\mathbf{x}), \tag{5.3}$$

where $h(\mathbf{x})$ and $f_i(\mathbf{x})$ with $i \in [n]$ are convex, the proximal mapping of h can be computed efficiently, and n is finite.

5.1.1 Accelerated Stochastic Coordinate Descent

To use the finiteness of n, we can solve (5.3) in its dual space in which n becomes the dimension of the dual variable and one coordinate update of the dual variable corresponds to one time of accessing the individual function. We first introduce accelerated stochastic coordinate descent (ASCD) [8, 11, 16] and later illustrate how to apply it to solve (5.3). ASCD is first proposed in [16], and later proximal versions are discovered in [8, 11]. With a little abuse of notation, without hindering the readers to understand the momentum technique, we still write the optimization model for ASCD as follows:

$$\min_{\mathbf{x} \in \mathcal{R}^n} F(\mathbf{x}) \equiv h(\mathbf{x}) + f(\mathbf{x}), \tag{5.4}$$

where $f(\mathbf{x})$ has L_c-coordinate Lipschitz continuous gradients (Definition A.13), $h(\mathbf{x})$ has coordinate separable structure, i.e., $h(\mathbf{x}) = \sum_{i=1}^{n} h_i(\mathbf{x}_i)$ with $\mathbf{x} = (\mathbf{x}_1^T, \cdots, \mathbf{x}_n^T)^T$, and $f(\mathbf{x})$ and $h_i(\mathbf{x})$ are convex.

Stochastic coordinate descent algorithms are efficient to solve (5.4). In each update, one random coordinate $\mathbf{x}_{i_k}$ is chosen to sufficiently reduce the objective value while keeping other coordinates fixed, reducing the per-iteration cost. More specifically, the following types of proximal subproblem are solved:

$$\boldsymbol{\delta} = \operatorname*{argmin}_{\boldsymbol{\delta}} \left(h_{i_k}(\mathbf{x}_{i_k}^k + \boldsymbol{\delta}) + \left\langle \nabla_{i_k} f(\mathbf{x}^k), \boldsymbol{\delta} \right\rangle + \frac{\theta_k}{2\gamma} \boldsymbol{\delta}^2 \right),$$

where $\nabla_{i_k} f(\mathbf{x})$ denotes the partial gradient of f with respect to $\mathbf{x}_{i_k}$. Fusing with the momentum technique, ASCD is shown in Algorithm 5.1.

We give the convergence result below. The proof is taken from [8].

Lemma 5.1 *If $\theta_0 \leq \frac{1}{n}$ and for all $k \geq 0$, $\theta_k \geq 0$ and is monotonically non-increasing, then $\mathbf{x}^k$ is a convex combination of $\mathbf{z}^0, \cdots, \mathbf{z}^k$, i.e., we have $\mathbf{x}^k = \sum_{i=0}^{k} e_{k,i} \mathbf{z}^i$, where $e_{0,0} = 1$, $e_{1,0} = 1 - n\theta_0$, $e_{1,1} = n\theta$, and for $k > 1$, we have*

$$e_{k+1,i} = \begin{cases} (1 - \theta_k) e_{k,i}, & i \leq k - 1, \\ n(1 - \theta_k)\theta_{k-1} + \theta_k - n\theta_k, & i = k, \\ n\theta_k, & i = k + 1. \end{cases} \tag{5.5}$$

Defining $\hat{h}_k = \sum_{i=0}^{k} e_{k,i} h(\mathbf{z}^i)$, we have

$$\mathbb{E}_{i_k}(\hat{h}_{k+1}) = (1 - \theta_k)\hat{h}_k + \theta_k \sum_{i_k=1}^{n} h_{i_k}(\mathbf{z}_{i_k}^{k+1}), \tag{5.6}$$

where $\mathbb{E}_{i_k}$ denotes that the expectation is taken only on the random number i_k under the condition that $\mathbf{x}^k$ and $\mathbf{z}^k$ are known.

Proof We prove $e_{k+1,i}$ first. When $k = 0$ and 1, we can check that it is right. We then prove (5.5). Since

$$\mathbf{x}^{k+1} \stackrel{a}{=} (1 - \theta_k)\mathbf{x}^k + \theta_k \mathbf{z}^k + n\theta_k(\mathbf{z}^{k+1} - \mathbf{z}^k)$$

$$= (1 - \theta_k)\sum_{i=0}^{k} e_{k,i}\mathbf{z}^i + \theta_k \mathbf{z}^k + n\theta_k(\mathbf{z}^{k+1} - \mathbf{z}^k)$$

$$= (1 - \theta_k)\sum_{i=0}^{k-1} e_{k,i}\mathbf{z}^i + \left[(1 - \theta_k)e_{k,k} + \theta_k - n\theta_k\right]\mathbf{z}^k + n\theta_k\mathbf{z}^{k+1},$$

where $\overset{a}{=}$ uses Step 5 of Algorithm 5.1. Comparing the results, we obtain (5.5). Next, we prove that the above is a convex combination. It is easy to prove that the weights sum to 1 (by induction), $0 \leq (1 - \theta_k)e_{k,j} \leq 1$, and $0 \leq n\theta_k \leq 1$. So $(1 - \theta_k)e_{k,k} + \theta_k - n\theta_k = n(1 - \theta_k)\theta_{k-1} + \theta_k - n\theta_k \leq 1$. On the other hand, we have

$$n(1 - \theta_k)\theta_{k-1} + \theta_k - n\theta_k \geq n(1 - \theta_k)\theta_k + \theta_k - n\theta_k = \theta_k(1 - n\theta_k) \geq 0.$$

For (5.6), we have

$$\mathbb{E}_{i_k}\hat{h}_{k+1}$$

$$\overset{a}{=} \sum_{i=0}^{k} e_{k+1,i}h(\mathbf{z}^i) + \mathbb{E}_{i_k}\left[n\theta_k h(\mathbf{z}^{k+1})\right]$$

$$= \sum_{i=0}^{k} e_{k+1,i}h(\mathbf{z}^i) + \frac{1}{n}\sum_{i_k} n\theta_k \left(h_{i_k}(\mathbf{z}_{i_k}^{k+1}) + \sum_{j\neq i_k} h_j(\mathbf{z}_j^k)\right)$$

$$= \sum_{i=0}^{k} e_{k+1,i}h(\mathbf{z}^i) + \theta_k \sum_{i_k} h_{i_k}(\mathbf{z}_{i_k}^{k+1}) + (n-1)\theta_k h(\mathbf{z}^k)$$

$$\overset{b}{=} \sum_{i=0}^{k-1} e_{k+1,i}h(\mathbf{z}^i) + [n(1 - \theta_k)\theta_{k-1} + \theta_k - n\theta_k]h(\mathbf{z}^k) + (n-1)\theta_k h(\mathbf{z}^k)$$

$$+ \theta_k \sum_{i_k} h_{i_k}(\mathbf{z}_{i_k}^{k+1})$$

$$= \sum_{i=0}^{k-1} e_{k+1,i}h(\mathbf{z}^i) + n(1 - \theta_k)\theta_{k-1}h(\mathbf{z}^k) + \theta_k \sum_{i_k} h_{i_k}(\mathbf{z}_{i_k}^{k+1})$$

$$\overset{c}{=} \sum_{i=0}^{k-1} e_{k,i}(1 - \theta_k)h(\mathbf{z}^i) + (1 - \theta_k)e_{k,k}h(\mathbf{z}^k) + \theta_k \sum_{i_k} h_{i_k}(\mathbf{z}_{i_k}^{k+1})$$

$$= \sum_{i=0}^{k} e_{k,i}(1 - \theta_k)h(\mathbf{z}^i) + \theta_k \sum_{i_k} h_{i_k}(\mathbf{z}_{i_k}^{k+1})$$

$$= (1 - \theta_k)\hat{h}_k + \theta_k \sum_{i_k} h_{i_k}(\mathbf{z}_{i_k}^{k+1}),$$

where in $\overset{a}{=}$ we use $e_{k+1,k+1} = n\theta_k$, in $\overset{b}{=}$ we use $e_{k+1,k} = n(1-\theta_k)\theta_{k-1} + \theta_k - n\theta_k$, and in $\overset{c}{=}$ we use $e_{k+1,i} = (1-\theta_k)e_{k,i}$ for $i \leq k-1$, and $e_{k,k} = n\theta_{k-1}$.   □

Theorem 5.1 *For Algorithm 5.1, setting $\theta_k = \frac{2}{2n+k}$, we have that $\mathbf{x}^k$ is a convex combination of $\mathbf{z}^0, \cdots, \mathbf{z}^k$, and*

$$\frac{\mathbb{E}[F(\mathbf{x}^{K+1})] - F(\mathbf{x}^*)}{\theta_K^2} + \frac{n^2 L_c}{2}\mathbb{E}\|\mathbf{z}^{K+1} - \mathbf{x}^*\|^2$$

$$\leq \frac{F(\mathbf{x}^0) - F(\mathbf{x}^*)}{\theta_{-1}^2} + \frac{n^2 L_c}{2}\|\mathbf{z}^0 - \mathbf{x}^*\|^2. \tag{5.7}$$

When $h(\mathbf{x})$ is strongly convex with modulus $0 \leq \mu \leq L_c$, setting $\theta_k = \frac{-\frac{\mu}{L_c}+\sqrt{\mu^2/L_c^2+4\mu/L_c}}{2n} \sim O\left(\frac{\sqrt{\mu/L_c}}{n}\right)$ which is denoted as θ instead, we have

$$\mathbb{E}[F(\mathbf{x}^{K+1})] - F(\mathbf{x}^*)$$

$$\leq (1-\theta)^{K+1}\left(F(\mathbf{x}^0) - F(\mathbf{x}^*) + \frac{n^2\theta^2 L_c + n\theta\mu}{2}\left\|\mathbf{z}^0 - \mathbf{x}^*\right\|^2\right). \tag{5.8}$$

Proof We can check that the setting of θ_k satisfies the assumptions in Lemma 5.1. We first consider the function value. By the optimality of $\mathbf{z}_{i_k}^{k+1}$ in Step 4, we have

$$n\theta_k(\mathbf{z}_{i_k}^{k+1} - \mathbf{z}_{i_k}^k) + \gamma\nabla_{i_k}f(\mathbf{y}^k) + \gamma\xi_{i_k}^k = 0, \tag{5.9}$$

where $\xi_{i_k}^k \in \partial h_{i_k}(\mathbf{z}_{i_k}^{k+1})$. By Steps 1 and 5, we have

$$\mathbf{x}^{k+1} = \mathbf{y}^k + n\theta_k(\mathbf{z}^{k+1} - \mathbf{z}^k). \tag{5.10}$$

Substituting (5.10) into (5.9), we have

$$\mathbf{x}_{i_k}^{k+1} - \mathbf{y}_{i_k}^k + \gamma\nabla_{i_k}f(\mathbf{y}^k) + \gamma\xi_{i_k}^k = 0. \tag{5.11}$$

Since f has L_c-coordinate Lipschitz continuous gradients on coordinate i_k (see (A.4)) and $\mathbf{x}^{k+1}$ and $\mathbf{y}^k$ only differ at the i_k-th entry, we have

$$f(\mathbf{x}^{k+1})$$

$$\overset{a}{\leq} f(\mathbf{y}^k) + \left\langle\nabla_{i_k}f(\mathbf{y}^k), \mathbf{x}_{i_k}^{k+1} - \mathbf{y}_{i_k}^k\right\rangle + \frac{L_c}{2}\left(\mathbf{x}_{i_k}^{k+1} - \mathbf{y}_{i_k}^k\right)^2$$

$$\overset{b}{=} f(\mathbf{y}^k) - \gamma\left\langle\nabla_{i_k}f(\mathbf{y}^k), \nabla_{i_k}f(\mathbf{y}^k) + \xi_{i_k}^k\right\rangle + \frac{L_c}{2}\left(\mathbf{x}_{i_k}^{k+1} - \mathbf{y}_{i_k}^k\right)^2$$

$$\overset{c}{=} f(\mathbf{y}^k) - \gamma \left\langle \nabla_{i_k} f(\mathbf{y}^k) + \boldsymbol{\xi}_{i_k}^k, \nabla_{i_k} f(\mathbf{y}^k) + \boldsymbol{\xi}_{i_k}^k \right\rangle + \frac{L_c}{2} \left(\mathbf{x}_{i_k}^{k+1} - \mathbf{y}_{i_k}^k \right)^2$$

$$+ \gamma \left\langle \boldsymbol{\xi}_{i_k}^k, \nabla_{i_k} f(\mathbf{y}^k) + \boldsymbol{\xi}_{i_k}^k \right\rangle$$

$$\overset{d}{=} f(\mathbf{y}^k) - \frac{\gamma}{2} \left(\frac{\mathbf{x}_{i_k}^{k+1} - \mathbf{y}_{i_k}^k}{\gamma} \right)^2 - \left\langle \boldsymbol{\xi}_{i_k}^k, \mathbf{x}_{i_k}^{k+1} - \mathbf{y}_{i_k}^k \right\rangle, \tag{5.12}$$

where $\overset{a}{\leq}$ uses Proposition A.7, in $\overset{b}{=}$ we use (5.11), in $\overset{c}{=}$ we insert $\gamma \langle \boldsymbol{\xi}_{i_k}^k, \nabla_{i_k} f(\mathbf{y}^k) + \boldsymbol{\xi}_{i_k}^k \rangle$, and $\overset{d}{=}$ uses $\gamma = \frac{1}{L_c}$.

Then we analyze $\left\| \mathbf{z}^{k+1} - \mathbf{x}^* \right\|^2$. We have

$$\frac{n^2}{2\gamma} \left\| \theta_k \mathbf{z}^{k+1} - \theta_k \mathbf{x}^* \right\|^2$$

$$= \frac{n^2}{2\gamma} \left\| \theta_k \mathbf{z}^k - \theta_k \mathbf{x}^* + \theta_k \mathbf{z}^{k+1} - \theta_k \mathbf{z}^k \right\|^2$$

$$= \frac{n^2}{2\gamma} \left\| \theta_k \mathbf{z}^k - \theta_k \mathbf{x}^* \right\|^2 + \frac{n^2}{2\gamma} \left(\theta_k \mathbf{z}_{i_k}^{k+1} - \theta_k \mathbf{z}_{i_k}^k \right)^2$$

$$+ \frac{n^2}{\gamma} \left\langle \theta_k \left(\mathbf{z}_{i_k}^{k+1} - \mathbf{z}_{i_k}^k \right), \theta_k \mathbf{z}_{i_k}^k - \theta_k \mathbf{x}_{i_k}^* \right\rangle$$

$$\overset{a}{=} \frac{n^2}{2\gamma} \left\| \theta_k \mathbf{z}^k - \theta_k \mathbf{x}^* \right\|^2 + \frac{1}{2\gamma} \left(\mathbf{x}_{i_k}^{k+1} - \mathbf{y}_{i_k}^k \right)^2$$

$$- n \left\langle \nabla_{i_k} f(\mathbf{y}^k) + \boldsymbol{\xi}_{i_k}^k, \theta_k \mathbf{z}_{i_k}^k - \theta_k \mathbf{x}_{i_k}^* \right\rangle, \tag{5.13}$$

where $\overset{a}{=}$ uses (5.9) and (5.10).

Then by taking expectation on (5.13), we have

$$\frac{n^2}{2\gamma} \mathbb{E}_{i_k} \left\| \theta_k \mathbf{z}^{k+1} - \theta_k \mathbf{x}^* \right\|^2$$

$$= \frac{n^2}{2\gamma} \left\| \theta_k \mathbf{z}^k - \theta_k \mathbf{x}^* \right\|^2 + \frac{1}{2\gamma n} \sum_{i_k=1}^{n} \left(\mathbf{x}_{i_k}^{k+1} - \mathbf{y}_{i_k}^k \right)^2 - \left\langle \nabla f(\mathbf{y}^k), \theta_k \mathbf{z}^k - \theta_k \mathbf{x}^* \right\rangle$$

$$- \sum_{i_k=1}^{n} \left\langle \boldsymbol{\xi}_{i_k}^k, \theta_k \mathbf{z}_{i_k}^k - \theta_k \mathbf{x}_{i_k}^* \right\rangle. \tag{5.14}$$

By Step 1, we have

$$-\left\langle \nabla f(\mathbf{y}^k), \theta_k \mathbf{z}^k - \theta_k \mathbf{x}^* \right\rangle = \left\langle \nabla f(\mathbf{y}^k), (1 - \theta_k)\mathbf{x}^k + \theta_k \mathbf{x}^* - \mathbf{y}^k \right\rangle$$

$$\overset{a}{\leq} (1 - \theta_k) f(\mathbf{x}^k) + \theta_k f(\mathbf{x}^*) - f(\mathbf{y}^k), \quad (5.15)$$

where $\overset{a}{\leq}$ uses the convexity of f. Taking expectation on (5.12) and adding (5.14) and (5.15), we have

$$\mathbb{E}_{i_k} f(\mathbf{x}^{k+1}) \leq (1 - \theta_k) f(\mathbf{x}^k) + \theta_k f(\mathbf{x}^*)$$

$$- \sum_{i_k=1}^{n} \left\langle \boldsymbol{\xi}_{i_k}^k, \theta_k \mathbf{z}_{i_k}^k - \theta_k \mathbf{x}_{i_k}^* + \frac{1}{n} \left(\mathbf{x}_{i_k}^{k+1} - \mathbf{y}_{i_k}^k \right) \right\rangle$$

$$+ \frac{n^2}{2\gamma} \left\| \theta_k \mathbf{z}^k - \theta_k \mathbf{x}^* \right\|^2 - \frac{n^2}{2\gamma} \mathbb{E}_{i_k} \left\| \theta_k \mathbf{z}^{k+1} - \theta_k \mathbf{x}^* \right\|^2$$

$$\overset{a}{=} (1 - \theta_k) f(\mathbf{x}^k) + \theta_k f(\mathbf{x}^*) - \sum_{i_k=1}^{n} \left\langle \boldsymbol{\xi}_{i_k}^k, \theta_k \mathbf{z}_{i_k}^{k+1} - \theta_k \mathbf{x}_{i_k}^* \right\rangle$$

$$+ \frac{n^2}{2\gamma} \left\| \theta_k \mathbf{z}^k - \theta_k \mathbf{x}^* \right\|^2 - \frac{n^2}{2\gamma} \mathbb{E}_{i_k} \left\| \theta_k \mathbf{z}^{k+1} - \theta_k \mathbf{x}^* \right\|^2,$$

where in $\overset{a}{=}$ we use (5.10).

Using the strong convexity of h_{i_k} ($\mu \geq 0$), we have

$$\theta_k \left\langle \boldsymbol{\xi}_{i_k}^k, \mathbf{x}_{i_k}^* - \mathbf{z}_{i_k}^{k+1} \right\rangle \leq \theta_k h_{i_k}(\mathbf{x}^*) - \theta_k h_{i_k}(\mathbf{z}_{i_k}^{k+1}) - \frac{\mu \theta_k}{2} \left(\mathbf{z}_{i_k}^{k+1} - \mathbf{x}_{i_k}^* \right)^2. \quad (5.16)$$

On the other hand, by analyzing the expectation we have

$$\mathbb{E}_{i_k} \left\| \mathbf{z}^{k+1} - \mathbf{x}^* \right\|^2 = \frac{1}{n} \sum_{i_k=1}^{n} \left[\left(\mathbf{z}_{i_k}^{k+1} - \mathbf{x}_{i_k}^* \right)^2 + \sum_{j \neq i_k} \left(\mathbf{z}_j^{k+1} - \mathbf{x}_j^* \right)^2 \right]$$

$$\overset{a}{=} \frac{1}{n} \sum_{i_k=1}^{n} \left[\left(\mathbf{z}_{i_k}^{k+1} - \mathbf{x}_{i_k}^* \right)^2 + \sum_{j \neq i_k} \left(\mathbf{z}_j^k - \mathbf{x}_j^* \right)^2 \right]$$

$$= \frac{1}{n} \sum_{i_k=1}^{n} \left(\mathbf{z}_{i_k}^{k+1} - \mathbf{x}_{i_k}^* \right)^2 + \frac{n-1}{n} \left\| \mathbf{z}^k - \mathbf{x}^* \right\|^2, \quad (5.17)$$

where in $\overset{a}{=}$ we use $\mathbf{z}_j^{k+1} = \mathbf{z}_j^k$, $j \neq i_k$. Similar to (5.17), we also have

$$
\mathbb{E}_{i_k} \left\| \mathbf{x}^{k+1} - \mathbf{y}^k \right\|^2 = \frac{1}{n} \sum_{i_k=1}^{n} \left[\left(\mathbf{x}_{i_k}^{k+1} - \mathbf{y}_{i_k}^k \right)^2 + \sum_{j \neq i_k} \left(\mathbf{x}_j^{k+1} - \mathbf{y}_j^k \right)^2 \right]
$$

$$
\overset{a}{=} \frac{1}{n} \sum_{i_k=1}^{n} \left(\mathbf{x}_{i_k}^{k+1} - \mathbf{y}_{i_k}^k \right)^2,
$$

where in $\overset{a}{=}$ we use $\mathbf{x}_j^{k+1} = \mathbf{y}_j^k$, $j \neq i_k$. Then we obtain

$$
\mathbb{E}_{i_k} f(\mathbf{x}^{k+1})
$$

$$
\overset{a}{\leq} (1 - \theta_k) f(\mathbf{x}^k) + \theta_k F(\mathbf{x}^*) - \theta_k \sum_{i_k=1}^{n} h_{i_k}(\mathbf{z}_{i_k}^{k+1}) - \sum_{i_k=1}^{n} \frac{\mu \theta_k}{2} \left(\mathbf{z}_{i_k}^{k+1} - \mathbf{x}_{i_k}^* \right)^2
$$

$$
+ \frac{n^2}{2\gamma} \left\| \theta_k \mathbf{z}^k - \theta_k \mathbf{x}^* \right\|^2 - \frac{n^2}{2\gamma} \mathbb{E}_{i_k} \left\| \theta_k \mathbf{z}^{k+1} - \theta_k \mathbf{x}^* \right\|^2
$$

$$
\overset{b}{=} (1 - \theta_k) f(\mathbf{x}^k) + \theta_k F(\mathbf{x}^*) - \theta_k \sum_{i_k=1}^{n} h_{i_k}(\mathbf{z}_{i_k}^{k+1})
$$

$$
+ \frac{n^2 \theta_k^2 + (n-1)\theta_k \mu \gamma}{2\gamma} \left\| \mathbf{z}^k - \mathbf{x}^* \right\|^2 - \frac{n^2 \theta_k^2 + n \theta_k \mu \gamma}{2\gamma} \mathbb{E}_{i_k} \left\| \mathbf{z}^{k+1} - \mathbf{x}^* \right\|^2
$$

$$
\overset{c}{=} (1 - \theta_k) f(\mathbf{x}^k) + \theta_k F(\mathbf{x}^*) + (1 - \theta_k) \hat{h}_k - \mathbb{E}_{i_k} \hat{h}_{k+1}
$$

$$
+ \frac{n^2 \theta_k^2 + (n-1)\theta_k \mu \gamma}{2\gamma} \left\| \mathbf{z}^k - \mathbf{x}^* \right\|^2 - \frac{n^2 \theta_k^2 + n \theta_k \mu \gamma}{2\gamma} \mathbb{E}_{i_k} \left\| \mathbf{z}^{k+1} - \mathbf{x}^* \right\|^2,
$$

$$
(5.18)
$$

where $\overset{a}{\leq}$ uses (5.16), $\overset{b}{=}$ uses (5.17), and $\overset{c}{=}$ uses Lemma 5.1.

For the generally convex case ($\mu = 0$), rearranging (5.18) and dividing both sides with θ_k^2, we have

$$
\frac{\mathbb{E}_{i_k} f(\mathbf{x}^{k+1}) + \mathbb{E}_{i_k} \hat{h}_{k+1} - F(\mathbf{x}^*)}{\theta_k^2} + \frac{n^2}{2\gamma} \mathbb{E}_{i_k} \left\| \mathbf{z}^{k+1} - \mathbf{x}^* \right\|^2
$$

$$
\leq \frac{1 - \theta_k}{\theta_k^2} \left(f(\mathbf{x}^k) + \hat{h}_k - F(\mathbf{x}^*) \right) + \frac{n^2}{2\gamma} \left\| \mathbf{z}^k - \mathbf{x}^* \right\|^2
$$

$$
\overset{a}{\leq} \frac{1}{\theta_{k-1}^2} \left(f(\mathbf{x}^k) + \hat{h}_k - F(\mathbf{x}^*) \right) + \frac{n^2}{2\gamma} \left\| \mathbf{z}^k - \mathbf{x}^* \right\|^2,
$$

where in $\overset{a}{\leq}$ we use $\frac{1-\theta_k}{\theta_k^2} \leq \frac{1}{\theta_{k-1}^2}$ when $k \geq -1$. Taking full expectation, we have

$$
\frac{\mathbb{E}f(\mathbf{x}^{k+1}) + \mathbb{E}\hat{h}_{k+1} - F(\mathbf{x}^*)}{\theta_k^2} + \frac{n^2}{2\gamma}\mathbb{E}\left\|\mathbf{z}^{k+1} - \mathbf{x}^*\right\|^2
$$

$$
\leq \frac{f(\mathbf{x}^0) + \hat{h}_0 - F(\mathbf{x}^*)}{\theta_{-1}^2} + \frac{n^2}{2\gamma}\left\|\mathbf{z}^0 - \mathbf{x}^*\right\|^2
$$

$$
\overset{a}{=} \frac{F(\mathbf{x}^0) - F(\mathbf{x}^*)}{\theta_{-1}^2} + \frac{n^2}{2\gamma}\|\mathbf{z}^0 - \mathbf{x}^*\|^2,
$$

where in $\overset{a}{=}$ we use $\hat{h}_0 = h(\mathbf{x}^0)$. Then using the convexity of $h(\mathbf{x})$ and that $\mathbf{x}^{k+1}$ is a convex combination of $\mathbf{z}^0, \cdots, \mathbf{z}^{k+1}$, we have

$$
h(\mathbf{x}^{k+1}) \leq \sum_{i=0}^{k+1} e_{k+1,i} h(\mathbf{z}^i) \overset{a}{=} \hat{h}_{k+1}, \tag{5.19}
$$

where $\overset{a}{=}$ uses Lemma 5.1. So we obtain (5.7).

For the strongly convex case ($\mu > 0$), (5.18) gives

$$
\mathbb{E}_{i_k} f(\mathbf{x}^{k+1}) + \mathbb{E}\hat{h}_{k+1} - F(\mathbf{x}^*) + \frac{n^2\theta^2 + n\theta\mu\gamma}{2\gamma}\mathbb{E}_{i_k}\left\|\mathbf{z}^{k+1} - \mathbf{x}^*\right\|^2
$$

$$
\leq (1-\theta)\left(f(\mathbf{x}^k) + \mathbb{E}\hat{h}_k - F(\mathbf{x}^*)\right) + \frac{n^2\theta^2 + (n-1)\theta\mu\gamma}{2\gamma}\left\|\mathbf{z}^k - \mathbf{x}^*\right\|^2.
$$

For the setting of θ, we have

$$
\theta^2 + (n-1)\theta\mu\gamma = (1-\theta)(n^2\theta^2 + n\theta\mu\gamma).
$$

Thus

$$
\mathbb{E}_{i_k} f(\mathbf{x}^{k+1}) + \mathbb{E}\hat{h}_{k+1} - F(\mathbf{x}^*) + \frac{n^2\theta^2 + n\theta\mu\gamma}{2\gamma}\mathbb{E}_{i_k}\left\|\mathbf{z}^{k+1} - \mathbf{x}^*\right\|^2
$$

$$
\leq (1-\theta)\left(f(\mathbf{x}^k) + \mathbb{E}\hat{h}_k - F(\mathbf{x}^*) + \frac{n^2\theta^2 + n\theta\mu\gamma}{2\gamma}\left\|\mathbf{z}^k - \mathbf{x}^*\right\|^2\right).
$$

Taking full expectation, expanding the result to $k = 0$, then using $h(\mathbf{x}^{k+1}) \leq \hat{h}_{k+1}$ (see (5.19)), $h(\mathbf{x}^0) = \hat{h}_0$, and $\|\mathbf{z}^{k+1} - \mathbf{x}^*\|^2 \geq 0$, we obtain (5.8).   □

An important application of ASCD is to solve ERM problems in the following form:

$$\min_{\mathbf{x} \in \mathbb{R}^d} P(\mathbf{x}) \equiv \frac{1}{n} \sum_{i=1}^{n} \phi_i(\mathbf{A}_i^T \mathbf{x}) + \frac{\lambda}{2} \|\mathbf{x}\|^2, \tag{5.20}$$

where $\lambda > 0$ and $\phi_i(\mathbf{A}_i^T \mathbf{x})$, $i = 1, \cdots, n$, are loss functions over training samples. Lots of machine learning problems can be formulated as (5.20), such as linear SVM, ridge regression, and logistic regression. For ASCD, we can solve (5.20) via its dual problem:

$$\min_{\mathbf{a} \in \mathbb{R}^n} D(\mathbf{a}) = \frac{1}{n} \sum_{i=1}^{n} \phi_i^*(-\mathbf{a}_i) + \frac{\lambda}{2} \left\| \frac{1}{\lambda n} \mathbf{A} \mathbf{a} \right\|^2, \tag{5.21}$$

where $\phi_i^*(\cdot)$ is the conjugate function (Definition A.21) of $\phi_i(\cdot)$ and $\mathbf{A} = [\mathbf{A}_1, \cdots, \mathbf{A}_n]$. Then (5.21) can be solved by Algorithm 5.1.

5.1.2   Background for Variance Reduction Methods

For stochastic gradient descent, due to the nonzero variance of the gradient, using a constant step size cannot guarantee the convergence of algorithms. Instead, it only guarantees a sublinear convergence rate on the strongly convex and L-smooth objective functions. For finite-sum objective functions, the way to solve the problem is called variance reduction (VR) [10] which reduces the variance to zero through the updates. The convergence rate can be accelerated to be linear for strongly convex and L-smooth objective functions. The first VR method might be SAG [17], which uses the sum of the latest individual gradients as an estimator. It requires $O(nd)$ memory storage and uses a biased gradient estimator. SDCA [18] also achieves a linear convergence rate. In the primal space, the algorithm is known as MISO [13], which is a majorization-minimization VR algorithm. SVRG [10] is a follow-up work of SAG [17] which reduces the memory costs to $O(d)$ and uses an unbiased gradient estimator. The main technique of SVRG [10] is frequently pre-storing a snapshot vector and bounding the variance by the distance from the snapshot vector and the latest variable. Later, SAGA [6] improves SAG by using an unbiased updates via the technique of SVRG [10].

Algorithm 5.2 Stochastic variance reduced gradient (SVRG) [10]

Input $\mathbf{x}_0^0$. Set epoch length m, $\tilde{\mathbf{x}}^0 = \mathbf{x}_0^0$, and step size γ.
1 **for** $s = 0$ **to** $S - 1$ **do**
2   **for** $k = 0$ **to** $m - 1$ **do**
3   Randomly sample $i_{k,s}$ from $[n]$,
5         $\tilde{\nabla} f(\mathbf{x}_k^s) = \nabla f_{i_{k,s}}(\mathbf{x}_k^s) - \nabla f_{i_{k,s}}(\tilde{\mathbf{x}}^s) + \frac{1}{n}\sum_{i=1}^{n} \nabla f_i(\tilde{\mathbf{x}}^s)$,
6         $\mathbf{x}_{k+1}^s = \mathbf{x}_k^s - \gamma \tilde{\nabla} f(\mathbf{x}_k^s)$,
7   **end for** k
8   Option I: $\mathbf{x}_0^{s+1} = \frac{1}{m}\sum_{k=0}^{m-1} \mathbf{x}_k^s$,
9   Option II: $\mathbf{x}_0^{s+1} = \mathbf{x}_m^s$,
10    $\tilde{\mathbf{x}}^{s+1} = \mathbf{x}_0^{s+1}$.
end for s
Output $\tilde{\mathbf{x}}^S$.

We introduce SVRG [10] as an example. We solve the following problem:

$$\min_{\mathbf{x} \in \mathbb{R}^d} f(\mathbf{x}) \equiv \frac{1}{n}\sum_{i=1}^{n} f_i(\mathbf{x}). \tag{5.22}$$

The proximal version can be obtained straightforwardly in Sect. 5.1.3. The algorithm of SVRG is Algorithm 5.2. We have the following theorem:

Theorem 5.2 *For Algorithm 5.2, if each $f_i(\mathbf{x})$ is μ-strongly convex and L-smooth and $\eta < \frac{1}{2L}$, then we have*

$$\mathbb{E} f(\tilde{\mathbf{x}}^s) - f(\mathbf{x}^*)$$

$$\leq \left(\frac{1}{\mu\eta(1 - 2L\eta)m} + \frac{2L\eta}{1 - 2L\eta} \right) \left(\mathbb{E} f(\tilde{\mathbf{x}}^{s-1}) - f(\mathbf{x}^*) \right), \ s \geq 1, \tag{5.23}$$

where $\mathbf{x}^* = \mathrm{argmin}_{\mathbf{x} \in \mathbb{R}^d} f(\mathbf{x})$. In other words, by setting $m = O(\frac{L}{\mu})$ and $\eta = O(\frac{1}{L})$, the IFO calls (Definition 5.1) to achieve an ϵ-accuracy solution is $O\left(\left(n + \frac{L}{\mu} \right) \log(1/\epsilon) \right)$. If each $f_i(\mathbf{x})$ is L-smooth (but might be nonconvex) and $f(\mathbf{x})$ is μ-strongly convex, by setting $\eta \leq \frac{\mu}{8(1+e)L^2}$ and $m = -\ln^{-1}(1 - \eta\mu/2) \sim O(\eta^{-1}\mu^{-1})$, we have

$$\mathbb{E} \|\mathbf{x}_k^s - \mathbf{x}^*\|^2 \leq (1 - \eta\mu/2)^{sm+k} \|\mathbf{x}_0^0 - \mathbf{x}^*\|^2.$$

In other words, the IFO calls to achieve an ϵ- accuracy solution is $O\left(\left(n + \frac{L^2}{\mu^2} \right) \log(1/\epsilon) \right)$.

The proof is mainly taken from [10] and [4].

Proof Let $\mathbb{E}_k$ denote the expectation taken only on the random number $i_{k,s}$ conditioned on $\mathbf{x}_k^s$. Then we have

$$\mathbb{E}_k \left(\tilde{\nabla} f_{i_{k,s}}(\mathbf{x}_k^s) \right) = \mathbb{E}_k \nabla f_{i_{k,s}}(\mathbf{x}_k^s) - \mathbb{E}_k \left(\nabla f_{i_{k,s}}(\tilde{\mathbf{x}}^s) - \frac{1}{n} \sum_{i=1}^n \nabla f_i(\tilde{\mathbf{x}}^s) \right)$$

$$= \nabla f(\mathbf{x}_k^s). \tag{5.24}$$

So $\tilde{\nabla} f_{i_{k,s}}(\mathbf{x}_k^s)$ is an unbiased estimator of $\nabla f(\mathbf{x}_k^s)$. Then

$$\mathbb{E}_k \| \tilde{\nabla} f(\mathbf{x}_k^s) \|^2$$

$$\overset{a}{\leq} 2\mathbb{E}_k \| \nabla f_{i_{k,s}}(\mathbf{x}_k^s) - \nabla f_{i_{k,s}}(\mathbf{x}^*) \|^2 + 2\mathbb{E}_k \| \nabla f_{i_{k,s}}(\tilde{\mathbf{x}}^s) - \nabla f_{i_{k,s}}(\mathbf{x}^*) - \nabla f(\tilde{\mathbf{x}}^s) \|^2$$

$$\overset{b}{\leq} 2\mathbb{E}_k \| \nabla f_{i_{k,s}}(\mathbf{x}_k^s) - \nabla f_{i_{k,s}}(\mathbf{x}^*) \|^2 + 2\mathbb{E}_k \| \nabla f_{i_{k,s}}(\tilde{\mathbf{x}}^s) - \nabla f_{i_{k,s}}(\mathbf{x}^*) \|^2, \tag{5.25}$$

where in $\overset{a}{\leq}$ we use $\|\mathbf{a} - \mathbf{b}\|^2 \leq 2\|\mathbf{a}\|^2 + 2\|\mathbf{b}\|^2$ and in $\overset{b}{\leq}$ we use Proposition A.2.

We first consider the case when each $f_i(\mathbf{x})$ is strongly convex. Since each $f_i(\mathbf{x})$ is convex and L-smooth, from (A.7) we have

$$\| \nabla f_i(\mathbf{x}) - \nabla f_i(\mathbf{y}) \|^2 \leq 2L \left(f_i(\mathbf{x}) - f_i(\mathbf{y}) + \langle \nabla f_i(\mathbf{y}), \mathbf{y} - \mathbf{x} \rangle \right). \tag{5.26}$$

Letting $\mathbf{x} = \mathbf{x}_k^s$ and $\mathbf{y} = \mathbf{x}^*$ in (5.26) and summing the result with $i = 1$ to n, we have

$$\mathbb{E}_k \left\| \nabla f_{i_{k,s}}(\mathbf{x}_k^s) - \nabla f_{i_{k,s}}(\mathbf{x}^*) \right\|^2 \leq 2L \left(f(\mathbf{x}_k^s) - f(\mathbf{x}^*) \right), \tag{5.27}$$

where we use $\nabla f(\mathbf{x}^*) = \mathbf{0}$. In the same way, we have

$$\mathbb{E}_k \left\| \nabla f_{i_{k,s}}(\tilde{\mathbf{x}}^s) - \nabla f_{i_{k,s}}(\mathbf{x}^*) \right\|^2 \leq 2L \left(f(\tilde{\mathbf{x}}^s) - f(\mathbf{x}^*) \right). \tag{5.28}$$

Plugging (5.27) and (5.28) into (5.25), we have

$$\mathbb{E}_k \left\| \tilde{\nabla} f(\mathbf{x}_k^s) \right\|^2 \leq 4L \left[(f(\mathbf{x}_k^s) - f(\mathbf{x}^*)) + (f(\tilde{\mathbf{x}}^s) - f(\mathbf{x}^*)) \right]. \tag{5.29}$$

On the other hand,

$$
\mathbb{E}_k \|\mathbf{x}^s_{k+1} - \mathbf{x}^*\|^2
$$

$$
= \|\mathbf{x}^s_k - \mathbf{x}^*\|^2 + 2\mathbb{E}_k \left\langle \mathbf{x}^s_{k+1} - \mathbf{x}^s_k, \mathbf{x}^s_k - \mathbf{x}^* \right\rangle + \mathbb{E}_k \|\mathbf{x}^s_{k+1} - \mathbf{x}^s_k\|^2
$$

$$
= \|\mathbf{x}^s_k - \mathbf{x}^*\|^2 - 2\eta \mathbb{E}_k \left\langle \tilde{\nabla} f(\mathbf{x}^s_k), \mathbf{x}^s_k - \mathbf{x}^* \right\rangle + \eta^2 \mathbb{E}_k \|\tilde{\nabla} f(\mathbf{x}^s_k)\|^2
$$

$$
\overset{a}{\leq} \|\mathbf{x}^s_k - \mathbf{x}^*\|^2 - 2\eta \left\langle \nabla f(\mathbf{x}^s_k), \mathbf{x}^s_k - \mathbf{x}^* \right\rangle
$$

$$
\quad + 4L\eta^2 \left[(f(\mathbf{x}^s_k) - f(\mathbf{x}^*)) + (f(\tilde{\mathbf{x}}^s) - f(\mathbf{x}^*)) \right]
$$

$$
\leq \|\mathbf{x}^s_k - \mathbf{x}^*\|^2 - 2\eta(1 - 2L\eta)[f(\mathbf{x}^s_k) - f(\mathbf{x}^*)] + 4L\eta^2[f(\tilde{\mathbf{x}}^s) - f(\mathbf{x}^*)],
$$

$$
\tag{5.30}
$$

where $\overset{a}{\leq}$ uses (5.24) and (5.29).

Suppose we choose Option I. By taking full expectation on (5.30) and telescoping the result with $k = 1$ to m, we have

$$
\mathbb{E}\|\mathbf{x}^s_m - \mathbf{x}^*\|^2 + 2\eta(1 - 2L\eta)m\mathbb{E}(f(\tilde{\mathbf{x}}^{s+1}) - f(\mathbf{x}^*))
$$

$$
\overset{a}{\leq} \mathbb{E}\|\mathbf{x}^s_m - \mathbf{x}^*\|^2 + 2\eta(1 - 2L\eta) \sum_{k=0}^{m-1} \mathbb{E}(f(\mathbf{x}^s_k) - f(\mathbf{x}^*))
$$

$$
\overset{b}{\leq} \mathbb{E}\|\mathbf{x}^s_0 - \mathbf{x}^*\|^2 + 4Lm\eta^2 \mathbb{E}\left(f(\tilde{\mathbf{x}}^s) - f(\mathbf{x}^*)\right)
$$

$$
\leq 2\left(\mu^{-1} + 2Lm\eta^2\right) \mathbb{E}\left(f(\tilde{\mathbf{x}}^s - f(\mathbf{x}^*)\right),
$$

where $\overset{a}{\leq}$ uses Option I of Algorithm 5.2 and $\overset{b}{\leq}$ is by telescoping (5.30). Using $\|\mathbf{x}^s_m - \mathbf{x}^*\|^2 \geq 0$, we can obtain (5.23).

Then we consider the case when $f(\mathbf{x})$ is strongly convex. From $f(\cdot)$ being L-smooth, by (A.5) we have

$$
f(\mathbf{x}^s_{k+1}) \leq f(\mathbf{x}^s_k) + \left\langle \nabla f(\mathbf{x}^s_k), \mathbf{x}^s_{k+1} - \mathbf{x}^s_k \right\rangle + \frac{L}{2} \|\mathbf{x}^s_{k+1} - \mathbf{x}^s_k\|^2, \tag{5.31}
$$

and from $f(\cdot)$ being μ-strongly convex, by (A.9) we have

$$
f(\mathbf{x}^s_k) \leq f(\mathbf{x}^*) + \left\langle \nabla f(\mathbf{x}^s_k), \mathbf{x}^s_k - \mathbf{x}^* \right\rangle - \frac{\mu}{2} \|\mathbf{x}^s_k - \mathbf{x}^*\|^2. \tag{5.32}
$$

Adding (5.31) and (5.32), we have

$$f(\mathbf{x}_{k+1}^s)$$

$$\leq f(\mathbf{x}^*) + \langle \nabla f(\mathbf{x}_k^s), \mathbf{x}_{k+1}^s - \mathbf{x}^* \rangle + \frac{L}{2} \|\mathbf{x}_{k+1}^s - \mathbf{x}_k^s\|^2 - \frac{\mu}{2} \|\mathbf{x}_k^s - \mathbf{x}^*\|^2$$

$$= f(\mathbf{x}^*) - \frac{1}{\eta} \langle \mathbf{x}_{k+1}^s - \mathbf{x}_k^s, \mathbf{x}_{k+1}^s - \mathbf{x}^* \rangle + \frac{L}{2} \|\mathbf{x}_{k+1}^s - \mathbf{x}_k^s\|^2$$

$$- \frac{\mu}{2} \|\mathbf{x}_k^s - \mathbf{x}^*\|^2 - \langle \tilde{\nabla} f(\mathbf{x}_k^s) - \nabla f(\mathbf{x}_k^s), \mathbf{x}_{k+1}^s - \mathbf{x}^* \rangle$$

$$= f(\mathbf{x}^*) - \frac{1}{2\eta} \|\mathbf{x}_{k+1}^s - \mathbf{x}^*\|^2 + \frac{1}{2\eta} \|\mathbf{x}_k^s - \mathbf{x}^*\|^2 - \left(\frac{1}{2\eta} - \frac{L}{2} \right) \|\mathbf{x}_{k+1}^s - \mathbf{x}_k^s\|^2$$

$$- \frac{\mu}{2} \|\mathbf{x}_k^s - \mathbf{x}^*\|^2 - \langle \tilde{\nabla} f(\mathbf{x}_k^s) - \nabla f(\mathbf{x}_k^s), \mathbf{x}_{k+1}^s - \mathbf{x}^* \rangle.$$

Then by rearranging terms and using $f(\mathbf{x}_{k+1}^s) - f(\mathbf{x}^*) \geq 0$, we have

$$\frac{1}{2} \|\mathbf{x}_{k+1}^s - \mathbf{x}^*\|^2$$

$$\leq \frac{1 - \eta\mu}{2} \|\mathbf{x}_k^s - \mathbf{x}^*\|^2 - \eta \langle \tilde{\nabla} f(\mathbf{x}_k^s) - \nabla f(\mathbf{x}_k^s), \mathbf{x}_{k+1}^s - \mathbf{x}^* \rangle$$

$$- \left(\frac{1}{2} - \frac{L\eta}{2} \right) \|\mathbf{x}_{k+1}^s - \mathbf{x}_k^s\|^2. \tag{5.33}$$

Considering the expectation taken on the random number of $i_{k,s}$, we have

$$-\eta \mathbb{E}_k \langle \tilde{\nabla} f(\mathbf{x}_k^s) - \nabla f(\mathbf{x}_k^s), \mathbf{x}_{k+1}^s - \mathbf{x}^* \rangle$$

$$\overset{a}{=} -\eta \mathbb{E}_k \langle \tilde{\nabla} f(\mathbf{x}_k^s) - \nabla f(\mathbf{x}_k^s), \mathbf{x}_{k+1}^s \rangle$$

$$\overset{b}{=} -\eta \mathbb{E}_k \langle \tilde{\nabla} f(\mathbf{x}_k^s) - \nabla f(\mathbf{x}_k^s), \mathbf{x}_{k+1}^s - \mathbf{x}_k^s \rangle$$

$$\leq \eta^2 \mathbb{E}_k \|\tilde{\nabla} f(\mathbf{x}_k^s) - \nabla f(\mathbf{x}_k^s)\|^2 + \frac{1}{4} \mathbb{E}_k \|\mathbf{x}_{k+1}^s - \mathbf{x}_k^s\|^2$$

$$\leq \eta^2 \mathbb{E}_k \left\| \nabla f_{i_{k,s}}(\mathbf{x}_k^s) - \nabla f_{i_{k,s}}(\tilde{\mathbf{x}}^s) - (\nabla f(\mathbf{x}_k^s) - \nabla f(\tilde{\mathbf{x}}^s)) \right\|^2 + \frac{1}{4} \mathbb{E}_k \|\mathbf{x}_{k+1}^s - \mathbf{x}_k^s\|^2$$

$$\overset{c}{\leq} \eta^2 \mathbb{E}_k \left\| \nabla f_{i_{k,s}}(\mathbf{x}_k^s) - \nabla f_{i_{k,s}}(\tilde{\mathbf{x}}^s) \right\|^2 + \frac{1}{4} \mathbb{E}_k \|\mathbf{x}_{k+1}^s - \mathbf{x}_k^s\|^2$$

$$\leq \eta^2 L^2 \|\mathbf{x}_k^s - \tilde{\mathbf{x}}^s\|^2 + \frac{1}{4} \mathbb{E}_k \|\mathbf{x}_{k+1}^s - \mathbf{x}_k^s\|^2$$

$$\leq 2\eta^2 L^2 \|\mathbf{x}_k^s - \mathbf{x}^*\|^2 + 2\eta^2 L^2 \|\tilde{\mathbf{x}}^s - \mathbf{x}^*\|^2 + \frac{1}{4} \mathbb{E}_k \|\mathbf{x}_{k+1}^s - \mathbf{x}_k^s\|^2, \tag{5.34}$$

where both $\overset{a}{=}$ and $\overset{b}{=}$ use (5.24) and $\overset{c}{\leq}$ uses Proposition A.2.

From the setting of η, we have $L\eta \leq \frac{1}{2}$. Substituting (5.34) into (5.33) after taking expectation on i_k, we have

$$
\frac{1}{2}\mathbb{E}_k \|\mathbf{x}_{k+1}^s - \mathbf{x}^*\|^2
$$

$$
\leq \frac{1 - \eta\mu}{2} \|\mathbf{x}_k^s - \mathbf{x}^*\|^2 + 2\eta^2 L^2 \|\mathbf{x}_k^s - \mathbf{x}^*\|^2 + 2\eta^2 L^2 \|\tilde{\mathbf{x}}^s - \mathbf{x}^*\|^2. \quad (5.35)
$$

Now we use induction to prove

$$
\mathbb{E}\|\mathbf{x}_k^s - \mathbf{x}^*\|^2 \leq (1 - \eta\mu/2)^{sm+k} \|\mathbf{x}_0^0 - \mathbf{x}^*\|^2.
$$

When $k = 0$, it is right. Suppose that at iteration $k \geq 0$, we have $\mathbb{E}\|\mathbf{x}_k^s - \mathbf{x}^*\|^2 \leq (1 - \eta\mu/2)^{sm+k} \|\mathbf{x}_0^0 - \mathbf{x}^*\|^2$. Now we consider $k + 1$. We consider Option II in Algorithm 5.2 and have

$$
\mathbb{E}\|\tilde{\mathbf{x}}^s - \mathbf{x}^*\|^2 = \mathbb{E}\|\mathbf{x}_0^s - \mathbf{x}^*\|^2 \leq (1 - \eta\mu/2)^{sm} \|\mathbf{x}_0^0 - \mathbf{x}^*\|^2
$$

$$
\overset{a}{\leq} e(1 - \eta\mu/2)^{sm+k} \|\mathbf{x}_0^0 - \mathbf{x}^*\|^2,
$$

where $\overset{a}{\leq}$ uses $k \leq m = -\ln^{-1}(1 - \eta\mu/2)$. Then we have

$$
4\eta^2 L^2 \mathbb{E}\|\mathbf{x}_k^s - \mathbf{x}^*\|^2 + 4\eta^2 L^2 \mathbb{E}\|\tilde{\mathbf{x}}^s - \mathbf{x}^*\|^2
$$

$$
\leq 4\eta^2 L^2 (1 + e)(1 - \eta\mu/2)^{sm+k} \|\mathbf{x}_0^0 - \mathbf{x}^*\|^2
$$

$$
\overset{a}{\leq} \eta\mu/2(1 - \eta\mu/2)^{sm+k} \|\mathbf{x}_0^0 - \mathbf{x}^*\|^2, \quad (5.36)
$$

where $\overset{a}{\leq}$ uses $\eta \leq \frac{\mu}{8(1+e)L^2}$.

Taking full expectation on (5.35) and substituting (5.36) into it, we can obtain that $\mathbb{E}\|\mathbf{x}_{k+1}^s - \mathbf{x}^*\|^2 \leq (1 - \eta\mu/2)^{sm+k+1} \|\mathbf{x}_0^0 - \mathbf{x}^*\|^2$. □

5.1.3   Accelerated Stochastic Variance Reduction Method

With the VR technique, we can fuse the momentum technique to achieve a faster rate. We show that the convergence rate can be improved to $O\left((n + \sqrt{n\kappa}) \log(1/\epsilon)\right)$ in the IC case, where $\kappa = \frac{L}{\mu}$. We introduce Katyusha [1], which is the first truly accelerated stochastic algorithm. The main technique in Katyusha [1] is introducing a "negative momentum" which restricts the extrapolation term to be not far from $\tilde{\mathbf{x}}$, the snapshot vector introduced by SVRG [10]. The algorithm is shown in Algorithm 5.3.

Algorithm 5.3 Katyusha [1]

Input θ_1, step size γ, $\mathbf{x}_0^0 = \mathbf{0}$, $\tilde{\mathbf{x}}^0 = \mathbf{0}$, $\mathbf{z}_0^0 = \mathbf{0}$, $\theta_2 = \frac{1}{2}$, m, and $\theta_3 = \frac{\mu\gamma}{\theta_1} + 1$.
for $s = 0$ **to** S **do**
    for $k = 0$ **to** $m - 1$
1    $\mathbf{y}_k^s = \theta_1 \mathbf{z}_k^s + \theta_2 \tilde{\mathbf{x}}^s + (1 - \theta_1 - \theta_2)\mathbf{x}_k^s$,
2   Randomly selected one (or b mini-batch in Sect. 6.2.1.1) sample(s), denoted as i_k^s,
3    $\tilde{\nabla}_k^s = \nabla f_{i_k^s}(\mathbf{y}_k^s) - \nabla f_{i_k^s}(\tilde{\mathbf{x}}^s) + \nabla f(\tilde{\mathbf{x}}^s)$,
4    $\boldsymbol{\delta}_k^s = \arg\min_{\boldsymbol{\delta}} \left(h(\mathbf{z}_k^s + \boldsymbol{\delta}) + \langle \tilde{\nabla}_k^s, \boldsymbol{\delta} \rangle + \frac{\theta_1}{2\gamma} \|\boldsymbol{\delta}\|^2 \right)$,
5    $\mathbf{z}_{k+1}^s = \mathbf{z}_k^s + \boldsymbol{\delta}_k^s$,
6    $\mathbf{x}_{k+1}^s = \theta_1 \mathbf{z}_{k+1}^s + \theta_2 \tilde{\mathbf{x}}^s + (1 - \theta_1 - \theta_2)\mathbf{x}_k^s$.
    end for k
    $\mathbf{x}_0^{s+1} = \mathbf{x}_m^s$,
    $\tilde{\mathbf{x}}^{s+1} = \left(\sum_{k=0}^{m-1} \theta_3^k \right)^{-1} \sum_{k=0}^{m-1} \theta_3^k \mathbf{x}_k^s$.
end for s
Output $\mathbf{x}_0^{S+1}$.

The proof is taken from [1]. We have the following lemma to bound the variance:

Lemma 5.2 *For* $f(\mathbf{x}) = \frac{1}{n} \sum_{i=1}^n f_i(\mathbf{x})$, *with each* f_i *with* $i \in [n]$ *being convex and* L-*smooth. For any* $\mathbf{u}$ *and* $\tilde{\mathbf{x}}$, *defining*

$$\tilde{\nabla} f(\mathbf{u}) = \nabla f_k(\mathbf{u}) - \nabla f_k(\tilde{\mathbf{x}}) + \frac{1}{n} \sum_{i=1}^n \nabla f_i(\tilde{\mathbf{x}}),$$

we have

$$\mathbb{E} \left\| \tilde{\nabla} f(\mathbf{u}) - \nabla f(\mathbf{u}) \right\|^2 \leq 2L \left(f(\tilde{\mathbf{x}}) - f(\mathbf{u}) + \langle \nabla f(\mathbf{u}), \mathbf{u} - \tilde{\mathbf{x}} \rangle \right), \quad (5.37)$$

where the expectation is taken on the random number k *under the condition that* $\mathbf{u}$ *and* $\tilde{\mathbf{x}}$ *are known.*

Proof

$$\mathbb{E} \left\| \tilde{\nabla} f(\mathbf{u}) - \nabla f(\mathbf{u}) \right\|^2$$

$$= \mathbb{E} \left(\left\| \nabla f_k(\mathbf{u}) - \nabla f_k(\tilde{\mathbf{x}}) - \left(\nabla f(\mathbf{u}) - \nabla f(\tilde{\mathbf{x}}) \right) \right\|^2 \right)$$

$$\overset{a}{\leq} \mathbb{E} \left\| \nabla f_k(\mathbf{u}) - \nabla f_k(\tilde{\mathbf{x}}) \right\|^2,$$

where in inequality $\overset{a}{\leq}$ we use

$$\mathbb{E}\left(\nabla f_k(\mathbf{u}) - \nabla f_k(\tilde{\mathbf{x}})\right) = \nabla f(\mathbf{u}) - \nabla f(\tilde{\mathbf{x}})$$

and Proposition A.2. Then by directly applying (A.7) we obtain (5.37). $\qquad\square$

Theorem 5.3 *Suppose that $h(\mathbf{x})$ is μ-strongly convex and $n \leq \frac{L}{4\mu}$. For Algorithm 5.3, if the step size $\gamma = \frac{1}{3L}$, $\theta_1 = \sqrt{\frac{n\mu}{L}}$, $\theta_3 = 1 + \frac{\mu\gamma}{\theta_1}$, and $m = n$, we have*

$$F(\tilde{\mathbf{x}}^{S+1}) - F(\mathbf{x}^*) \leq \theta_3^{-Sn}\left[\frac{1}{4n\gamma}\|\mathbf{z}_0^0 - \mathbf{x}^*\|^2 + \left(1 + \frac{1}{n}\right)\left(F(\mathbf{x}_0^0) - F(\mathbf{x}^*)\right)\right].$$

Proof Because $n \leq \frac{L}{4\mu}$ and $\theta_1 = \sqrt{\frac{n\mu}{L}}$, we have

$$\theta_1 \leq \frac{1}{2}, \tag{5.38}$$

$$1 - \theta_1 - \theta_2 \geq 0.$$

For Step 1,

$$\mathbf{y}_k^s = \theta_1 \mathbf{z}_k^s + \theta_2 \tilde{\mathbf{x}}^s + (1 - \theta_1 - \theta_2)\mathbf{x}_k^s. \tag{5.39}$$

Together with Step 6, we have

$$\mathbf{x}_{k+1}^s = \mathbf{y}_k^s + \theta_1(\mathbf{z}_{k+1}^s - \mathbf{z}_k^s). \tag{5.40}$$

Through the optimality of $\mathbf{z}_{k+1}^s$ in Step 4 of Algorithm 5.3, there exists $\boldsymbol{\xi}_{k+1}^s \in \partial h(\mathbf{z}_{k+1}^s)$ satisfying

$$\theta_1(\mathbf{z}_{k+1}^s - \mathbf{z}_k^s) + \gamma \tilde{\nabla}_k^s + \gamma \boldsymbol{\xi}_{k+1}^s = \mathbf{0}.$$

From (5.39) and (5.40), we have

$$\mathbf{x}_{k+1}^s - \mathbf{y}_k^s + \gamma \tilde{\nabla}_k^s + \gamma \boldsymbol{\xi}_{k+1}^s = \mathbf{0}. \tag{5.41}$$

For $f(\cdot)$ is L-smooth, we have

$$f(\mathbf{x}_{k+1}^s) \leq f(\mathbf{y}_k^s) + \langle \nabla f(\mathbf{y}_k^s), \mathbf{x}_{k+1}^s - \mathbf{y}_k^s \rangle + \frac{L}{2}\|\mathbf{x}_{k+1}^s - \mathbf{y}_k^s\|^2$$

$$= f(\mathbf{y}_k^s) - \gamma \langle \nabla f(\mathbf{y}_k^s), \tilde{\nabla}_k^s + \boldsymbol{\xi}_{k+1}^s \rangle + \frac{L}{2}\|\mathbf{x}_{k+1}^s - \mathbf{y}_k^s\|^2$$

$$\overset{a}{=} f(\mathbf{y}_k^s) - \gamma \langle \tilde{\nabla}_k^s + \boldsymbol{\xi}_{k+1}^s, \tilde{\nabla}_k^s + \boldsymbol{\xi}_{k+1}^s \rangle$$

$$+ \frac{L}{2} \left\| \mathbf{x}_{k+1}^s - \mathbf{y}_k^s \right\|^2 + \gamma \langle \tilde{\nabla}_k^s + \boldsymbol{\xi}_{k+1}^s - \nabla f(\mathbf{y}_k^s), \tilde{\nabla}_k^s + \boldsymbol{\xi}_{k+1}^s \rangle$$

$$\overset{b}{=} f(\mathbf{y}_k^s) - \gamma \left(1 - \frac{\gamma L}{2} \right) \left\| \frac{1}{\gamma} \left(\mathbf{x}_{k+1}^s - \mathbf{y}_k^s \right) \right\|^2 - \langle \tilde{\nabla}_k^s - \nabla f(\mathbf{y}_k^s), \mathbf{x}_{k+1}^s - \mathbf{y}_k^s \rangle$$

$$- \langle \boldsymbol{\xi}_{k+1}^s, \mathbf{x}_{k+1}^s - \mathbf{y}_k^s \rangle, \tag{5.42}$$

where in $\overset{a}{=}$ we add and subtract the term $\gamma \langle \tilde{\nabla}_k^s + \boldsymbol{\xi}_{k+1}^s, \tilde{\nabla}_k^s + \boldsymbol{\xi}_{k+1}^s \rangle$ and $\overset{b}{=}$ uses the equality (5.41).

For the last but one term of (5.42), we have

$$\mathbb{E}_k \langle \tilde{\nabla}_k^s - \nabla f(\mathbf{y}_k^s), \mathbf{y}_k^s - \mathbf{x}_{k+1}^s \rangle$$

$$\overset{a}{\leq} \frac{\gamma}{2C_3} \mathbb{E}_k \left\| \tilde{\nabla}_k^s - \nabla f(\mathbf{y}_k^s) \right\|^2 + \frac{\gamma C_3}{2} \mathbb{E}_k \left\| \frac{1}{\gamma} \left(\mathbf{x}_{k+1}^s - \mathbf{y}_k^s \right) \right\|^2$$

$$\overset{b}{\leq} \frac{\gamma L}{C_3} \left(f(\tilde{\mathbf{x}}^s) - f(\mathbf{y}_k^s) + \langle \nabla f(\mathbf{y}_k^s), \mathbf{y}_k^s - \tilde{\mathbf{x}}^s \rangle \right) + \frac{\gamma C_3}{2} \mathbb{E}_k \left\| \frac{1}{\gamma} \left(\mathbf{x}_{k+1}^s - \mathbf{y}_k^s \right) \right\|^2, \tag{5.43}$$

where we use $\mathbb{E}_k$ to denote that expectation is taken on the random number i_k^s (step k and epoch s) under the condition that $\mathbf{y}_k^s$ is known, in $\overset{a}{\leq}$ we use the Cauchy–Schwartz inequality, and $\overset{b}{\leq}$ uses (5.37). C_3 is an absolute constant determined later.

Taking expectation for (5.42) on the random number i_k^s and adding (5.43), we obtain

$$\mathbb{E}_k f(\mathbf{x}_{k+1}^s)$$

$$\leq f(\mathbf{y}_k^s) - \gamma \left(1 - \frac{\gamma L}{2} - \frac{C_3}{2} \right) \mathbb{E}_k \left\| \frac{1}{\gamma} \left(\mathbf{x}_{k+1}^s - \mathbf{y}_k^s \right) \right\|^2$$

$$- \mathbb{E}_k \langle \boldsymbol{\xi}_{k+1}^s, \mathbf{x}_{k+1}^s - \mathbf{y}_k^s \rangle + \frac{\gamma L}{C_3} \left(f(\tilde{\mathbf{x}}^s) - f(\mathbf{y}_k^s) + \langle \nabla f(\mathbf{y}_k^s), \mathbf{y}_k^s - \tilde{\mathbf{x}}^s \rangle \right). \tag{5.44}$$

On the other hand, we analyze $\| \mathbf{z}_k^s - \mathbf{x}^* \|^2$. Setting $a = 1 - \theta_1 - \theta_2$, we have

$$\left\| \theta_1 \mathbf{z}_{k+1}^s - \theta_1 \mathbf{x}^* \right\|^2$$

$$= \left\| \mathbf{x}_{k+1}^s - a \mathbf{x}_k^s - \theta_2 \tilde{\mathbf{x}}^s - \theta_1 \mathbf{x}^* \right\|^2$$

$$= \left\| \mathbf{y}_k^s - a \mathbf{x}_k^s - \theta_2 \tilde{\mathbf{x}}^s - \theta_1 \mathbf{x}^* - (\mathbf{y}_k^s - \mathbf{x}_{k+1}^s) \right\|^2$$

$$= \left\| \mathbf{y}_k^s - a \mathbf{x}_k^s - \theta_2 \tilde{\mathbf{x}}^s - \theta_1 \mathbf{x}^* \right\|^2 + \left\| \mathbf{y}_k^s - \mathbf{x}_{k+1}^s \right\|^2$$

$$
-2\gamma \langle \boldsymbol{\xi}_{k+1}^{s} + \tilde{\nabla}_{k}^{s}, \mathbf{y}_{k}^{s} - a\mathbf{x}_{k}^{s} - \theta_{2}\tilde{\mathbf{x}}^{s} - \theta_{1}\mathbf{x}^{*} \rangle
$$

$$
\overset{a}{=} \left\| \theta_{1}\mathbf{z}_{k}^{s} - \theta_{1}\mathbf{x}^{*} \right\|^{2} + \left\| \mathbf{y}_{k}^{s} - \mathbf{x}_{k+1}^{s} \right\|^{2}
$$

$$
-2\gamma \left\langle \boldsymbol{\xi}_{k+1}^{s} + \tilde{\nabla}_{k}^{s}, \mathbf{y}_{k}^{s} - a\mathbf{x}_{k}^{s} - \theta_{2}\tilde{\mathbf{x}}^{s} - \theta_{1}\mathbf{x}^{*} \right\rangle, \tag{5.45}
$$

where $\overset{a}{=}$ uses Step 1 of Algorithm 5.3.

For the last term of (5.45), we have

$$
\mathbb{E}_{k} \left\langle \tilde{\nabla}_{k}^{s}, a\mathbf{x}_{k}^{s} + \theta_{2}\tilde{\mathbf{x}}^{s} + \theta_{1}\mathbf{x}^{*} - \mathbf{y}_{k}^{s} \right\rangle
$$

$$
\overset{a}{=} \left\langle \nabla f(\mathbf{y}_{k}^{s}), a\mathbf{x}_{k}^{s} + \theta_{1}\mathbf{x}^{*} - (1 - \theta_{2})\mathbf{y}_{k}^{s} \right\rangle + \theta_{2} \left\langle \nabla f(\mathbf{y}_{k}^{s}), \tilde{\mathbf{x}}^{s} - \mathbf{y}_{k}^{s} \right\rangle
$$

$$
\overset{b}{\leq} af(\mathbf{x}_{k}^{s}) + \theta_{1}f(\mathbf{x}^{*}) - (1 - \theta_{2})f(\mathbf{y}_{k}^{s}) + \theta_{2} \left\langle \nabla f(\mathbf{y}_{k}^{s}), \tilde{\mathbf{x}}^{s} - \mathbf{y}_{k}^{s} \right\rangle, \tag{5.46}
$$

where in $\overset{a}{\leq}$ we use $\mathbb{E}_{k}(\tilde{\nabla}_{k}^{s}) = \nabla f(\mathbf{y}_{k}^{s})$ and that $a\mathbf{x}_{k}^{s} + \theta_{2}\tilde{\mathbf{x}}^{s} + \theta_{1}\mathbf{x}^{*} - \mathbf{y}_{k}^{s}$ is constant, because the expectation is taken only on the random number $i_{k,s}$, and in inequality $\overset{b}{\leq}$ we use the convexity of $f(\cdot)$ and so for any vector $\mathbf{u}$,

$$
\langle \nabla f(\mathbf{y}_{k}^{s}), \mathbf{u} - \mathbf{y}_{k}^{s} \rangle \leq f(\mathbf{u}) - f(\mathbf{y}_{k}^{s}).
$$

Then dividing (5.45) by 2γ and taking expectation on the random number i_{k}^{s}, we have

$$
\frac{1}{2\gamma} \mathbb{E}_{k} \left\| \theta_{1}\mathbf{z}_{k+1}^{s} - \theta_{1}\mathbf{x}^{*} \right\|^{2}
$$

$$
\leq \frac{1}{2\gamma} \left\| \theta_{1}\mathbf{z}_{k}^{s} - \theta_{1}\mathbf{x}^{*} \right\|^{2} + \frac{\gamma}{2} \mathbb{E}_{k} \left\| \frac{1}{\gamma} \left(\mathbf{y}_{k}^{s} - \mathbf{x}_{k+1}^{s} \right) \right\|^{2}
$$

$$
- \mathbb{E}_{k} \left\langle \boldsymbol{\xi}_{k+1}^{s} + \tilde{\nabla}_{s}^{k}, \mathbf{y}_{k}^{s} - a\mathbf{x}_{k}^{s} - \theta_{2}\tilde{\mathbf{x}}^{s} - \theta_{1}\mathbf{x}^{*} \right\rangle
$$

$$
\overset{a}{\leq} \frac{1}{2\gamma} \left\| \theta_{1}\mathbf{z}_{k}^{s} - \theta_{1}\mathbf{x}^{*} \right\|^{2} + \frac{\gamma}{2} \mathbb{E}_{k} \left\| \frac{1}{\gamma} \left(\mathbf{y}_{k}^{s} - \mathbf{x}_{k+1}^{s} \right) \right\|^{2}
$$

$$
- \mathbb{E}_{k} \left\langle \boldsymbol{\xi}_{k+1}^{s}, \mathbf{y}_{k}^{s} - a\mathbf{x}_{k}^{s} - \theta_{2}\tilde{\mathbf{x}}^{s} - \theta_{1}\mathbf{x}^{*} \right\rangle
$$

$$
+ af(\mathbf{x}_{k}^{s}) + \theta_{1}f(\mathbf{x}^{*}) - (1 - \theta_{2})f(\mathbf{y}_{k}^{s}) + \theta_{2} \left\langle \nabla f(\mathbf{y}_{k}^{s}), \tilde{\mathbf{x}}^{s} - \mathbf{y}_{k}^{s} \right\rangle, \tag{5.47}
$$

where $\overset{a}{\leq}$ uses (5.46). Adding (5.47) and (5.44), we obtain that

$$
\mathbb{E}_{k} f(\mathbf{x}_{k+1}^{s}) + \frac{1}{2\gamma} \mathbb{E}_{k} \left\| \theta_{1}\mathbf{z}_{k+1}^{s} - \theta_{1}\mathbf{x}^{*} \right\|^{2}
$$

$$
\leq af(\mathbf{x}_{k}^{s}) + \theta_{1}f(\mathbf{x}^{*}) + \theta_{2}f(\mathbf{y}_{k}^{s}) + \theta_{2}\langle \nabla f(\mathbf{y}_{k}^{s}), \tilde{\mathbf{x}}^{s} - \mathbf{y}_{k}^{s} \rangle
$$

$$-\gamma \left(\frac{1}{2} - \frac{\gamma L}{2} - \frac{C_3}{2} \right) \mathbb{E}_k \left\| \frac{1}{\gamma} \left(\mathbf{y}_k^s - \mathbf{x}_{k+1}^s \right) \right\|^2$$

$$-\mathbb{E}_k \langle \boldsymbol{\xi}_{k+1}^s, \mathbf{x}_{k+1}^s - a\mathbf{x}_k^s - \theta_2 \tilde{\mathbf{x}}^s - \theta_1 \mathbf{x}^* \rangle$$

$$+\frac{\gamma L}{C_3} \left(f(\tilde{\mathbf{x}}^s) - f(\mathbf{y}_k^s) + \langle \nabla f(\mathbf{y}_k^s), \mathbf{y}_k^s - \tilde{\mathbf{x}}^s \rangle \right) + \frac{1}{2\gamma} \left\| \theta_1 \mathbf{z}_k^s - \theta_1 \mathbf{x}^* \right\|^2$$

$$\overset{a}{\le} af(\mathbf{x}_k^s) + \theta_1 f(\mathbf{x}^*) + \theta_2 f(\tilde{\mathbf{x}}^s) + \frac{1}{2\gamma} \left\| \theta_1 \mathbf{z}_k^s - \theta_1 \mathbf{x}^* \right\|^2$$

$$-\mathbb{E}_k \langle \boldsymbol{\xi}_{k+1}^s, \mathbf{x}_{k+1}^s - a\mathbf{x}_k^s - \theta_2 \tilde{\mathbf{x}}^s - \theta_1 \mathbf{x}^* \rangle, \tag{5.48}$$

where in $\overset{a}{\le}$ we set $C_3 = \frac{\gamma L}{\theta_2}$. For the last term of (5.48), we have

$$-\langle \boldsymbol{\xi}_{k+1}^s, \mathbf{x}_{k+1}^s - a\mathbf{x}_k^s - \theta_2 \tilde{\mathbf{x}}^s - \theta_1 \mathbf{x}^* \rangle$$

$$= -\langle \boldsymbol{\xi}_{k+1}^s, \theta_1 \mathbf{z}_{k+1}^s - \theta_1 \mathbf{x}^* \rangle$$

$$\le \theta_1 h(\mathbf{x}^*) - \theta_1 h(\mathbf{z}_{k+1}^s) - \frac{\mu \theta_1}{2} \left\| \mathbf{z}_{k+1}^s - \mathbf{x}^* \right\|^2$$

$$\overset{a}{\le} \theta_1 h(\mathbf{x}^*) - h(\mathbf{x}_{k+1}^s) + \theta_2 h(\tilde{\mathbf{x}}^s) + ah(\mathbf{x}_k^s) - \frac{\mu \theta_1}{2} \left\| \mathbf{z}_{k+1}^s - \mathbf{x}^* \right\|^2, \tag{5.49}$$

where in $\overset{a}{\le}$ we use $\mathbf{x}_{k+1}^s = a\mathbf{x}_k^s + \theta_2 \tilde{\mathbf{x}}^s + \theta_1 \mathbf{z}_{k+1}^s$ and the convexity of $h(\cdot)$. Substituting (5.49) into (5.48), we obtain

$$\mathbb{E}_k F(\mathbf{x}_{k+1}^s) + \frac{1 + \frac{\mu \gamma}{\theta_1}}{2\gamma} \left\| \theta_1 \mathbf{z}_{k+1}^s - \theta_1 \mathbf{x}^* \right\|^2$$

$$\le aF(\mathbf{x}_k^s) + \theta_1 F(\mathbf{x}^*) + \theta_2 F(\tilde{\mathbf{x}}^s) + \frac{1}{2\gamma} \left\| \theta_1 \mathbf{z}_k^s - \theta_1 \mathbf{x}^* \right\|^2. \tag{5.50}$$

Set $\theta_1 = \sqrt{\frac{n\mu}{L}}$ and $\theta_3 = \frac{\mu \gamma}{\theta_1} + 1 \le \frac{L\gamma}{\theta_1} + 1 \le \frac{1}{3}\sqrt{\frac{\mu}{Ln}} + 1$. Taking expectation on (5.50) for the first $k - 1$ iterations, then multiplying it with θ_3^k, and telescoping the results with k from 0 to $m - 1$, we have

$$\sum_{k=1}^{m} \theta_3^{k-1} \left(F(\mathbf{x}_k^s) - F(\mathbf{x}^*) \right) - a \sum_{k=0}^{m-1} \theta_3^k \left(F(\mathbf{x}_k^s) - F(\mathbf{x}^*) \right)$$

$$-\theta_2 \sum_{k=0}^{m-1} \left[\theta_3^k \left(F(\tilde{\mathbf{x}}^s) - F(\mathbf{x}^*) \right) \right] + \frac{\theta_3^m}{2\gamma} \left\| \theta_1 \mathbf{z}_m^s - \theta_1 \mathbf{x}^* \right\|^2 \le \frac{1}{2\gamma} \left\| \theta_1 \mathbf{z}_0^s - \theta_1 \mathbf{x}^* \right\|^2. \tag{5.51}$$

By rearranging the terms of (5.51), we have

$$[\theta_1 + \theta_2 - (1 - 1/\theta_3)] \sum_{k=1}^{m} \theta_3^k \left(F(\mathbf{x}_k^s) - F(\mathbf{x}^*) \right) + \theta_3^m a \left(F(\mathbf{x}_m^s) - F(\mathbf{x}^*) \right)$$

$$+ \frac{\theta_3^m}{2\gamma} \left\| \theta_1 \mathbf{z}_m^s - \theta_1 \mathbf{x}^* \right\|^2$$

$$\leq \theta_2 \sum_{k=0}^{m-1} \left[\theta_3^k \left(F(\tilde{\mathbf{x}}^s) - F(\mathbf{x}^*) \right) \right] + a \left(F(\mathbf{x}_0^s) - F(\mathbf{x}^*) \right) + \frac{1}{2\gamma} \left\| \theta_1 \mathbf{z}_0^s - \theta_1 \mathbf{x}^* \right\|^2 .$$

From the definition $\tilde{\mathbf{x}}^{s+1} = \left(\sum_{j=0}^{m-1} \theta_3^j \right)^{-1} \sum_{j=0}^{m-1} \theta_3^j \mathbf{x}_j^s$, we have

$$[\theta_1 + \theta_2 - (1 - 1/\theta_3)]\theta_3 \left(\sum_{k=0}^{m-1} \theta_3^k \right) \left(F(\tilde{\mathbf{x}}^{s+1}) - F(\mathbf{x}^*) \right) + \theta_3^m a \left(F(\mathbf{x}_m^s) - F(\mathbf{x}^*) \right)$$

$$+ \frac{\theta_3^m}{2\gamma} \left\| \theta_1 \mathbf{z}_m^s - \theta_1 \mathbf{x}^* \right\|^2$$

$$\leq \theta_2 \left(\sum_{k=0}^{m-1} \theta_3^k \right) \left(F(\tilde{\mathbf{x}}^s) - F(\mathbf{x}^*) \right) + a \left(F(\mathbf{x}_0^s) - F(\mathbf{x}^*) \right)$$

$$+ \frac{1}{2\gamma} \left\| \theta_1 \mathbf{z}_0^s - \theta_1 \mathbf{x}^* \right\|^2 . \tag{5.52}$$

Since

$$\theta_2 \left(\theta_3^{m-1} - 1 \right) + (1 - 1/\theta_3)$$

$$\leq \frac{1}{2} \left[\left(1 + \frac{1}{3} \sqrt{\frac{\mu}{nL}} \right)^{m-1} - 1 \right] + \frac{\frac{1}{3}\sqrt{\frac{\mu}{nL}}}{\theta_3}$$

$$\overset{a}{\leq} \frac{1}{2} \frac{1}{2} \sqrt{\frac{n\mu}{L}} + \frac{\frac{1}{3}\sqrt{\frac{\mu}{nL}}}{\theta_3} \overset{b}{\leq} \frac{7}{12} \sqrt{\frac{n\mu}{L}} \leq \theta_1, \tag{5.53}$$

where in $\overset{a}{\leq}$ we use $\mu n \leq L/4 < L$, $m - 1 = n - 1 \leq n$, and the fact that

$$g(x) = (1 + x)^c \leq 1 + \frac{3}{2} cx \quad \text{when } c \geq 1 \text{ and } x \leq \frac{1}{c}, \tag{5.54}$$

and in $\overset{b}{\leq}$ we use $\theta_3 n \geq \theta_3 \geq 1$. To prove (5.54), we can use Taylor expansion at point $x = 0$ to obtain

$$(1+x)^c = 1 + cx + \frac{c(c-1)}{2}\xi^2 \leq 1 + cx + \frac{c(c-1)}{2}\frac{1}{c}x \leq 1 + \frac{3}{2}cx,$$

where $\xi \in [0, x]$.

Equation (5.53) indicates that $\theta_1 + \theta_2 - (1 - 1/\theta_3) \geq \theta_2\theta_3^{m-1}$. This and (5.52) gives

$$\theta_2\theta_3^m \left(\sum_{k=0}^{m-1} \theta_3^k\right) \left(F(\tilde{\mathbf{x}}^{s+1}) - F(\mathbf{x}^*)\right) + \theta_3^m a \left(F(\mathbf{x}_m^s) - F(\mathbf{x}^*)\right)$$

$$+ \frac{\theta_3^m}{2\gamma}\left\|\theta_1\mathbf{z}_m^s - \theta_1\mathbf{x}^*\right\|^2$$

$$\leq \theta_2 \left(\sum_{k=0}^{m-1} \theta_3^k\right) \left(F(\tilde{\mathbf{x}}^s) - F(\mathbf{x}^*)\right) + a \left(F(\mathbf{x}_0^s) - F(\mathbf{x}^*)\right)$$

$$+ \frac{1}{2\gamma}\left\|\theta_1\mathbf{z}_0^s - \theta_1\mathbf{x}^*\right\|^2.$$

By expanding the above inequality from $s = S, \cdots, 0$, we have

$$\theta_2 \left(\sum_{k=0}^{m-1} \theta_3^k\right) \left(F(\tilde{\mathbf{x}}^{S+1}) - F(\mathbf{x}^*)\right) + (1 - \theta_1 - \theta_2) \left(F(\mathbf{x}_m^S) - F(\mathbf{x}^*)\right)$$

$$+ \frac{\theta_1^2}{2\gamma}\left\|\mathbf{z}_0^{S+1} - \mathbf{x}^*\right\|^2$$

$$\leq \theta_3^{-Sm} \left\{\left[\theta_2 \left(\sum_{k=0}^{m-1} \theta_3^k\right) + (1 - \theta_1 - \theta_2)\right] \left(F(\mathbf{x}_0^0) - F(\mathbf{x}^*)\right)\right.$$

$$\left. + \frac{\theta_1^2}{2\gamma}\left\|\mathbf{z}_0^0 - \mathbf{x}^*\right\|^2\right\}.$$

Since $\theta_3^k \geq 1$, we have $\sum_{k=0}^{m-1} \theta_3^k \geq n$. Then using $\theta_2 = \frac{1}{2}$ and $\theta_1 \leq \frac{1}{2}$ (see (5.38)), we have

$$F(\tilde{\mathbf{x}}^{S+1}) - F(\mathbf{x}^*) \leq \theta_3^{-Sn}\left[\left(1 + \frac{1}{n}\right)\left(F(\mathbf{x}_0^0) - F(\mathbf{x}^*)\right) + \frac{1}{4n\gamma}\left\|\mathbf{z}_0^0 - \mathbf{x}^*\right\|^2\right].$$

This ends the proof. $\qquad\square$

5.1.4 Black-Box Acceleration

In this section, we introduce black-box acceleration methods. In general, the methods solve (5.3) by constructing a series of subproblems known as "mediator" [15], which can be solved efficiently to a high accuracy. The main advantages of these methods are listed as follows:

1. The black-box methods make acceleration easier, because we only need to concern the method to solve subproblems. In most time, the subproblems have a good condition number and so are easy to be solved. Specifically, for general problem of (5.3), the subproblems can be solved by arbitrary vanilla VR methods. For specific forms of (5.3), one is allowed to design appropriate methods according to the characteristic of functions to solve them without considering acceleration techniques.
2. The black-box methods make the acceleration technique more general. For different properties of objectives, no matter strongly convex or not and smooth or not, the black-box methods are able to give a universal accelerated algorithm.

The first stochastic black-box acceleration might be Acc-SDCA [19]. Its convergence rate for the IC case is $O\left((n + \sqrt{n\kappa})\log(\kappa)\log^2(1/\epsilon)\right)$. Later, Lin et al. proposed a generic acceleration, called Catalyst [12], which achieved a convergence rate $O\left((n + \sqrt{n\kappa})\log^2(1/\epsilon)\right)$, outperforming Acc-SDCA by a factor of $\log(\kappa)$. Allen-Zhu et al. [2] designed a black-box acceleration by gradually decreasing the condition number of the subproblem, achieving $O(\log(1/\epsilon))$ faster rate than Catalyst [12] on some general objective functions. In the following, we introduce Catalyst [12] as an example. The algorithm of Catalyst is shown in Algorithm 5.4. The main theorem for Algorithm 5.4 is as follows:

Theorem 5.4 *For problem* (5.3), *suppose that $F(\mathbf{x})$ is μ-strongly convex and set $\alpha_0 = \sqrt{q}$ with $q = \frac{\mu}{\mu+\kappa}$ and*

$$\epsilon_k = \frac{2}{9}(F(\mathbf{x}_0) - F^*)(1 - \rho)^k \quad with \ \rho \leq \sqrt{q}.$$

Then Algorithm 5.4 generates iterates $\{\mathbf{x}_k\}_{k \geq 0}$ such that

$$F(\mathbf{x}_k) - F^* \leq C(1 - \rho)^{k+1}(F(\mathbf{x}_0) - F^*) \quad with \ C = \frac{8}{(\sqrt{q} - \rho)^2}.$$

We leave the proof to Sect. 5.2 and first introduce how to use it to obtain an accelerated algorithm. We have the following theorem:

Algorithm 5.4 Catalyst [12]

Input $\mathbf{x}^0$, parameters κ and α_0, sequence $\{\epsilon_k\}_{k\geq 0}$, and optimization method $\mathcal{M}$.

1 Initialize $q = \mu/(\mu + \kappa)$ and $\mathbf{y}_0 = \mathbf{x}_0$.

2 **for** $k = 0$ to K **do**

3   Applying $\mathcal{M}$ to solve: $\mathbf{x}_k = \mathrm{argmin}_{\mathbf{x}\in\mathbb{R}^p}\left\{G_k(\mathbf{x}) \equiv F(\mathbf{x}) + \frac{\kappa}{2}\|\mathbf{x} - \mathbf{y}_{k-1}\|^2\right\}$
     to the accuracy satisfying $G_k(\mathbf{x}_k) - G_k^* \leq \epsilon_k$,

4   Compute $\alpha_k \in (0, 1)$ from equation $\alpha_k^2 = (1 - \alpha_k)\alpha_{k-1}^2 + q\alpha_k$,

5    $\mathbf{y}_k = \mathbf{x}_k + \beta_k(\mathbf{x}_k - \mathbf{x}_{k-1})$ with $\beta_k = \frac{\alpha_{k-1}(1-\alpha_{k-1})}{\alpha_{k-1}^2+\alpha_k}$.

   end for k

   Output $\mathbf{x}^{K+1}$.

Theorem 5.5 *For (5.3), assume that each $f_i(\mathbf{x})$ is convex and L-smooth and $h(\mathbf{x})$ is μ-strongly convex satisfying $\mu \leq L/n$.[1] For Algorithm 5.4, if setting $\kappa = \frac{L}{n-1}$ and solving Step 3 by SVRG [10], then one can obtain an ϵ-accuracy solution satisfying $F(\mathbf{x}) - F(\mathbf{x}^*) \leq \epsilon$ with IFO calls of $O\left(\sqrt{nL/\mu}\log^2(1/\epsilon)\right)$.*

Proof The subproblem in Step 3 is $\left(\mu + \frac{L}{n-1}\right)$-strongly convex and $\left(L + \frac{L}{n-1}\right)$-smooth. From Theorem 5.2, applying SVRG to solve the subproblem needs
$$O\left(\left(n + \frac{L+\frac{L}{n-1}}{\mu+\frac{L}{n-1}}\right)\log(1/\epsilon)\right) = O\left(n\log(1/\epsilon)\right) \text{ IFOs. So the total complexity is}$$
$O\left(\sqrt{nL/\mu}\log^2(1/\epsilon)\right)$.   □

The black-box algorithms, e.g., Catalyst [12], are proposed earlier than Katyusha [1], discussed in Sect. 5.1.3. From the convergence results, Catalyst [12] is $O(\log(1/\epsilon))$ times lower than Katyusha [1]. However, the black-box algorithms are more flexible and easier to obtain an accelerated rate. For example, in the next section we apply Catalyst to the INC case.

5.2 The Individually Nonconvex Case

In this section, we consider solving (5.3) by allowing nonconvexity of $f_i(\mathbf{x})$ but $f(\mathbf{x})$, the sum of $f_i(\mathbf{x})$, is still convex. One important application of it is principal component analysis [9]. It is also the core technique to obtain faster rate in the NC case.

Notice that the convexity of $f(\mathbf{x})$ guarantees an achievable global minimum of (5.3). As shown in Theorem 5.2, to reach a minimizer of $F(\mathbf{x}) - F^* \leq \epsilon$, vanilla SVRG needs IFO calls of $O\left(\left(n + \frac{L^2}{\mu^2}\right)\log^2(1/\epsilon)\right)$. We will show that the convergence rate can be improved to be $O\left(\left(n + n^{3/4}\sqrt{\frac{L}{\mu}}\right)\log^2(1/\epsilon)\right)$ by

---

[1] If $n \geq O(L/\mu)$, $O(n)$ is the dominant dependency in the convergence rate, thus the momentum technique cannot achieve a faster rate by order.

acceleration. Compared with the computation costs in the IC case, the costs of NC cases are $\Omega(n^{1/4})$ times larger, which is caused by the nonconvexity of individuals. We still use Catalyst [12] to obtain the result. We first prove Theorem 5.4 in the context of INC. The proof is directly taken from [12], which is based on the estimate sequence in [14] (see Sect. 2.1) and further takes the error of inexact solver into account.

Proof of Theorem 5.4 in the Context of INC Define the estimate sequence as follows:

1. $\phi_0(\mathbf{x}) \equiv F(\mathbf{x}_0) + \frac{\gamma_0}{2}\|\mathbf{x} - \mathbf{x}_0\|^2$;
2. For $k \geq 0$,

$$\phi_{k+1}(\mathbf{x}) = (1 - \alpha_k)\phi_k(\mathbf{x}) + \alpha_k \left[F(\mathbf{x}_{k+1}) + \langle \kappa(\mathbf{y}_k - \mathbf{x}_{k+1}), \mathbf{x} - \mathbf{x}_{k+1} \rangle \right]$$
$$+ \frac{\mu}{2}\|\mathbf{x} - \mathbf{x}_{k+1}\|^2,$$

where $\gamma_0 \geq 0$ will be defined in Step 3. One can find that the main difference of estimate sequence defined here with the original one in [15] is replacing $\nabla f(\mathbf{y}_k)$ with $\kappa(\mathbf{y}_k - \mathbf{x}_{k+1})$ (see (2.5)).

Step 1: For all $k \geq 0$, we can have

$$\phi_k(\mathbf{x}) = \phi_k^* + \frac{\gamma_k}{2}\|\mathbf{x} - \mathbf{v}_k\|^2, \tag{5.55}$$

where $\phi_0^* = F(\mathbf{x}^0)$ and $\mathbf{v}_0 = \mathbf{x}_0$ when $k = 0$, and γ_k, $\mathbf{v}_k$, and ϕ_k^* satisfy:

$$\gamma_k = (1 - \alpha_{k-1})\gamma_{k-1} + \alpha_{k-1}\mu, \tag{5.56}$$

$$\mathbf{v}_k = \frac{1}{\gamma_k} \left[(1 - \alpha_{k-1})\gamma_{k-1}\mathbf{v}_{k-1} + \alpha_{k-1}\mu\mathbf{x}_k - \alpha_{k-1}\kappa(\mathbf{y}_{k-1} - \mathbf{x}_k) \right], \tag{5.57}$$

$$\phi_k^* = (1 - \alpha_{k-1})\phi_{k-1}^* + \alpha_{k-1}F(\mathbf{x}_k) - \frac{\alpha_{k-1}^2}{2\gamma_k}\|\kappa(\mathbf{y}_{k-1} - \mathbf{x}_k)\|^2$$

$$+ \frac{\alpha_{k-1}(1 - \alpha_{k-1})\gamma_{k-1}}{\gamma_k}\left(\frac{\mu}{2}\|\mathbf{x}_k - \mathbf{v}_{k-1}\|^2 + \langle \kappa(\mathbf{y}_{k-1} - \mathbf{x}_k), \mathbf{v}_{k-1} - \mathbf{x}_k \rangle \right), \tag{5.58}$$

when $k > 0$.

Proof of Step 1: we need to check

$$\phi_k^* + \frac{\gamma_k}{2}\|\mathbf{x} - \mathbf{v}_k\|^2$$

$$= (1 - \alpha_{k-1})\phi_{k-1}(\mathbf{x})$$

$$+ \alpha_{k-1}\left(F(\mathbf{x}_k) + \langle \kappa(\mathbf{y}_{k-1} - \mathbf{x}_k), \mathbf{x} - \mathbf{x}_k \rangle] + \frac{\mu}{2}\|\mathbf{x} - \mathbf{x}_k\|^2 \right).$$

Suppose that at iteration $k - 1$, (5.55) is right. Then we need to prove

$$\phi_k^* + \frac{\gamma_k}{2}\|\mathbf{x} - \mathbf{v}_k\|^2$$

$$= (1 - \alpha_{k-1})\left(\phi_{k-1}^* + \frac{\gamma_{k-1}}{2}\|\mathbf{x} - \mathbf{v}_{k-1}\|^2\right)$$

$$+ \alpha_{k-1}\left(F(\mathbf{x}_k) + \langle \kappa(\mathbf{y}_{k-1} - \mathbf{x}_k), \mathbf{x} - \mathbf{x}_k \rangle + \frac{\mu}{2}\|\mathbf{x} - \mathbf{x}_k\|^2\right). \quad (5.59)$$

Both sides of the equation are simple quadratic forms. By comparing the coefficient of $\|\mathbf{x}\|^2$, we have $\gamma_k = (1 - \alpha_{k-1})\gamma_{k-1} + \alpha_{k-1}\mu$. Then by computing the gradient at $\mathbf{x} = \mathbf{0}$ on both sides of (5.59), we can obtain

$$\gamma_k \mathbf{v}_k = (1 - \alpha_{k-1})\gamma_{k-1}\mathbf{v}_{k-1} - \alpha_{k-1}\kappa(\mathbf{y}_{k-1} - \mathbf{x}_k) + \alpha_{k-1}\mu\mathbf{x}_k.$$

For ϕ_k^*, we can set $\mathbf{x} = \mathbf{x}_k$ in (5.59) and obtain:

$$\phi_k^* = -\frac{\gamma_k}{2}\|\mathbf{x}_k - \mathbf{v}_k\|^2 + (1 - \alpha_{k-1})\left(\phi_{k-1}^* + \frac{\gamma_{k-1}}{2}\|\mathbf{x}_k - \mathbf{v}_{k-1}\|^2\right) + \alpha_{k-1}F(\mathbf{x}_k).$$

Then by substituting (5.56) and (5.57) to the above, we can obtain (5.58).

Step 2: For Algorithm 5.4, we can have

$$F(\mathbf{x}_k) \leq \phi_k^* + \xi_k, \quad (5.60)$$

where $\xi_0 = 0$ and $\xi_k = (1 - \alpha_{k-1})\left(\xi_{k-1} + \epsilon_k - (\kappa + \mu)\langle \mathbf{x}_k - \mathbf{x}_k^*, \mathbf{x}_{k-1} - \mathbf{x}_k \rangle\right)$.

Proof of Step 2: Suppose that $\mathbf{x}_k^*$ is the optimal solution in Step 3 of Algorithm 5.4. By the $(\mu + \kappa)$-strongly convexity of $G_k(\mathbf{x})$, we have

$$G_k(\mathbf{x}) \geqslant G_k^* + \frac{\kappa + \mu}{2}\|\mathbf{x} - \mathbf{x}_k^*\|^2.$$

Then

$$F(\mathbf{x}) \geq G_k^* + \frac{\kappa + \mu}{2}\|\mathbf{x} - \mathbf{x}_k^*\|^2 - \frac{\kappa}{2}\|\mathbf{x} - \mathbf{y}_{k-1}\|^2$$

$$\overset{a}{=} G_k(\mathbf{x}_k) - \epsilon_k + \frac{\kappa + \mu}{2}\|(\mathbf{x} - \mathbf{x}_k) + (\mathbf{x}_k - \mathbf{x}_k^*)\|^2 - \frac{\kappa}{2}\|\mathbf{x} - \mathbf{y}_{k-1}\|^2$$

$$\geq F(\mathbf{x}_k) + \frac{\kappa}{2}\|\mathbf{x}_k - \mathbf{y}_{k-1}\|^2 - \epsilon_k + \frac{\kappa + \mu}{2}\|\mathbf{x} - \mathbf{x}_k\|^2 - \frac{\kappa}{2}\|\mathbf{x} - \mathbf{y}_{k-1}\|^2$$

$$+ (\kappa + \mu)\langle \mathbf{x}_k - \mathbf{x}_k^*, \mathbf{x} - \mathbf{x}_k \rangle$$

$$= F(\mathbf{x}_k) + \kappa \langle \mathbf{y}_{k-1} - \mathbf{x}_k, \mathbf{x} - \mathbf{x}_k \rangle - \epsilon_k + \frac{\mu}{2}\|\mathbf{x} - \mathbf{x}_k\|^2$$

$$+ (\kappa + \mu)\langle \mathbf{x}_k - \mathbf{x}_k^*, \mathbf{x} - \mathbf{x}_k \rangle, \quad (5.61)$$

where in $\overset{a}{=}$ we use that $G_k(\mathbf{x}_k)$ is an ϵ-accuracy solution of Step 3.

Now we prove (5.60). When $k = 0$, we have $F(\mathbf{x}_0) = \phi_0^*$. Suppose that for $k-1$ with $k \geq 1$, (5.60) is right, i.e., $F(\mathbf{x}_{k-1}) \leq \phi_{k-1}^* + \xi_{k-1}$. Then

$$
\begin{aligned}
\phi_{k-1}^* &\geq F(\mathbf{x}_{k-1}) - \xi_{k-1} \\
&\overset{a}{\geq} F(\mathbf{x}_k) + \langle \kappa(\mathbf{y}_{k-1} - \mathbf{x}_k), \mathbf{x}_{k-1} - \mathbf{x}_k \rangle + (\kappa + \mu)\langle \mathbf{x}_k - \mathbf{x}_k^*, \mathbf{x}_{k-1} - \mathbf{x}_k \rangle \\
&\quad - \epsilon_k - \xi_{k-1} \\
&= F(\mathbf{x}_k) + \langle \kappa(\mathbf{y}_{k-1} - \mathbf{x}_k), \mathbf{x}_{k-1} - \mathbf{x}_k \rangle - \xi_k/(1 - \alpha_{k-1}), \quad (5.62)
\end{aligned}
$$

where $\overset{a}{\geq}$ uses (5.61). Then from (5.58), we have

$$
\begin{aligned}
\phi_k^* &= (1 - \alpha_{k-1})\phi_{k-1}^* + \alpha_{k-1}F(\mathbf{x}_k) - \frac{\alpha_{k-1}^2}{2\gamma_k}\|\kappa(\mathbf{y}_{k-1} - \mathbf{x}_k)\|^2 \\
&\quad + \frac{\alpha_{k-1}(1 - \alpha_{k-1})\gamma_{k-1}}{\gamma_k}\left(\frac{\mu}{2}\|\mathbf{x}_k - \mathbf{v}_{k-1}\|^2 + \langle \kappa(\mathbf{y}_{k-1} - \mathbf{x}_k), \mathbf{v}_{k-1} - \mathbf{x}_k \rangle\right) \\
&\overset{a}{\geq} (1 - \alpha_{k-1})F(\mathbf{x}_k) + (1 - \alpha_{k-1})\langle \kappa(\mathbf{y}_{k-1} - \mathbf{x}_k), \mathbf{x}_{k-1} - \mathbf{x}_k \rangle - \xi_k + \alpha_{k-1}F(\mathbf{x}_k) \\
&\quad - \frac{\alpha_{k-1}^2}{2\gamma_k}\|\kappa(\mathbf{y}_{k-1} - \mathbf{x}_k)\|^2 + \frac{\alpha_{k-1}(1 - \alpha_{k-1})\gamma_{k-1}}{\gamma_k}\langle \kappa(\mathbf{y}_{k-1} - \mathbf{x}_k), \mathbf{v}_{k-1} - \mathbf{x}_k \rangle \\
&= F(\mathbf{x}_k) + (1 - \alpha_{k-1})\left\langle \kappa(\mathbf{y}_{k-1} - \mathbf{x}_k), \mathbf{x}_{k-1} - \mathbf{x}_k + \frac{\alpha_{k-1}\gamma_{k-1}}{\gamma_k}(\mathbf{v}_{k-1} - \mathbf{x}_k) \right\rangle \\
&\quad - \frac{\alpha_{k-1}^2}{2\gamma_k}\|\kappa(\mathbf{y}_{k-1} - \mathbf{x}_k)\|^2 - \xi_k \\
&\overset{b}{=} F(\mathbf{x}_k) + (1 - \alpha_{k-1})\left\langle \kappa(\mathbf{y}_{k-1} - \mathbf{x}_k), \mathbf{x}_{k-1} - \mathbf{y}_{k-1} + \frac{\alpha_{k-1}\gamma_{k-1}}{\gamma_k}(\mathbf{v}_{k-1} - \mathbf{y}_{k-1}) \right\rangle \\
&\quad + \left(1 - \frac{(\kappa + 2\mu)\alpha_{k-1}^2}{2\gamma_k}\right)\kappa\|\mathbf{y}_{k-1} - \mathbf{x}_k\|^2 - \xi_k,
\end{aligned}
$$

where $\overset{a}{\geq}$ uses (5.62) and $\overset{b}{=}$ uses (5.56).

Step 3: Set

$$
\mathbf{x}_{k-1} - \mathbf{y}_{k-1} + \frac{\alpha_{k-1}\gamma_{k-1}}{\gamma_k}(\mathbf{v}_{k-1} - \mathbf{y}_{k-1}) = \mathbf{0}, \quad (5.63)
$$

$$
\gamma_0 = \frac{\alpha_0[(\kappa + \mu)\alpha_0 - \mu]}{1 - \alpha_0},
$$

$$
\gamma_k = (\kappa + \mu)\alpha_{k-1}^2, \quad k \geq 1. \quad (5.64)
$$

By Step 4 of Algorithm 5.4, we can have

$$\mathbf{y}_k = \mathbf{x}_k + \frac{\alpha_{k-1}(1 - \alpha_{k-1})}{\alpha_{k-1}^2 + \alpha_k}(\mathbf{x}_k - \mathbf{x}_{k-1}). \tag{5.65}$$

Proof of Step 3: Suppose that at iteration $k - 1$, (5.65) is right, then from (5.57) in Step 1, we have

$$\mathbf{v}_k = \frac{1}{\gamma_k}[(1 - \alpha_{k-1})\gamma_{k-1}\mathbf{v}_{k-1} + \alpha_{k-1}\mu\mathbf{x}_k - \alpha_{k-1}\kappa(\mathbf{y}_{k-1} - \mathbf{x}_k)]$$

$$\overset{a}{=} \frac{1}{\gamma_k}\left\{\frac{1 - \alpha_{k-1}}{\alpha_{k-1}}[(\gamma_k + \alpha_{k-1}\gamma_{k-1})\mathbf{y}_{k-1} - \gamma_k\mathbf{x}_{k-1}]\right.$$

$$\left. + \alpha_{k-1}\mu\mathbf{x}_k - \alpha_{k-1}\kappa(\mathbf{y}_{k-1} - \mathbf{x}_k)\right\}$$

$$\overset{b}{=} \frac{1}{\alpha_{k-1}}[\mathbf{x}_k - (1 - \alpha_{k-1})\mathbf{x}_{k-1}], \tag{5.66}$$

where $\overset{a}{=}$ uses (5.63) and in $\overset{b}{=}$ we use

$$(1 - \alpha_{k-1})(\gamma_k + \alpha_{k-1}\gamma_{k-1}) \overset{c}{=} (1 - \alpha_{k-1})(\gamma_{k-1} + \alpha_{k-1}\mu)$$

$$\overset{d}{=} \gamma_k - \mu\alpha_{k-1}^2 \overset{e}{=} \kappa\alpha_{k-1}^2,$$

in which both $\overset{c}{=}$ and $\overset{d}{=}$ use (5.56) and $\overset{e}{=}$ uses (5.64).

Plugging (5.66) into (5.63) with $k - 1$ being replaced by k, we obtain (5.65).

Step 4: Set $\lambda_k = \prod_{i=0}^{k-1}(1 - \alpha_i)$, we can have

$$\frac{1}{\lambda_k}\left(F(\mathbf{x}_k) - F^* + \frac{\gamma_k}{2}\|\mathbf{x}^* - \mathbf{v}_k\|^2\right)$$

$$\leq \phi_0(\mathbf{x}^*) - F^* + \sum_{i=1}^{k}\frac{\epsilon_i}{\lambda_i} + \sum_{i=1}^{k}\frac{\sqrt{2\epsilon_i\gamma_i}}{\lambda_i}\|\mathbf{x}^* - \mathbf{v}_i\|. \tag{5.67}$$

Proof of Step 4: From the definition of $\phi_k(\mathbf{x})$, we have

$$\phi_k(\mathbf{x}^*) = (1 - \alpha_{k-1})\phi_{k-1}(\mathbf{x}^*)$$

$$+ \alpha_{k-1}\left(F(\mathbf{x}_k) + \langle\kappa(\mathbf{y}_{k-1} - \mathbf{x}_k), \mathbf{x}^* - \mathbf{x}_k\rangle + \frac{\mu}{2}\|\mathbf{x}^* - \mathbf{x}_k\|^2\right)$$

$$\overset{a}{\leq} (1 - \alpha_{k-1})\phi_{k-1}(\mathbf{x}^*)$$

$$+ \alpha_{k-1}\left[F(\mathbf{x}^*) + \epsilon_k - (\kappa + \mu)\langle\mathbf{x}_k - \mathbf{x}_k^*, \mathbf{x}^* - \mathbf{x}_k\rangle\right],$$

where $\overset{a}{\le}$ uses (5.61). Rearranging terms and using the definition of $\xi_k = (1 - \alpha_{k-1})\left(\xi_{k-1} + \epsilon_k - (\kappa + \mu)\langle \mathbf{x}_k - \mathbf{x}_k^*, \mathbf{x}_{k-1} - \mathbf{x}_k \rangle\right)$, we have

$$
\begin{aligned}
\phi_k(&\mathbf{x}^*) + \xi_k - F^* \\
&\le (1 - \alpha_{k-1})(\phi_{k-1}(\mathbf{x}^*) + \xi_{k-1} - F^*) + \epsilon_k \\
&\quad -(\kappa + \mu)\langle \mathbf{x}_k - \mathbf{x}_k^*, (1 - \alpha_{k-1})\mathbf{x}_{k-1} + \alpha_{k-1}\mathbf{x}^* - \mathbf{x}_k \rangle \\
&\overset{a}{\le} (1 - \alpha_{k-1})(\phi_{k-1}(\mathbf{x}^*) + \xi_{k-1} - F^*) + \epsilon_k + \sqrt{2\epsilon_k \gamma_k}\|\mathbf{x}^* - \mathbf{v}_k\|, \quad (5.68)
\end{aligned}
$$

where $F^* = F(\mathbf{x}^*)$ and in $\overset{a}{=}$ we use

$$
\begin{aligned}
-(\kappa + \mu)&\langle \mathbf{x}_k - \mathbf{x}_k^*, (1 - \alpha_{k-1})\mathbf{x}_{k-1} + \alpha_{k-1}\mathbf{x}^* - \mathbf{x}_k \rangle \\
&\overset{a}{=} -\alpha_{k-1}(\kappa + \mu)\langle \mathbf{x}_k - \mathbf{x}_k^*, \mathbf{x}^* - \mathbf{v}_k \rangle \\
&\le \alpha_{k-1}(\kappa + \mu)\|\mathbf{x}_k - \mathbf{x}_k^*\|\|\mathbf{x}^* - \mathbf{v}_k\| \\
&\overset{b}{\le} \alpha_{k-1}\sqrt{2(\kappa + \mu)\epsilon_k}\|\mathbf{x}^* - \mathbf{v}_k\| \overset{c}{=} \sqrt{2\epsilon_k \gamma_k}\|\mathbf{x}^* - \mathbf{v}_k\|,
\end{aligned}
$$

where $\overset{a}{=}$ uses (5.66), $\overset{b}{\le}$ uses (A.10), and $\overset{c}{=}$ uses (5.64).

Then dividing λ_k on both sides of (5.68) and telescoping the result with $k = 1$ to k, we have

$$
\frac{1}{\lambda_k}\left(\phi_k(\mathbf{x}^*) + \xi_k - F^*\right) \le \phi_0(\mathbf{x}^*) - F^* + \sum_{i=1}^{k}\frac{\epsilon_i}{\lambda_i} + \sum_{i=1}^{k}\frac{\sqrt{2\epsilon_i \gamma_i}}{\lambda_i}\|\mathbf{x}^* - \mathbf{v}_i\|.
$$

Using the definition of $\phi_k(\mathbf{x}^*)$, we have

$$
\begin{aligned}
\phi_k(\mathbf{x}^*) + \xi_k - F^* &= \phi_k^* + \xi_k - F^* + \frac{\gamma_k}{2}\|\mathbf{x}^* - \mathbf{v}_k\|^2 \\
&\overset{a}{\ge} F(\mathbf{x}_k) - F^* + \frac{\gamma_k}{2}\|\mathbf{x}^* - \mathbf{v}_k\|^2,
\end{aligned}
$$

where $\overset{a}{\ge}$ uses (5.60). So we obtain (5.67).

Step 5: By setting $u_i = \sqrt{\frac{\gamma_i}{2\lambda_i}}\|\mathbf{x}^* - \mathbf{v}_i\|$, $\alpha_i = 2\sqrt{\frac{\epsilon_i}{\lambda_i}}$, and $S_k = \phi_0(\mathbf{x}^*) - F^* + \sum_{i=1}^{k}\frac{\epsilon_i}{\lambda_i}$, (2.37) of Lemma 2.8 is validated due to (5.67) and $F(\mathbf{x}) - F^* \ge 0$. So by Lemma 2.8 we have

$$
F(\mathbf{x}_k) - F^* \overset{a}{\le} \lambda_k\left(S_k + \sum_{i=1}^{k}\alpha_i u_i\right) \le \lambda_k\left(\sqrt{S_k} + 2\sum_{i=1}^{k}\sqrt{\frac{\epsilon_i}{\lambda_i}}\right)^2, \quad (5.69)
$$

where $\overset{a}{\le}$ uses (5.67).

Step 6: We prove Theorem 5.4. By setting $\alpha_0 = \sqrt{q} = \sqrt{\frac{\mu}{\mu+\kappa}}$, we can have $\alpha_k = \sqrt{q}$ for $k \geq 0$, and then $\lambda_k = (1 - \sqrt{q})^k$ and $\gamma_0 = \mu$. Thus by the μ-strong convexity of $F(\cdot)$, we have $\frac{\gamma_0}{2}\|\mathbf{x}_0 - \mathbf{x}^*\|^2 \leq F(\mathbf{x}_0) - F^*$. Then

$$
\sqrt{S_k} + 2\sum_{i=1}^{k}\sqrt{\frac{\epsilon_i}{\lambda_i}}
$$

$$
= \sqrt{F(\mathbf{x}_0) - F^* + \frac{\gamma_0}{2}\|\mathbf{x}_0 - \mathbf{x}_*\|^2 + \sum_{i=1}^{k}\frac{\epsilon_i}{\lambda_i} + 2\sum_{i=1}^{k}\sqrt{\frac{\epsilon_i}{\lambda_i}}}
$$

$$
\leq \sqrt{F(\mathbf{x}_0) - F^* + \frac{\gamma_0}{2}\|\mathbf{x}_0 - \mathbf{x}_*\|^2 + 3\sum_{i=1}^{k}\sqrt{\frac{\epsilon_i}{\lambda_i}}}
$$

$$
\leq \sqrt{2(F(\mathbf{x}_0) - F^*)} + 3\sum_{i=1}^{k}\sqrt{\frac{\epsilon_i}{\lambda_i}}
$$

$$
= \sqrt{2(F(\mathbf{x}_0) - F^*)}\left[1 + \sum_{i=1}^{k}\left(\sqrt{\frac{1-\rho}{1-\sqrt{q}}}\right)^i\right]
$$

$$
= \sqrt{2(F(\mathbf{x}_0) - F^*)}\frac{\eta^{k+1} - 1}{\eta - 1}
$$

$$
\leq \sqrt{2(F(\mathbf{x}_0) - F^*)}\frac{\eta^{k+1}}{\eta - 1},
$$

where we set $\eta = \sqrt{\frac{1-\rho}{1-\sqrt{q}}}$. Then from (5.69), we have

$$
F(\mathbf{x}_k) - F^*
$$

$$
\leq 2\lambda_k(F(\mathbf{x}_0) - F^*)\left(\frac{\eta^{k+1}}{\eta - 1}\right)^2
$$

$$
\overset{a}{\leq} 2\left(\frac{\eta}{\eta - 1}\right)^2(1 - \rho)^k(F(\mathbf{x}_0) - F^*)
$$

$$
= 2\left(\frac{\sqrt{1-\rho}}{\sqrt{1-\rho} - \sqrt{1-\sqrt{q}}}\right)^2(1 - \rho)^k(F(\mathbf{x}_0) - F^*)
$$

$$
= 2\left(\frac{1}{\sqrt{1-\rho} - \sqrt{1-\sqrt{q}}}\right)^2(1 - \rho)^{k+1}(F(\mathbf{x}_0) - F^*),
$$

where in $\overset{a}{\leq}$ we use $\lambda_k = \prod_{i=0}^{k-1}(1 - \alpha_i) \leq \left(1 - \sqrt{q}\right)^k$. Using that $\sqrt{1 - x} + \frac{x}{2}$ is monotonically decreasing, we have $\sqrt{1 - \rho} + \frac{\rho}{2} \geq \sqrt{1 - \sqrt{q}} + \frac{\sqrt{q}}{2}$. So we have

$$F(\mathbf{x}_k) - F^* \leq \frac{8}{(\sqrt{q} - \rho)^2}(1 - \rho)^{k+1}\left(F(\mathbf{x}_0) - F^*\right). \tag{5.70}$$

This ends the proof. $\square$

With a special setting of the parameters for Algorithm 5.4, we are able to give the convergence result for the INC case.

Theorem 5.6 *For (5.3), assume that each $f_i(\mathbf{x})$ is L-smooth, $f(\mathbf{x})$ is convex, and $h(\mathbf{x})$ is μ-strongly convex, satisfying $\frac{L}{\mu} \geq n^{1/2}$, then for Algorithm 5.4, by setting $\kappa = \frac{L}{\sqrt{n-1}}$ and solving Step 3 by SVRG [10] by running $O(n \log(1/\epsilon))$ steps, one can obtain an ϵ-accuracy solution satisfying $F(\mathbf{x}) - F^* \leq \epsilon$ with IFO calls of $O(n^{3/4}\sqrt{L/\mu} \log^2(1/\epsilon))$.*

5.3 The Nonconvex Case

In this section, we consider a hard case when $f(\mathbf{x})$ is nonconvex. We only consider the problem where $h(\mathbf{x}) = \mathbf{0}$ and focus on the IFO complexity to achieve an approximate first-order stationary point satisfying $\|\nabla f(\mathbf{x})\| \leq \epsilon$. In the IC case, SVRG [10] has already achieved nearly optimal rate (ignoring some constants) when n is sufficiently large ($n \geq \frac{L}{\mu}$). However, for the NC case SVRG is not the optimal. We will introduce SPIDER [7] which can find an approximate first-order stationary point in an optimal (ignoring some constants) $O\left(\frac{n^{1/2}}{\epsilon^2}\right)$ rate. Next, we will show that if further assume that the objective function has Lipschitz continuous Hessians, the momentum technique can ensure a faster rate when n is much smaller than $\kappa = L/\mu$.

5.3.1 SPIDER

The Stochastic Path-Integrated Differential Estimator (SPIDER) [7] technique is a radical VR method which is used to track quantities using reduced stochastic oracles. Let us consider an arbitrary deterministic vector quantity $Q(\mathbf{x})$. Assume that we observe a sequence $\hat{\mathbf{x}}_{0:K}$ and we want to dynamically track $Q(\hat{\mathbf{x}}^k)$ for $k = 0, 1, \cdots, K$. Further assume that we have an initial estimate $\tilde{Q}(\hat{\mathbf{x}}^0) \approx Q(\hat{\mathbf{x}}^0)$ and an unbiased estimate $\boldsymbol{\xi}_k(\hat{\mathbf{x}}_{0:k})$ of $Q(\hat{\mathbf{x}}^k) - Q(\hat{\mathbf{x}}^{k-1})$ such that for each $k = 1, \cdots, K$,

$$\mathbb{E}\left[\boldsymbol{\xi}_k(\hat{\mathbf{x}}_{0:k}) \mid \hat{\mathbf{x}}_{0:k}\right] = Q(\hat{\mathbf{x}}^k) - Q(\hat{\mathbf{x}}^{k-1}).$$

Then we can integrate (in the discrete sense) the stochastic differential estimate as

$$\tilde{Q}(\hat{\mathbf{x}}_{0:K}) \equiv \tilde{Q}(\hat{\mathbf{x}}^0) + \sum_{k=1}^{K} \boldsymbol{\xi}_k(\hat{\mathbf{x}}_{0:k}). \tag{5.71}$$

We call estimator $\tilde{Q}(\hat{\mathbf{x}}_{0:K})$ the *Stochastic Path-Integrated Differential Estimator*, or SPIDER for brevity. We have

Proposition 5.1 *The martingale (Definition A.4) variance bound has*

$$\mathbb{E} \left\| \tilde{Q}(\hat{\mathbf{x}}_{0:K}) - Q(\hat{\mathbf{x}}^K) \right\|^2 = \mathbb{E} \left\| \tilde{Q}(\hat{\mathbf{x}}^0) - Q(\hat{\mathbf{x}}^0) \right\|^2$$

$$+ \sum_{k=1}^{K} \mathbb{E} \left\| \boldsymbol{\xi}_k(\hat{\mathbf{x}}_{0:k}) - (Q(\hat{\mathbf{x}}^k) - Q(\hat{\mathbf{x}}^{k-1})) \right\|^2. \tag{5.72}$$

Proposition 5.1 can be easily proven using the property of square-integrable martingales.

Now, let $\mathcal{B}_i$ map any $\mathbf{x} \in \mathbb{R}^d$ to a random estimate $\mathcal{B}_i(\mathbf{x})$, where $\mathcal{B}(\mathbf{x})$ is the true value to be estimated. At each step k, let S_* be a subset that samples $|S_*|$ elements in $[n]$ with replacement and let the stochastic estimator $\mathcal{B}_{S_*} = (1/|S_*|) \sum_{i \in S_*} \mathcal{B}_i$ satisfy

$$\mathbb{E} \left\| \mathcal{B}_i(\mathbf{x}) - \mathcal{B}_i(\mathbf{y}) \right\|^2 \le L_{\mathcal{B}}^2 \|\mathbf{x} - \mathbf{y}\|^2, \tag{5.73}$$

and $\left\| \mathbf{x}^k - \mathbf{x}^{k-1} \right\| \le \epsilon_1$ for all $k = 1, \cdots, K$. Finally, we set our estimator $\mathcal{V}^k$ of $\mathcal{B}(\mathbf{x}^k)$ as

$$\mathcal{V}^k = \mathcal{B}_{S_*}(\mathbf{x}^k) - \mathcal{B}_{S_*}(\mathbf{x}^{k-1}) + \mathcal{V}^{k-1}.$$

Applying Proposition 5.1 immediately concludes the following lemma, which gives an error bound of the estimator $\mathcal{V}^k$ in terms of the second moment of $\left\| \mathcal{V}^k - \mathcal{B}(\mathbf{x}^k) \right\|$:

Lemma 5.3 *Under the condition (5.73) we have that for all $k = 1, \cdots, K$,*

$$\mathbb{E} \left\| \mathcal{V}^k - \mathcal{B}(\mathbf{x}^k) \right\|^2 \le \frac{k L_{\mathcal{B}}^2 \epsilon_1^2}{|S_*|} + \mathbb{E} \left\| \mathcal{V}^0 - \mathcal{B}(\mathbf{x}^0) \right\|^2. \tag{5.74}$$

Proof For any $k > 0$, we have from Proposition 5.1 (by applying $\tilde{Q} = \mathcal{V}$)

$$\mathbb{E}_k \left\| \mathcal{V}^k - \mathcal{B}(\mathbf{x}^k) \right\|^2 = \mathbb{E}_k \left\| \mathcal{B}_{S_*}(\mathbf{x}^k) - \mathcal{B}(\mathbf{x}^k) - \mathcal{B}_{S_*}(\mathbf{x}^{k-1}) + \mathcal{B}(\mathbf{x}^{k-1}) \right\|^2$$

$$+ \left\| \mathcal{V}^{k-1} - \mathcal{B}(\mathbf{x}^{k-1}) \right\|^2. \tag{5.75}$$

Then

$$\mathbb{E}_k \left\| \mathcal{B}_{S_*}(\mathbf{x}^k) - \mathcal{B}(\mathbf{x}^k) - \mathcal{B}_{S_*}(\mathbf{x}^{k-1}) + \mathcal{B}(\mathbf{x}^{k-1}) \right\|^2$$

$$\stackrel{a}{=} \frac{1}{|S_*|} \mathbb{E} \left\| \mathcal{B}_i(\mathbf{x}^k) - \mathcal{B}(\mathbf{x}^k) - \mathcal{B}_i(\mathbf{x}^{k-1}) + \mathcal{B}(\mathbf{x}^{k-1}) \right\|^2$$

$$\stackrel{b}{\leq} \frac{1}{|S_*|} \mathbb{E} \left\| \mathcal{B}_i(\mathbf{x}^k) - \mathcal{B}_i(\mathbf{x}^{k-1}) \right\|^2$$

$$\stackrel{c}{\leq} \frac{1}{|S_*|} L_{\mathcal{B}}^2 \mathbb{E} \left\| \mathbf{x}^k - \mathbf{x}^{k-1} \right\|^2 \leq \frac{L_{\mathcal{B}}^2 \epsilon_1^2}{|S_*|}, \tag{5.76}$$

where in $\stackrel{a}{=}$ we use that S_* is randomly sampled from $[n]$ with replacement, so the variance reduces by $\frac{1}{|S_*|}$ times. In $\stackrel{b}{\leq}$ and $\stackrel{c}{\leq}$ we use Proposition A.2 and (5.73), respectively.

Combining (5.75) and (5.76), we have

$$\mathbb{E}_k \left\| \mathcal{V}^k - \mathcal{B}(\mathbf{x}^k) \right\|^2 \leq \frac{L_{\mathcal{B}}^2 \epsilon_1^2}{|S_*|} + \left\| \mathcal{V}^{k-1} - \mathcal{B}(\mathbf{x}^{k-1}) \right\|^2. \tag{5.77}$$

Telescoping the above display for $k' = k - 1, \cdots, 0$ and using the iterated law of expectation (Proposition A.4), we have

$$\mathbb{E} \left\| \mathcal{V}^k - \mathcal{B}(\mathbf{x}^k) \right\|^2 \leq \frac{k L_{\mathcal{B}}^2 \epsilon_1^2}{|S_*|} + \mathbb{E} \left\| \mathcal{V}^0 - \mathcal{B}(\mathbf{x}^0) \right\|^2. \tag{5.78}$$

$\square$

The algorithm using SPIDER to solve (5.2) is shown in Algorithm 5.5. We have the following theorem.

Theorem 5.7 *For the optimization problem (5.22) in the online case ($n = \infty$), assume that each $f_i(\mathbf{x})$ is L-smooth and $\mathbb{E} \|\nabla f_i(\mathbf{x}) - \nabla f(\mathbf{x})\|^2 \leq \sigma^2$. Set the parameters $S_1, S_2, \eta,$ and q as*

$$S_1 = \frac{2\sigma^2}{\epsilon^2}, \quad S_2 = \frac{2\sigma}{\epsilon n_0}, \quad \eta = \frac{\epsilon}{L n_0}, \quad \eta_k = \min\left(\frac{\epsilon}{L n_0 \|\mathbf{v}^k\|}, \frac{1}{2L n_0}\right), \quad q = \frac{\sigma n_0}{\epsilon}, \tag{5.79}$$

and set $K = \lfloor (4L\Delta n_0)\epsilon^{-2} \rfloor + 1$. Then running Algorithm 5.5 with OPTION II for K iterations outputs an $\tilde{\mathbf{x}}$ satisfying

$$\mathbb{E} \|\nabla f(\tilde{\mathbf{x}})\| \leq 5\epsilon, \tag{5.80}$$

Algorithm 5.5 SPIDER for searching first-order stationary point (SPIDER-SFO)

1: Input $\mathbf{x}^0$, q, S_1, S_2, n_0, ϵ, and $\tilde{\epsilon}$.
2: **for** $k = 0$ to K **do**
3:     **if** mod $(k, q) = 0$ **then**
4:         Draw S_1 samples (or compute the full gradient for the finite-sum case) and let $\mathbf{v}^k = \nabla f_{S_1}(\mathbf{x}^k)$.
5:     **else**
6:         Draw S_2 samples and let $\mathbf{v}^k = \nabla f_{S_2}(\mathbf{x}^k) - \nabla f_{S_2}(\mathbf{x}^{k-1}) + \mathbf{v}^{k-1}$.
7:     **end if**

8: **OPTION I**   ⋄ for convergence rates in high probability
9:     **if** $\|\mathbf{v}^k\| \le 2\tilde{\epsilon}$ **then**
10:         return $\mathbf{x}^k$.
11:     **else**
12:         $\mathbf{x}^{k+1} = \mathbf{x}^k - \eta \cdot (\mathbf{v}^k/\|\mathbf{v}^k\|)$, where $\eta = \dfrac{\epsilon}{Ln_0}$.
13:     **end if**

14: **OPTION II**   ⋄ for convergence rates in expectation
15:     $\mathbf{x}^{k+1} = \mathbf{x}^k - \eta_k \mathbf{v}^k$, where $\eta_k = \min\left(\dfrac{\epsilon}{Ln_0\|\mathbf{v}^k\|}, \dfrac{1}{2Ln_0}\right)$.
16: **end for**

17: **OPTION I**: Return $\mathbf{x}^K$.   ⋄ however, this line is *not* reached with high probability

18: **OPTION II**: Return $\tilde{\mathbf{x}}$ chosen uniformly at random from $\{\mathbf{x}^k\}_{k=0}^{K-1}$.

---

where $\Delta = f(\mathbf{x}^0) - f^*$ ($f^* = \inf_{\mathbf{x}} f(\mathbf{x})$). The gradient cost is bounded by $24L\Delta\sigma \cdot \epsilon^{-3} + 2\sigma^2\epsilon^{-2} + 4\sigma n_0^{-1}\epsilon^{-1}$ for any choice of $n_0 \in [1, 2\sigma/\epsilon]$. Treating Δ, L, and σ as positive constants, the stochastic gradient complexity is $O(\epsilon^{-3})$.

To prove Theorem 5.7, we first prepare the following lemmas.

Lemma 5.4 *Set the parameters* S_1, S_2, η, and q as in (5.79), and $k_0 = \lfloor k/q \rfloor \cdot q$. We have

$$\mathbb{E}_{k_0}\left[\left\|\mathbf{v}^k - \nabla f(\mathbf{x}^k)\right\|^2 \middle| \mathbf{x}_{0:k_0}\right] \le \epsilon^2, \tag{5.81}$$

where $\mathbb{E}_{k_0}$ denotes the conditional expectation over the randomness of $\mathbf{x}_{(k_0+1):k}$.

Proof For $k = k_0$, we have

$$\mathbb{E}_{k_0}\left\|\mathbf{v}^{k_0} - \nabla f(\mathbf{x}^{k_0})\right\|^2 = \mathbb{E}_{k_0}\left\|\nabla f_{S_1}(\mathbf{x}^{k_0}) - \nabla f(\mathbf{x}^{k_0})\right\|^2 \le \frac{\sigma^2}{S_1} = \frac{\epsilon^2}{2}. \tag{5.82}$$

From Line 15 of Algorithm 5.5 we have that for all $k \geq 0$,

$$\left\| \mathbf{x}^{k+1} - \mathbf{x}^k \right\| = \min \left(\frac{\epsilon}{Ln_0 \|\mathbf{v}^k\|}, \frac{1}{2Ln_0} \right) \|\mathbf{v}^k\| \leq \frac{\epsilon}{Ln_0}. \tag{5.83}$$

Applying Lemma 5.3 with $\epsilon_1 = \epsilon/(Ln_0)$, $S_2 = 2\sigma/(\epsilon n_0)$, and $K = k - k_0 \leq q = \sigma n_0/\epsilon$, we have

$$\mathbb{E}_{k_0} \left\| \mathbf{v}^k - \nabla f(\mathbf{x}^k) \right\|^2 \leq \frac{\sigma n_0}{\epsilon} \cdot L^2 \cdot \left(\frac{\epsilon}{Ln_0} \right)^2 \cdot \frac{\epsilon n_0}{2\sigma} + \mathbb{E}_{k_0} \left\| \mathbf{v}^{k_0} - \nabla f(\mathbf{x}^{k_0}) \right\|^2 \overset{a}{\leq} \epsilon^2,$$

where $\overset{a}{\leq}$ uses (5.82). The proof is completed. $\qquad\qquad\square$

Lemma 5.5 *Setting $k_0 = \lfloor k/q \rfloor \cdot q$, we have*

$$\mathbb{E}_{k_0} \left[f(\mathbf{x}^{k+1}) - f(\mathbf{x}^k) \right] \leq -\frac{\epsilon}{4Ln_0} \mathbb{E}_{k_0} \left\| \mathbf{v}^k \right\| + \frac{3\epsilon^2}{4n_0 L}. \tag{5.84}$$

Proof We have

$$\|\nabla f(\mathbf{x}) - \nabla f(\mathbf{y})\|^2 = \|\mathbb{E}_i \left(\nabla f_i(\mathbf{x}) - \nabla f_i(\mathbf{y}) \right)\|^2 \leq \mathbb{E}_i \|\nabla f_i(\mathbf{x}) - \nabla f_i(\mathbf{y})\|^2$$
$$\leq L^2 \|\mathbf{x} - \mathbf{y}\|^2.$$

So $f(\mathbf{x})$ is L-smooth, then

$$f(\mathbf{x}^{k+1}) \leq f(\mathbf{x}^k) + \left\langle \nabla f(\mathbf{x}^k), \mathbf{x}^{k+1} - \mathbf{x}^k \right\rangle + \frac{L}{2} \left\| \mathbf{x}^{k+1} - \mathbf{x}^k \right\|^2$$
$$= f(\mathbf{x}^k) - \eta_k \left\langle \nabla f(\mathbf{x}^k), \mathbf{v}^k \right\rangle + \frac{L\eta_k^2}{2} \left\| \mathbf{v}^k \right\|^2$$
$$= f(\mathbf{x}^k) - \eta_k \left(1 - \frac{\eta_k L}{2} \right) \left\| \mathbf{v}^k \right\|^2 - \eta_k \left\langle \nabla f(\mathbf{x}^k) - \mathbf{v}^k, \mathbf{v}^k \right\rangle$$
$$\overset{a}{\leq} f(\mathbf{x}^k) - \eta_k \left(\frac{1}{2} - \frac{\eta_k L}{2} \right) \left\| \mathbf{v}^k \right\|^2 + \frac{\eta_k}{2} \left\| \mathbf{v}^k - \nabla f(\mathbf{x}^k) \right\|^2, \tag{5.85}$$

where in $\overset{a}{\leq}$ we apply the Cauchy–Schwartz inequality. Since

$$\eta_k = \min \left(\frac{\epsilon}{Ln_0 \|\mathbf{v}^k\|}, \frac{1}{2Ln_0} \right) \leq \frac{1}{2Ln_0} \leq \frac{1}{2L},$$

we have

$$\eta_k \left(\frac{1}{2} - \frac{\eta_k L}{2} \right) \left\| \mathbf{v}^k \right\|^2 \geq \frac{1}{4} \eta_k \left\| \mathbf{v}^k \right\|^2$$

$$= \frac{\epsilon^2}{8 n_0 L} \min \left(2 \left\| \frac{\mathbf{v}^k}{\epsilon} \right\|, \left\| \frac{\mathbf{v}^k}{\epsilon} \right\|^2 \right) \overset{a}{\geq} \frac{\epsilon \left\| \mathbf{v}^k \right\| - 2\epsilon^2}{4 n_0 L},$$

where in $\overset{a}{\geq}$ we use $V(x) = \min \left(|x|, \frac{x^2}{2} \right) \geq |x| - 2$ for all x. Hence

$$f(\mathbf{x}^{k+1}) \leq f(\mathbf{x}^k) - \frac{\epsilon \left\| \mathbf{v}^k \right\|}{4 L n_0} + \frac{\epsilon^2}{2 n_0 L} + \frac{\eta_k}{2} \left\| \mathbf{v}^k - \nabla f(\mathbf{x}^k) \right\|^2$$

$$\overset{a}{\leq} f(\mathbf{x}^k) - \frac{\epsilon \left\| \mathbf{v}^k \right\|}{4 L n_0} + \frac{\epsilon^2}{2 n_0 L} + \frac{1}{4 L n_0} \left\| \mathbf{v}^k - \nabla f(\mathbf{x}^k) \right\|^2, \quad (5.86)$$

where $\overset{a}{\leq}$ uses $\eta_k \leq \frac{1}{2 L n_0}$.

Taking expectation on the above display and using Lemma 5.4, we have

$$\mathbb{E}_{k_0} f(\mathbf{x}^{k+1}) - \mathbb{E}_{k_0} f(\mathbf{x}^k) \leq - \frac{\epsilon}{4 L n_0} \mathbb{E}_{k_0} \left\| \mathbf{v}^k \right\| + \frac{3\epsilon^2}{4 L n_0}. \quad (5.87)$$

$\square$

Lemma 5.6 *For all $k \geq 0$, we have*

$$\mathbb{E} \left\| \nabla f(\mathbf{x}^k) \right\| \leq \mathbb{E} \left\| \mathbf{v}^k \right\| + \epsilon. \quad (5.88)$$

Proof By taking the total expectation on (5.81), we have

$$\mathbb{E} \left\| \mathbf{v}^k - \nabla f(\mathbf{x}^k) \right\|^2 \leq \epsilon^2. \quad (5.89)$$

Then by Jensen's inequality (Proposition A.3),

$$\left(\mathbb{E} \left\| \mathbf{v}^k - \nabla f(\mathbf{x}^k) \right\| \right)^2 \leq \mathbb{E} \left\| \mathbf{v}^k - \nabla f(\mathbf{x}^k) \right\|^2 \leq \epsilon^2.$$

So using the triangle inequality,

$$\mathbb{E} \left\| \nabla f(\mathbf{x}^k) \right\| = \mathbb{E} \left\| \mathbf{v}^k - (\mathbf{v}^k - \nabla f(\mathbf{x}^k)) \right\|$$

$$\leq \mathbb{E} \left\| \mathbf{v}^k \right\| + \mathbb{E} \left\| \mathbf{v}^k - \nabla f(\mathbf{x}^k) \right\| \leq \mathbb{E} \left\| \mathbf{v}^k \right\| + \epsilon. \quad (5.90)$$

This completes our proof. $\square$

Now, we are ready to prove Theorem 5.7.

Proof of Theorem 5.7 Taking full expectation on (5.84) and telescoping the results from $k = 0$ to $K - 1$, we have

$$\frac{\epsilon}{4Ln_0} \sum_{k=0}^{K-1} \mathbb{E} \left\| \mathbf{v}^k \right\| \le f(\mathbf{x}^0) - \mathbb{E} f(\mathbf{x}^K) + \frac{3K\epsilon^2}{4Ln_0} \overset{a}{\le} \Delta + \frac{3K\epsilon^2}{4Ln_0}, \qquad (5.91)$$

where $\overset{a}{\le}$ uses $\mathbb{E} f(\mathbf{x}^K) \ge f^*$.

Dividing both sides of (5.91) by $\frac{\epsilon}{4Ln_0} K$ and using $K = \lfloor \frac{4L\Delta n_0}{\epsilon^2} \rfloor + 1 \ge \frac{4L\Delta n_0}{\epsilon^2}$, we have

$$\frac{1}{K} \sum_{k=0}^{K-1} \mathbb{E} \left\| \mathbf{v}^k \right\| \le \Delta \cdot \frac{4Ln_0}{\epsilon} \frac{1}{K} + 3\epsilon \le 4\epsilon. \qquad (5.92)$$

Then from the choice of $\tilde{\mathbf{x}}$ in Line 17 of Algorithm 5.5, we have

$$\mathbb{E} \| \nabla f(\tilde{\mathbf{x}}) \| = \frac{1}{K} \sum_{k=0}^{K-1} \mathbb{E} \left\| \nabla f(\mathbf{x}^k) \right\| \overset{a}{\le} \frac{1}{K} \sum_{k=0}^{K-1} \mathbb{E} \left\| \mathbf{v}^k \right\| + \epsilon \overset{b}{\le} 5\epsilon, \qquad (5.93)$$

where $\overset{a}{\le}$ and $\overset{b}{\le}$ use (5.88) and (5.92), respectively.

To compute the gradient cost, note that in each q iteration we access for one time of S_1 stochastic gradients and for q times of $2S_2$ stochastic gradients, hence the cost is

$$\left\lceil K \cdot \frac{1}{q} \right\rceil S_1 + 2K S_2 \overset{a}{\le} 3K \cdot S_2 + S_1$$

$$\le \left[3 \left(\frac{4Ln_0\Delta}{\epsilon^2} \right) + 2 \right] \frac{2\sigma}{\epsilon n_0} + \frac{2\sigma^2}{\epsilon^2}$$

$$= \frac{24L\sigma\Delta}{\epsilon^3} + \frac{4\sigma}{n_0\epsilon} + \frac{2\sigma^2}{\epsilon^2}, \qquad (5.94)$$

where $\overset{a}{\le}$ uses $S_1 = q S_2$. This concludes a gradient cost of $24L\Delta\sigma\epsilon^{-3} + 2\sigma^2\epsilon^{-2} + 4\sigma n_0^{-1}\epsilon^{-1}$. $\qquad\qquad\qquad\qquad\qquad\qquad\qquad\qquad\qquad\qquad\qquad\qquad\qquad\qquad\qquad\qquad\Box$

Theorem 5.8 *For the optimization problem (5.22) in the finite-sum case ($n < \infty$), assume that each $f_i(\mathbf{x})$ is L-smooth, set the parameters S_2, η_k, and q as*

$$S_2 = \frac{n^{1/2}}{n_0}, \quad \eta = \frac{\epsilon}{Ln_0}, \quad \eta_k = \min \left(\frac{\epsilon}{Ln_0\|\mathbf{v}^k\|}, \frac{1}{2Ln_0} \right), \quad q = n_0 n^{1/2}, \quad (5.95)$$

set $K = \lfloor (4L\Delta n_0)\epsilon^{-2} \rfloor + 1$, and let $S_1 = n$, i.e., we obtain the full gradient in Line 4. Then running Algorithm 5.5 with OPTION II for K iterations outputs an $\tilde{\mathbf{x}}$ satisfying

$$\mathbb{E}\|\nabla f(\tilde{\mathbf{x}})\| \leq 5\epsilon.$$

The gradient cost is bounded by $n + 12(L\Delta) \cdot n^{1/2}\epsilon^{-2} + 2n_0^{-1}n^{1/2}$ for any choice of $n_0 \in [1, n^{1/2}]$. Treating Δ, L, and σ as positive constants, the stochastic gradient complexity is $O(n + n^{1/2}\epsilon^{-2})$.

Proof For $k = k_0$, we have

$$\mathbb{E}_{k_0} \left\| \mathbf{v}^{k_0} - \nabla f(\mathbf{x}^{k_0}) \right\|^2 = \mathbb{E}_{k_0} \left\| \nabla f(\mathbf{x}^{k_0}) - \nabla f(\mathbf{x}^{k_0}) \right\|^2 = 0. \qquad (5.96)$$

For $k \neq k_0$, applying Lemma 5.3 with $\epsilon_1 = \frac{\epsilon}{Ln_0}$, $S_2 = \frac{n^{1/2}}{\epsilon n_0}$, and $K = k - k_0 \leq q = n_0 n^{1/2}$, we have

$$\mathbb{E}_{k_0} \left\| \mathbf{v}^k - \nabla f(\mathbf{x}^k) \right\|^2 \leq n_0 n^{1/2} \cdot L^2 \cdot \left(\frac{\epsilon}{Ln_0} \right)^2 \cdot \frac{n_0}{n^{1/2}} + \mathbb{E}_{k_0} \left\| \mathbf{v}^{k_0} - \nabla f(\mathbf{x}^{k_0}) \right\|^2$$

$$\overset{a}{=} \epsilon^2,$$

where $\overset{a}{=}$ uses (5.96). So Lemma 5.4 holds for all k. Then from the same technique of the online case ($n = \infty$), we can also obtain (5.83), (5.84), and (5.93). The gradient cost analysis is computed as

$$\left\lceil K \cdot \frac{1}{q} \right\rceil S_1 + 2K S_2 \overset{a}{\leq} 3K \cdot S_2 + S_1$$

$$\leq \left[3\left(\frac{4Ln_0\Delta}{\epsilon^2} \right) + 2 \right] \frac{n^{1/2}}{n_0} + n$$

$$= \frac{12(L\Delta) \cdot n^{1/2}}{\epsilon^2} + \frac{2n^{1/2}}{n_0} + n, \qquad (5.97)$$

where $\overset{a}{\leq}$ uses $S_1 = q S_2$. This concludes a gradient cost of $n + 12(L\Delta) \cdot n^{1/2}\epsilon^{-2} + 2n_0^{-1}n^{1/2}$. $\square$

5.3.2 Momentum Acceleration

When computing a first-order stationary point, SPIDER is actually (nearly) optimal, if only with the gradient-smoothness condition under certain regimes. Thus only with the condition of smoothness of the gradient, it is hard to apply the momentum techniques to accelerate algorithms. However, one can obtain a faster rate with an additional assumption on Hessian:

Assumption 5.1 *Each* $f_i(\mathbf{x})$ *has* ρ-*Lipschitz continuous Hessians (Definition A.14).*

The technique to accelerate nonconvex algorithms is briefly described as follows:

- Run an efficient Negative Curvature Search (NC-Search) iteration to find an δ-approximate negative Hessian direction $\mathbf{w}_1$ using stochastic gradients,[2] e.g., the shift-and-invert technique in [9].
- If NC-Search finds a $\mathbf{w}_1$, update $\mathbf{x}^{k+1} \leftarrow \mathbf{x}^k \pm (\delta/\rho)\mathbf{w}_1$.
- If not, solve the INC problem:

$$\mathbf{x}^{k+1} = \underset{\mathbf{x}}{\operatorname{argmin}} \left(f(\mathbf{x}) + \frac{\Omega(\delta)}{2} \|\mathbf{x} - \mathbf{x}^k\|^2 \right),$$

using a momentum acceleration technique, e.g., Catalyst [12] described in Sect. 5.1.4. If $\|\mathbf{x}^{k+1} - \mathbf{x}^k\| \geq \Omega(\delta)$, return to Step 1, otherwise output $\mathbf{x}^{k+1}$.

We informally list the convergence result as follows:

Theorem 5.9 *Suppose solving NC-Search by [9] and INC blocks by Catalyst described in Theorem 5.6, the total stochastic gradient complexity to achieve an* ϵ-*accuracy solution satisfying* $\|\nabla f(\mathbf{x}_k)\| \leq \epsilon$ *and* $\lambda_{\min}(\nabla^2 f(\mathbf{x}_k)) \geq -\sqrt{\epsilon}$ *is* $\tilde{O}(n^{3/4}\epsilon^{-1.75})$.

The proof of Theorem 5.9 is lengthy, so we omit the proof here. We mention that when n is large, e.g., $n \geq \epsilon^{-1}$, the above method might not be faster than SPIDER [7]. Thus the existing lowest complexity to find a first-order stationary point is $\tilde{O}(\min(n^{3/4}\epsilon^{-1.75}, n^{1/2}\varepsilon^{-2}, \varepsilon^{-3}))$. In fact, for the problem of searching a stationary point in nonconvex (stochastic) optimization, neither the upper nor the lower bounds have been well studied up to now. However, it is a very hot topic recently and has aroused a lot of attention in both the optimization and the machine learning communities due to the empirical practicability for nonconvex models. Interested readers may refer to [3, 7, 9] for some latest developments.

---

[2]A task that given a point $\mathbf{x} \in \mathbb{R}^d$, decides if $\lambda_{\min}(\nabla^2 f(\mathbf{x})) \geq -2\delta$ or finds a unit vector $\mathbf{w}_1$ such that $\mathbf{w}_1^T \nabla^2 f(\mathbf{x})\mathbf{w}_1 \leq -\delta$ (for numerical reasons, one has to leave some room).

Algorithm 5.6 Inner loop of Acc-SADMM

---

for $k = 0$ to $m - 1$ **do**

Update dual variable: $\lambda_s^k = \tilde{\lambda}_s^k + \frac{\beta\theta_2}{\theta_{1,s}} \left(A_1 x_{s,1}^k + A_2 x_{s,2}^k - \tilde{b}_s \right)$,

Update $x_{s,1}^{k+1}$ by (5.99),

Update $x_{s,2}^{k+1}$ by (5.100),

Update dual variable: $\tilde{\lambda}_s^{k+1} = \lambda_s^k + \beta \left(A_1 x_{s,1}^{k+1} + A_2 x_{s,2}^{k+1} - b \right)$,

Update y_s^{k+1} by $y_s^{k+1} = x_s^{k+1} + (1 - \theta_{1,s} - \theta_2)(x_s^{k+1} - x_s^k)$.

end for k

5.4 Constrained Problem

We extend the stochastic acceleration methods to solve the constrained problem in this section. As a simple example, we consider the convex finite-sum problem with linear constraints:

$$\min_{x_1, x_2} \ h_1(x_1) + f_1(x_1) + h_2(x_2) + \frac{1}{n} \sum_{i=1}^{n} f_{2,i}(x_2), \qquad (5.98)$$

$$s.t. \ A_1 x_1 + A_2 x_2 = b,$$

where $f_1(x_1)$ and $f_{2,i}(x_2)$ with $i \in [n]$ are convex and have Lipschitz continuous gradients, and $h_1(x_1)$ and $h_2(x_2)$ are also convex and their proximal mappings can be solved efficiently. We use L_1 to denote the Lipschitz constant of $f_1(x_1)$, and L_2 to denote the Lipschitz constant of $f_{2,i}(x_2)$ with $i \in [n]$, and $f_2(x) = \frac{1}{n}\sum_{i=1}^{n} f_{2,i}(x)$. We show that by fusing the VR technique and momentum, the convergence rate can be improved to be *non-ergodic* $O(1/K)$.

We list the notations and variables in Table 5.1. The algorithm has double loops: in the inner loop, we update primal variables $x_{s,1}^k$ and $x_{s,2}^k$ through extrapolation terms $y_{s,1}^k$ and $y_{s,2}^k$ and the dual variable λ_s^k; in the outer loop, we maintain snapshot vectors $\tilde{x}_{s+1,1}, \tilde{x}_{s+1,2}$, and $\tilde{b}_{s+1}$, and then assign the initial value to the extrapolation terms $y_{s+1,1}^0$ and $y_{s+1,2}^0$. The whole algorithm is shown in Algorithm 5.7. In the

Table 5.1 Notations and variables

Notation	Meaning	Variable	Meaning
$\langle x, y \rangle_G$, $\|x\|_G$	$x^T G y$, $\sqrt{x^T G x}$	$y_{s,1}^k, y_{s,2}^k$	Extrapolation variables
$F_i(x_i)$	$h_i(x_i) + f_i(x_i)$	$x_{s,1}^k, x_{s,2}^k$	Primal variables
x	$(x_1^T, x_2^T)^T$	$\tilde{\lambda}_s^k, \lambda_s^k, \hat{\lambda}^k$	Dual and temporary variables
y	$(y_1^T, y_2^T)^T$	$\tilde{x}_{s,1}, \tilde{x}_{s,2}, \tilde{b}_s$	Snapshot vectors
$F(x)$	$F_1(x_1) + F_2(x_2)$		used for VR
A	$[A_1, A_2]$	$(x_1^*, x_2^*, \lambda^*)$	KKT point of (5.98)
$\mathcal{I}_{k,s}$	Mini-batch indices	b	Batch size

process of solving primal variables, we linearize both the smooth term $f_i(\mathbf{x}_i)$ and the augmented term $\frac{\beta}{2}\|\mathbf{A}_1\mathbf{x}_1 + \mathbf{A}_2\mathbf{x}_2 - \mathbf{b} + \frac{\lambda}{\beta}\|^2$. The update rules of $\mathbf{x}_1$ and $\mathbf{x}_2$ can be written as

$$
\begin{aligned}
\mathbf{x}_{s,1}^{k+1} = \operatorname*{argmin}_{\mathbf{x}_1} h_1(\mathbf{x}_1) + \left\langle \nabla f_1(\mathbf{y}_{s,1}^k), \mathbf{x}_1 \right\rangle \\
+ \left\langle \frac{\beta}{\theta_{1,s}} \left(\mathbf{A}_1\mathbf{y}_{s,1}^k + \mathbf{A}_2\mathbf{y}_{s,2}^k - \mathbf{b} \right) + \lambda_s^k, \mathbf{A}_1\mathbf{x}_1 \right\rangle \\
+ \left(\frac{L_1}{2} + \frac{\beta\|\mathbf{A}_1^T\mathbf{A}_1\|}{2\theta_{1,s}} \right) \left\| \mathbf{x}_1 - \mathbf{y}_{s,1}^k \right\|^2
\end{aligned} \tag{5.99}
$$

and

$$
\begin{aligned}
\mathbf{x}_{s,2}^{k+1} = \operatorname*{argmin}_{\mathbf{x}_2} h_2(\mathbf{x}_2) + \left\langle \tilde{\nabla} f_2(\mathbf{y}_{s,1}^k), \mathbf{x}_2 \right\rangle \\
+ \left\langle \frac{\beta}{\theta_{1,s}} \left(\mathbf{A}_1\mathbf{x}_{s,1}^{k+1} + \mathbf{A}_2\mathbf{y}_{s,2}^k - \mathbf{b} \right) + \lambda_s^k, \mathbf{A}_2\mathbf{x}_2 \right\rangle \\
+ \left(\frac{\left(1 + \frac{1}{b\theta_2}\right) L_2}{2} + \frac{\beta\|\mathbf{A}_2^T\mathbf{A}_2\|}{2\theta_{1,s}} \right) \left\| \mathbf{x}_2 - \mathbf{y}_{s,2}^k \right\|^2 ,
\end{aligned} \tag{5.100}
$$

where $\tilde{\nabla} f_2(\mathbf{y}_{s,2}^k)$ is defined as

$$
\tilde{\nabla} f_2(\mathbf{y}_{s,2}^k) = \frac{1}{b} \sum_{i_{k,s} \in \mathcal{I}_{k,s}} \left(\nabla f_{2,i_{k,s}}(\mathbf{y}_{s,2}^k) - \nabla f_{2,i_{k,s}}(\tilde{\mathbf{x}}_{s,2}) + \nabla f_2(\tilde{\mathbf{x}}_{s,2}) \right),
$$

in which $\mathcal{I}_{k,s}$ is a mini-batch of indices randomly chosen from $[n]$ with a size of b.

Now, we give the convergence result. The main property of Acc-SADMM (Algorithm 5.7) in the inner loop is shown below.

Lemma 5.7 *For Algorithm 5.6, in any epoch with fixed s (for simplicity we drop the subscript s throughout the proof unless necessary), we have*

$$
\begin{aligned}
& \mathbb{E}_{i_k} L(\mathbf{x}_1^{k+1}, \mathbf{x}_2^{k+1}, \lambda^*) - \theta_2 L(\tilde{\mathbf{x}}_1, \tilde{\mathbf{x}}_2, \lambda^*) - (1 - \theta_2 - \theta_1) L(\mathbf{x}_1^k, \mathbf{x}_2^k, \lambda^*) \\
& \leq \frac{\theta_1}{2\beta} \left(\left\| \hat{\lambda}^k - \lambda^* \right\|^2 - \mathbb{E}_{i_k} \left\| \hat{\lambda}^{k+1} - \lambda^* \right\|^2 \right) \\
& \quad + \frac{1}{2} \left\| \mathbf{y}_1^k - (1 - \theta_1 - \theta_2)\mathbf{x}_1^k - \theta_2\tilde{\mathbf{x}}_1 - \theta_1\mathbf{x}_1^* \right\|_{\mathbf{G}_1}^2 \\
& \quad - \frac{1}{2} \mathbb{E}_{i_k} \left\| \mathbf{x}_1^{k+1} - (1 - \theta_1 - \theta_2)\mathbf{x}_1^k - \theta_2\tilde{\mathbf{x}}_1 - \theta_1\mathbf{x}_1^* \right\|_{\mathbf{G}_1}^2
\end{aligned}
$$

Algorithm 5.7 Accelerated stochastic alternating direction method of multiplier (Acc-SADMM)

Input: epoch length $m > 2$, β, $\tau = 2$, $c = 2$, $\mathbf{x}_0^0 = \mathbf{0}$, $\tilde{\boldsymbol{\lambda}}_0^0 = \mathbf{0}$, $\tilde{\mathbf{x}}_0 = \mathbf{x}_0^0$, $\mathbf{y}_0^0 = \mathbf{x}_0^0$, $\theta_{1,s} = \frac{1}{c + \tau s}$,

and $\theta_2 = \frac{m - \tau}{\tau(m-1)}$.

for $s = 0$ to $S - 1$ **do**

Do inner loop, as stated in Algorithm 5.6,

Set primal variables: $\mathbf{x}_{s+1}^0 = \mathbf{x}_s^m$,

Update $\tilde{\mathbf{x}}_{s+1}$ by $\tilde{\mathbf{x}}_{s+1} = \frac{1}{m}\left(\left[1 - \frac{(\tau - 1)\theta_{1,s+1}}{\theta_2}\right]\mathbf{x}_s^m + \left[1 + \frac{(\tau - 1)\theta_{1,s+1}}{(m-1)\theta_2}\right]\sum_{k=1}^{m-1}\mathbf{x}_s^k\right)$,

Update dual variable: $\tilde{\boldsymbol{\lambda}}_{s+1}^0 = \boldsymbol{\lambda}_s^{m-1} + \beta(1 - \tau)(\mathbf{A}_1\mathbf{x}_{s,1}^m + \mathbf{A}_2\mathbf{x}_{s,2}^m - \mathbf{b})$,

Update dual snapshot variable: $\tilde{\mathbf{b}}_{s+1} = \mathbf{A}_1\tilde{\mathbf{x}}_{s+1,1} + \mathbf{A}_2\tilde{\mathbf{x}}_{s+1,2}$,

Update extrapolation terms $\mathbf{y}_{s+1}^0$ through

$$\mathbf{y}_{s+1}^0 = (1 - \theta_2)\mathbf{x}_s^m + \theta_2\tilde{\mathbf{x}}_{s+1} + \frac{\theta_{1,s+1}}{\theta_{1,s}}\left[(1 - \theta_{1,s})\mathbf{x}_s^m - (1 - \theta_{1,s} - \theta_2)\mathbf{x}_s^{m-1} - \theta_2\tilde{\mathbf{x}}_s\right].$$

end for s

Output:

$$\hat{\mathbf{x}}_S = \frac{1}{(m-1)(\theta_{1,s} + \theta_2) + 1}\mathbf{x}_S^m + \frac{\theta_{1,s} + \theta_2}{(m-1)(\theta_{1,s} + \theta_2) + 1}\sum_{k=1}^{m-1}\mathbf{x}_S^k. \qquad (5.101)$$

$$+ \frac{1}{2}\left\|\mathbf{y}_2^k - (1 - \theta_1 - \theta_2)\mathbf{x}_2^k - \theta_2\tilde{\mathbf{x}}_2 - \theta_1\mathbf{x}_2^*\right\|_{\mathbf{G}_2}^2$$

$$- \frac{1}{2}\mathbb{E}_{i_k}\left\|\mathbf{x}_2^{k+1} - (1 - \theta_1 - \theta_2)\mathbf{x}_2^k - \theta_2\tilde{\mathbf{x}}_2 - \theta_1\mathbf{x}_2^*\right\|_{\mathbf{G}_2}^2, \qquad (5.102)$$

where $\mathbb{E}_{i_k}$ *denotes that the expectation is taken over the random samples in the mini-batch* $\mathcal{I}_{k,s}$, $L(\mathbf{x}_1, \mathbf{x}_2, \boldsymbol{\lambda}) = F_1(\mathbf{x}_1) + F_2(\mathbf{x}_2) + \langle \boldsymbol{\lambda}, \mathbf{A}_1\mathbf{x}_1 + \mathbf{A}_2\mathbf{x}_2 - \mathbf{b}\rangle$ *is the Lagrangian function,* $\hat{\boldsymbol{\lambda}}^k = \tilde{\boldsymbol{\lambda}}^k + \frac{\beta(1 - \theta_1)}{\theta_1}(\mathbf{A}\mathbf{x}^k - \mathbf{b})$, $\mathbf{G}_1 = \left(L_1 + \frac{\beta\|\mathbf{A}_1^T\mathbf{A}_1\|}{\theta_1}\right)\mathbf{I} - \frac{\beta\mathbf{A}_1^T\mathbf{A}_1}{\theta_1}$, *and*

$\mathbf{G}_2 = \left(\left(1 + \frac{1}{b\theta_2}\right)L_2 + \frac{\beta\|\mathbf{A}_2^T\mathbf{A}_2\|}{\theta_1}\right)\mathbf{I}$. *Other notations can be found in Table 5.1.*

Proof Step 1: We first analyze $\mathbf{x}_1$. By the optimality of $\mathbf{x}_1^{k+1}$ in (5.99) and the convexity of $F_1(\cdot)$, we can obtain

$$F_1(\mathbf{x}_1^{k+1})$$

$$\leq (1 - \theta_1 - \theta_2)F_1(\mathbf{x}_1^k) + \theta_2 F_1(\tilde{\mathbf{x}}_1) + \theta_1 F_1(\mathbf{x}_1^*)$$

$$- \left\langle \mathbf{A}_1^T\bar{\boldsymbol{\lambda}}(\mathbf{x}_1^{k+1}, \mathbf{y}_2^k), \mathbf{x}_1^{k+1} - (1 - \theta_1 - \theta_2)\mathbf{x}_1^k - \theta_2\tilde{\mathbf{x}}_1 - \theta_1\mathbf{x}_1^*\right\rangle$$

$$+ \frac{L_1}{2}\left\|\mathbf{x}_1^{k+1} - \mathbf{y}_1^k\right\|^2$$

$$- \left\langle \mathbf{x}_1^{k+1} - \mathbf{y}_1^k, \mathbf{x}_1^{k+1} - (1 - \theta_1 - \theta_2)\mathbf{x}_1^k - \theta_2\tilde{\mathbf{x}}_1 - \theta_1\mathbf{x}^*\right\rangle_{\mathbf{G}_1}. \qquad (5.103)$$

We prove (5.103) below.

By the same proof technique of Lemma 5.2 for Katyusha (Algorithm 5.3), we can bound the variance through

$$\mathbb{E}_{i_k} \left\| \nabla f_2(\mathbf{y}_2^k) - \tilde{\nabla} f_2(\mathbf{y}_2^k) \right\|^2 \le \frac{2L_2}{b} \left[f_2(\tilde{\mathbf{x}}_2) - f_2(\mathbf{y}_2^k) - \langle \nabla f_2(\mathbf{y}_2^k), \tilde{\mathbf{x}}_2 - \mathbf{y}_2^k \rangle \right].$$
(5.104)

Set

$$\bar{\lambda}(\mathbf{x}_1, \mathbf{x}_2) = \lambda^k + \frac{\beta}{\theta_1} \left(\mathbf{A}_1 \mathbf{x}_1 + \mathbf{A}_2 \mathbf{x}_2 - \mathbf{b} \right).$$

For the optimality solution of $\mathbf{x}_1^{k+1}$ in (5.99), we have

$$\left(L_1 + \frac{\beta \left\| \mathbf{A}_1^T \mathbf{A}_1 \right\|}{\theta_1} \right) \left(\mathbf{x}_1^{k+1} - \mathbf{y}_1^k \right) + \nabla f_1(\mathbf{y}_1^k) + \mathbf{A}_1^T \bar{\lambda}(\mathbf{y}_1^k, \mathbf{y}_2^k) \in -\partial h_1(\mathbf{x}_1^{k+1}).$$
(5.105)

Since f_1 is L_1-smooth, we have

$$f_1(\mathbf{x}_1^{k+1}) \le f_1(\mathbf{y}_1^k) + \left\langle \nabla f_1(\mathbf{y}_1^k), \mathbf{x}_1^{k+1} - \mathbf{y}_1^k \right\rangle + \frac{L_1}{2} \left\| \mathbf{x}_1^{k+1} - \mathbf{y}_1^k \right\|^2$$

$$\overset{a}{\le} f_1(\mathbf{u}_1) + \left\langle \nabla f_1(\mathbf{y}_1^k), \mathbf{x}_1^{k+1} - \mathbf{u}_1 \right\rangle + \frac{L_1}{2} \left\| \mathbf{x}_1^{k+1} - \mathbf{y}_1^k \right\|^2$$

$$\overset{b}{\le} f_1(\mathbf{u}_1) - \left\langle \partial h_1(\mathbf{x}_1^{k+1}), \mathbf{x}_1^{k+1} - \mathbf{u}_1 \right\rangle - \langle \mathbf{A}_1^T \bar{\lambda}(\mathbf{y}_1^k, \mathbf{y}_2^k), \mathbf{x}_1^{k+1} - \mathbf{u}_1 \rangle$$

$$- \left(L_1 + \frac{\beta \left\| \mathbf{A}_1^T \mathbf{A}_1 \right\|}{\theta_1} \right) \left\langle \mathbf{x}_1^{k+1} - \mathbf{y}_1^k, \mathbf{x}_1^{k+1} - \mathbf{u}_1 \right\rangle + \frac{L_1}{2} \left\| \mathbf{x}_1^{k+1} - \mathbf{y}_1^k \right\|^2,$$

where $\mathbf{u}_1$ is an arbitrary variable. In the inequality $\overset{a}{\le}$ we use the fact that $f_1(\cdot)$ is convex and so $f_1(\mathbf{y}_1^k) \le f_1(\mathbf{u}_1) + \langle \nabla f_1(\mathbf{y}_1^k), \mathbf{y}_1^k - \mathbf{u}_1 \rangle$. The inequality $\overset{b}{\le}$ uses (5.105). Then the convexity of $h_1(\cdot)$ gives $h_1(\mathbf{x}_1^{k+1}) \le h_1(\mathbf{u}_1) + \langle \partial h_1(\mathbf{x}_1^{k+1}), \mathbf{x}_1^{k+1} - \mathbf{u}_1 \rangle$. So we have

$$F_1(\mathbf{x}_1^{k+1}) \le F_1(\mathbf{u}_1) - \left\langle \mathbf{A}_1^T \bar{\lambda}(\mathbf{y}_1^k, \mathbf{y}_2^k), \mathbf{x}_1^{k+1} - \mathbf{u}_1 \right\rangle + \frac{L_1}{2} \left\| \mathbf{x}_1^{k+1} - \mathbf{y}_1^k \right\|^2$$

$$- \left(L_1 + \frac{\beta \left\| \mathbf{A}_1^T \mathbf{A}_1 \right\|}{\theta_1} \right) \left\langle \mathbf{x}_1^{k+1} - \mathbf{y}_1^k, \mathbf{x}_1^{k+1} - \mathbf{u}_1 \right\rangle.$$

Setting $\mathbf{u}_1$ be $\mathbf{x}_1^k$, $\tilde{\mathbf{x}}_1$, and $\mathbf{x}_1^*$, respectively, then multiplying the three inequalities by $(1 - \theta_1 - \theta_2)$, θ_2, and θ_1, respectively, and adding them, we have

$$F_1(\mathbf{x}_1^{k+1})$$

$$\leq (1 - \theta_1 - \theta_2) F_1(\mathbf{x}_1^k) + \theta_2 F_1(\tilde{\mathbf{x}}_1) + \theta_1 F_1(\mathbf{x}_1^*) + \frac{L_1}{2} \left\| \mathbf{x}_1^{k+1} - \mathbf{y}_1^k \right\|^2$$

$$- \left\langle \mathbf{A}_1^T \bar{\boldsymbol{\lambda}}(\mathbf{y}_1^k, \mathbf{y}_2^k), \mathbf{x}_1^{k+1} - (1 - \theta_1 - \theta_2)\mathbf{x}_1^k - \theta_2 \tilde{\mathbf{x}}_1 - \theta_1 \mathbf{x}_1^* \right\rangle$$

$$- \left(L_1 + \frac{\beta \left\| \mathbf{A}_1^T \mathbf{A}_1 \right\|}{\theta_1} \right) \left\langle \mathbf{x}_1^{k+1} - \mathbf{y}_1^k, \mathbf{x}_1^{k+1} - (1 - \theta_1 - \theta_2)\mathbf{x}_1^k - \theta_2 \tilde{\mathbf{x}}_1 - \theta_1 \mathbf{x}_1^* \right\rangle$$

$$\overset{a}{=} (1 - \theta_1 - \theta_2) F_1(\mathbf{x}_1^k) + \theta_2 F_1(\tilde{\mathbf{x}}_1) + \theta_1 F_1(\mathbf{x}_1^*) + \frac{L_1}{2} \left\| \mathbf{x}_1^{k+1} - \mathbf{y}_1^k \right\|^2$$

$$- \left\langle \mathbf{A}_1^T \bar{\boldsymbol{\lambda}}(\mathbf{x}_1^{k+1}, \mathbf{y}_2^k), \mathbf{x}_1^{k+1} - (1 - \theta_1 - \theta_2)\mathbf{x}_1^k - \theta_2 \tilde{\mathbf{x}}_1 - \theta_1 \mathbf{x}_1^* \right\rangle$$

$$- \left\langle \mathbf{x}_1^{k+1} - \mathbf{y}_1^k, \mathbf{x}_1^{k+1} - (1 - \theta_1 - \theta_2)\mathbf{x}_1^k - \theta_2 \tilde{\mathbf{x}}_1 - \theta_1 \mathbf{x}^* \right\rangle_{\mathbf{G}_1}, \tag{5.106}$$

where in the equality $\overset{a}{\leq}$ we replace $\mathbf{A}_1^T \bar{\boldsymbol{\lambda}}(\mathbf{y}_1^k, \mathbf{y}_2^k)$ with $\mathbf{A}_1^T \bar{\boldsymbol{\lambda}}(\mathbf{x}_1^{k+1}, \mathbf{y}_2^k) - \frac{\beta \mathbf{A}_1^T \mathbf{A}_1}{\theta_1}(\mathbf{x}_1^{k+1} - \mathbf{y}_1^k)$.

Step 2: We next analyze $\mathbf{x}_2$. By the optimality of $\mathbf{x}_2^{k+1}$ in (5.100) and the convexity of $F_2(\cdot)$, we can obtain

$$\mathbb{E}_{i_k} F_2(\mathbf{x}_2^{k+1})$$

$$\leq -\mathbb{E}_{i_k} \left\langle \mathbf{A}_2^T \bar{\boldsymbol{\lambda}}(\mathbf{x}_1^{k+1}, \mathbf{y}_2^k) + \left(\alpha L_2 + \frac{\beta \left\| \mathbf{A}_2^T \mathbf{A}_2 \right\|}{\theta_1} \right) (\mathbf{x}_2^{k+1} - \mathbf{y}_2^k), \right.$$

$$\left. \mathbf{x}_2^{k+1} - \theta_2 \tilde{\mathbf{x}}_2 \right\rangle$$

$$- \mathbb{E}_{i_k} \left\langle \mathbf{A}_2^T \bar{\boldsymbol{\lambda}}(\mathbf{x}_1^{k+1}, \mathbf{y}_2^k) + \left(\alpha L_2 + \frac{\beta \left\| \mathbf{A}_2^T \mathbf{A}_2 \right\|}{\theta_1} \right) (\mathbf{x}_2^{k+1} - \mathbf{y}_2^k), \right.$$

$$\left. -(1 - \theta_2 - \theta_1)\mathbf{x}_2^k - \theta_1 \mathbf{x}_2^* \right\rangle$$

$$+ (1 - \theta_2 - \theta_1) F_2(\mathbf{x}_2^k) + \theta_1 F_2(\mathbf{x}_2^*) + \theta_2 F_2(\tilde{\mathbf{x}}_2)$$

$$+ \mathbb{E}_{i_k} \left(\frac{\left(1 + \frac{1}{b\theta_2} \right) L_2}{2} \left\| \mathbf{x}_2^{k+1} - \mathbf{y}_2^k \right\|^2 \right). \tag{5.107}$$

We prove (5.107) below.

For the optimality of $\mathbf{x}_2^{k+1}$ in (5.100), we have

$$\left(\alpha L_2 + \frac{\beta \left\|\mathbf{A}_2^T \mathbf{A}_2\right\|}{\theta_1}\right)\left(\mathbf{x}_2^{k+1} - \mathbf{y}_2^k\right) + \tilde{\nabla} f_2(\mathbf{y}_2^k) + \mathbf{A}_2^T \bar{\lambda}(\mathbf{x}_1^{k+1}, \mathbf{y}_2^k)$$

$$\in -\partial h_2(\mathbf{x}_2^{k+1}), \tag{5.108}$$

where we set $\alpha = 1 + \frac{1}{b\theta_2}$. Since f_2 is L_2-smooth, we have

$$f_2(\mathbf{x}_2^{k+1}) \le f_2(\mathbf{y}_2^k) + \left\langle \nabla f_2(\mathbf{y}_2^k), \mathbf{x}_2^{k+1} - \mathbf{y}_2^k \right\rangle + \frac{L_2}{2}\left\|\mathbf{x}_2^{k+1} - \mathbf{y}_2^k\right\|^2. \tag{5.109}$$

We first consider $\langle \nabla f_2(\mathbf{y}_2^k), \mathbf{x}_2^{k+1} - \mathbf{y}_2^k \rangle$ and have

$$\left\langle \nabla f_2(\mathbf{y}_2^k), \mathbf{x}_2^{k+1} - \mathbf{y}_2^k \right\rangle$$

$$\overset{a}{=} \left\langle \nabla f_2(\mathbf{y}_2^k), \mathbf{u}_2 - \mathbf{y}_2^k + \mathbf{x}_2^{k+1} - \mathbf{u}_2 \right\rangle$$

$$\overset{b}{=} \left\langle \nabla f_2(\mathbf{y}_2^k), \mathbf{u}_2 - \mathbf{y}_2^k \right\rangle - \theta_3 \left\langle \nabla f_2(\mathbf{y}_2^k), \mathbf{y}_2^k - \tilde{\mathbf{x}}_2 \right\rangle + \left\langle \nabla f_2(\mathbf{y}_2^k), \mathbf{z}^{k+1} - \mathbf{u}_2 \right\rangle$$

$$= \left\langle \nabla f_2(\mathbf{y}_2^k), \mathbf{u}_2 - \mathbf{y}_2^k \right\rangle - \theta_3 \left\langle \nabla f_2(\mathbf{y}_2^k), \mathbf{y}_2^k - \tilde{\mathbf{x}}_2 \right\rangle$$

$$+ \left\langle \tilde{\nabla} f_2(\mathbf{y}_2^k), \mathbf{z}^{k+1} - \mathbf{u}_2 \right\rangle + \left\langle \nabla f_2(\mathbf{y}_2^k) - \tilde{\nabla} f_2(\mathbf{y}_2^k), \mathbf{z}^{k+1} - \mathbf{u}_2 \right\rangle, \tag{5.110}$$

where in the equality $\overset{a}{=}$ we introduce an arbitrary variable $\mathbf{u}_2$ (we will set it to be $\mathbf{x}_2^k$, $\tilde{\mathbf{x}}_2$, and $\mathbf{x}_2^*$) and in the equality $\overset{b}{=}$ we set

$$\mathbf{z}^{k+1} = \mathbf{x}_2^{k+1} + \theta_3(\mathbf{y}_2^k - \tilde{\mathbf{x}}_2), \tag{5.111}$$

in which θ_3 is an absolute constant determined later. For $\langle \tilde{\nabla} f_2(\mathbf{y}_2^k), \mathbf{z}^{k+1} - \mathbf{u}_2 \rangle$, we have

$$\left\langle \tilde{\nabla} f_2(\mathbf{y}_2^k), \mathbf{z}^{k+1} - \mathbf{u}_2 \right\rangle$$

$$\overset{a}{=} -\left\langle \partial h_2(\mathbf{x}_2^{k+1}) + \mathbf{A}_2^T \bar{\lambda}(\mathbf{x}_1^{k+1}, \mathbf{y}_2^k) + \left(\alpha L_2 + \frac{\beta \left\|\mathbf{A}_2^T \mathbf{A}_2\right\|}{\theta_1}\right)\right.$$

$$\left.\left(\mathbf{x}_2^{k+1} - \mathbf{y}_2^k\right), \mathbf{z}^{k+1} - \mathbf{u}_2 \right\rangle$$

$$\overset{b}{=} -\left\langle \partial h_2(\mathbf{x}_2^{k+1}), \mathbf{x}_2^{k+1} + \theta_3(\mathbf{y}_2^k - \tilde{\mathbf{x}}_2) - \mathbf{u}_2 \right\rangle$$

$$-\left\langle \mathbf{A}_2^T \bar{\lambda}(\mathbf{x}_1^{k+1}, \mathbf{y}_2^k) + \left(\alpha L_2 + \frac{\beta \|\mathbf{A}_2^T \mathbf{A}_2\|}{\theta_1}\right)(\mathbf{x}_2^{k+1} - \mathbf{y}_2^k), \mathbf{z}^{k+1} - \mathbf{u}_2\right\rangle$$

$$= -\left\langle \partial h_2(\mathbf{x}_2^{k+1}), \mathbf{x}_2^{k+1} + \theta_3(\mathbf{y}_2^k - \mathbf{x}_2^{k+1} + \mathbf{x}_2^{k+1} - \tilde{\mathbf{x}}_2) - \mathbf{u}_2\right\rangle$$

$$-\left\langle \mathbf{A}_2^T \bar{\lambda}(\mathbf{x}_1^{k+1}, \mathbf{y}_2^k) + \left(\alpha L_2 + \frac{\beta \|\mathbf{A}_2^T \mathbf{A}_2\|}{\theta_1}\right)(\mathbf{x}_2^{k+1} - \mathbf{y}_2^k), \mathbf{z}^{k+1} - \mathbf{u}_2\right\rangle$$

$$\overset{c}{\leq} h_2(\mathbf{u}_2) - h_2(\mathbf{x}_2^{k+1}) + \theta_3 h_2(\tilde{\mathbf{x}}_2) - \theta_3 h_2(\mathbf{x}_2^{k+1}) - \theta_3\left\langle \partial h_2(\mathbf{x}_2^{k+1}), \mathbf{y}_2^k - \mathbf{x}_2^{k+1}\right\rangle$$

$$-\left\langle \mathbf{A}_2^T \bar{\lambda}(\mathbf{x}_1^{k+1}, \mathbf{y}_2^k) + \left(\alpha L_2 + \frac{\beta \|\mathbf{A}_2^T \mathbf{A}_2\|}{\theta_1}\right)(\mathbf{x}_2^{k+1} - \mathbf{y}_2^k), \mathbf{z}^{k+1} - \mathbf{u}_2\right\rangle$$

$$\overset{d}{=} h_2(\mathbf{u}_2) - h_2(\mathbf{x}_2^{k+1}) + \theta_3 h_2(\tilde{\mathbf{x}}_2) - \theta_3 h_2(\mathbf{x}_2^{k+1})$$

$$-\left\langle \mathbf{A}_2^T \bar{\lambda}(\mathbf{x}_1^{k+1}, \mathbf{y}_2^k) + \left(\alpha L_2 + \frac{\beta \|\mathbf{A}_2^T \mathbf{A}_2\|}{\theta_1}\right)(\mathbf{x}_2^{k+1} - \mathbf{y}_2^k), \mathbf{z}^{k+1} - \mathbf{u}_2\right\rangle$$

$$-\theta_3\left\langle \mathbf{A}_2^T \bar{\lambda}(\mathbf{x}_1^{k+1}, \mathbf{y}_2^k) + \left(\alpha L_2 + \frac{\beta \|\mathbf{A}_2^T \mathbf{A}_2\|}{\theta_1}\right)(\mathbf{x}_2^{k+1} - \mathbf{y}_2^k)\right.$$

$$\left. + \tilde{\nabla} f_2(\mathbf{y}_2^k), \mathbf{x}_2^{k+1} - \mathbf{y}_2^k\right\rangle, \tag{5.112}$$

where in the equalities $\overset{a}{=}$ and $\overset{b}{=}$, we use (5.108) and (5.111), respectively. The inequality $\overset{d}{=}$ uses (5.108) again. The inequality $\overset{c}{\leq}$ uses the convexity of h_2:

$$\left\langle \partial h_2(\mathbf{x}_2^{k+1}), \mathbf{w} - \mathbf{x}_2^{k+1}\right\rangle \leq h_2(\mathbf{w}) - h_2(\mathbf{x}_2^{k+1}), \quad \mathbf{w} = \mathbf{u}_2, \tilde{\mathbf{x}}_2.$$

Rearranging terms in (5.112) and using $\tilde{\nabla} f_2(\mathbf{y}_2^k) = \nabla f_2(\mathbf{y}_2^k) + \left(\tilde{\nabla} f_2(\mathbf{y}_2^k) - \nabla f_2(\mathbf{y}_2^k)\right)$, we have

$$\left\langle \tilde{\nabla} f_2(\mathbf{y}_2^k), \mathbf{z}^{k+1} - \mathbf{u}_2\right\rangle$$

$$= h_2(\mathbf{u}_2) - h_2(\mathbf{x}_2^{k+1}) + \theta_3 h_2(\tilde{\mathbf{x}}_2) - \theta_3 h_2(\mathbf{x}_2^{k+1})$$

$$-\left\langle \mathbf{A}_2^T \bar{\lambda}(\mathbf{x}_1^{k+1}, \mathbf{y}_2^k) + \left(\alpha L_2 + \frac{\beta \|\mathbf{A}_2^T \mathbf{A}_2\|}{\theta_1}\right)(\mathbf{x}_2^{k+1} - \mathbf{y}_2^k),\right.$$

$$\left. \theta_3(\mathbf{x}_2^{k+1} - \mathbf{y}_2^k) + \mathbf{z}^{k+1} - \mathbf{u}_2\right\rangle$$

$$-\theta_3\left\langle \nabla f_2(\mathbf{y}_2^k) + \left(\tilde{\nabla} f_2(\mathbf{y}_2^k) - \nabla f_2(\mathbf{y}_2^k)\right), \mathbf{x}_2^{k+1} - \mathbf{y}_2^k\right\rangle. \tag{5.113}$$

Substituting (5.113) in (5.110), we obtain

$$
(1 + \theta_3) \left\langle \nabla f_2(\mathbf{y}_2^k), \mathbf{x}_2^{k+1} - \mathbf{y}_2^k \right\rangle
$$

$$
= \left\langle \nabla f_2(\mathbf{y}_2^k), \mathbf{u}_2 - \mathbf{y}_2^k \right\rangle - \theta_3 \left\langle \nabla f_2(\mathbf{y}_2^k), \mathbf{y}_2^k - \tilde{\mathbf{x}}_2 \right\rangle + h_2(\mathbf{u}_2) - h_2(\mathbf{x}_2^{k+1})
$$

$$
+ \theta_3 h_2(\tilde{\mathbf{x}}_2) - \theta_3 h_2(\mathbf{x}_2^{k+1})
$$

$$
- \left\langle \mathbf{A}_2^T \bar{\lambda}(\mathbf{x}_1^{k+1}, \mathbf{y}_2^k) + \left(\alpha L_2 + \frac{\beta \left\| \mathbf{A}_2^T \mathbf{A}_2 \right\|}{\theta_1} \right) \left(\mathbf{x}_2^{k+1} - \mathbf{y}_2^k \right), \right.
$$

$$
\left. \mathbf{z}^{k+1} - \mathbf{u}_2 + \theta_3(\mathbf{x}_2^{k+1} - \mathbf{y}_2^k) \right\rangle
$$

$$
+ \left\langle \nabla f_2(\mathbf{y}_2^k) - \tilde{\nabla} f_2(\mathbf{y}_2^k), \theta_3(\mathbf{x}_2^{k+1} - \mathbf{y}_2^k) + \mathbf{z}^{k+1} - \mathbf{u}_2 \right\rangle. \tag{5.114}
$$

Multiplying (5.109) by $(1 + \theta_3)$ and then adding (5.114), we can eliminate the term $\langle \nabla f_2(\mathbf{y}_2^k), \mathbf{x}_2^{k+1} - \mathbf{y}_2^k \rangle$ and obtain

$$
(1 + \theta_3) F_2(\mathbf{x}_2^{k+1})
$$

$$
\leq (1 + \theta_3) f_2(\mathbf{y}_2^k) + \left\langle \nabla f_2(\mathbf{y}_2^k), \mathbf{u}_2 - \mathbf{y}_2^k \right\rangle - \theta_3 \langle \nabla f_2(\mathbf{y}_2^k), \mathbf{y}_2^k - \tilde{\mathbf{x}}_2 \rangle
$$

$$
+ h_2(\mathbf{u}_2) + \theta_3 h_2(\tilde{\mathbf{x}}_2)
$$

$$
- \left\langle \mathbf{A}_2^T \bar{\lambda}(\mathbf{x}_1^{k+1}, \mathbf{y}_2^k) + \left(\alpha L_2 + \frac{\beta \left\| \mathbf{A}_2^T \mathbf{A}_2 \right\|}{\theta_1} \right) \left(\mathbf{x}_2^{k+1} - \mathbf{y}_2^k \right), \right.
$$

$$
\left. \mathbf{z}^{k+1} - \mathbf{u}_2 + \theta_3(\mathbf{x}_2^{k+1} - \mathbf{y}_2^k) \right\rangle
$$

$$
+ \left\langle \nabla f_2(\mathbf{y}_2^k) - \tilde{\nabla} f_2(\mathbf{y}_2^k), \theta_3(\mathbf{x}_2^{k+1} - \mathbf{y}_2^k) + \mathbf{z}^{k+1} - \mathbf{u}_2 \right\rangle
$$

$$
+ \frac{(1 + \theta_3) L_2}{2} \left\| \mathbf{x}_2^{k+1} - \mathbf{y}_2^k \right\|^2
$$

$$
\overset{a}{\leq} F_2(\mathbf{u}_2) - \theta_3 \left\langle \nabla f(\mathbf{y}_2^k), \mathbf{y}_2^k - \tilde{\mathbf{x}}_2 \right\rangle + \theta_3 f_2(\mathbf{y}_2^k) + \theta_3 h_2(\tilde{\mathbf{x}}_2)
$$

$$
- \left\langle \mathbf{A}_2^T \bar{\lambda}(\mathbf{x}_1^{k+1}, \mathbf{y}_2^k) + \left(\alpha L_2 + \frac{\beta \left\| \mathbf{A}_2^T \mathbf{A}_2 \right\|}{\theta_1} \right) \left(\mathbf{x}_2^{k+1} - \mathbf{y}_2^k \right), \right.
$$

$$
\left. \mathbf{z}^{k+1} - \mathbf{u}_2 + \theta_3(\mathbf{x}_2^{k+1} - \mathbf{y}_2^k) \right\rangle
$$

$$+\left\langle \nabla f(\mathbf{y}_2^k) - \tilde{\nabla} f_2(\mathbf{y}_2^k), \theta_3(\mathbf{x}_2^{k+1} - \mathbf{y}_2^k) + \mathbf{z}^{k+1} - \mathbf{u}_2 \right\rangle$$

$$+\frac{(1+\theta_3)L_2}{2}\left\| \mathbf{x}_2^{k+1} - \mathbf{y}_2^k \right\|^2, \tag{5.115}$$

where the inequality $\overset{a}{\leq}$ uses the convexity of f_2: $\langle \nabla f_2(\mathbf{y}_2^k), \mathbf{u}_2 - \mathbf{y}_2^k \rangle \leq f_2(\mathbf{u}_2) - f_2(\mathbf{y}_2^k)$.

We now consider the term $\left\langle \nabla f_2(\mathbf{y}_2^k) - \tilde{\nabla} f_2(\mathbf{y}_2^k), \theta_3(\mathbf{x}_2^{k+1} - \mathbf{y}_2^k) + \mathbf{z}^{k+1} - \mathbf{u}_2 \right\rangle$. We will set $\mathbf{u}_2$ to be $\mathbf{x}_2^k$ and $\mathbf{x}_2^*$, which do not depend on $\mathcal{I}_{k,s}$. So we obtain

$$\mathbb{E}_{i_k}\left\langle \nabla f_2(\mathbf{y}_2^k) - \tilde{\nabla} f_2(\mathbf{y}^k), \theta_3(\mathbf{x}_2^{k+1} - \mathbf{y}_2^k) + \mathbf{z}^{k+1} - \mathbf{u}_2 \right\rangle$$

$$= \mathbb{E}_{i_k}\left\langle \nabla f_2(\mathbf{y}_2^k) - \tilde{\nabla} f_2(\mathbf{y}_2^k), \theta_3\mathbf{z}^{k+1} + \mathbf{z}^{k+1} \right\rangle$$

$$- \mathbb{E}_{i_k}\left\langle \nabla f_2(\mathbf{y}_2^k) - \tilde{\nabla} f_2(\mathbf{y}_2^k), \theta_3^2(\mathbf{y}_2^k - \tilde{\mathbf{x}}_2) + \theta_3\mathbf{y}_2^k + \mathbf{u}_2 \right\rangle$$

$$\overset{a}{=} (1+\theta_3)\mathbb{E}_{i_k}\langle \nabla f_2(\mathbf{y}_2^k) - \tilde{\nabla} f_2(\mathbf{y}_2^k), \mathbf{z}^{k+1} \rangle$$

$$\overset{b}{=} (1+\theta_3)\mathbb{E}_{i_k}\langle \nabla f_2(\mathbf{y}_2^k) - \tilde{\nabla} f_2(\mathbf{y}_2^k), \mathbf{x}_2^{k+1} \rangle$$

$$\overset{c}{=} (1+\theta_3)\mathbb{E}_{i_k}\langle \nabla f_2(\mathbf{y}_2^k) - \tilde{\nabla} f_2(\mathbf{y}_2^k), \mathbf{x}_2^{k+1} - \mathbf{y}_2^k \rangle$$

$$\overset{d}{\leq} \mathbb{E}_{i_k}\left(\frac{\theta_3 b}{2L_2}\left\| \nabla f_2(\mathbf{y}_2^k) - \tilde{\nabla} f_2(\mathbf{y}_2^k) \right\|^2 \right) + \mathbb{E}_{i_k}\left(\frac{(1+\theta_3)^2 L_2}{2\theta_3 b}\left\| \mathbf{x}_2^{k+1} - \mathbf{y}_2^k \right\|^2 \right)$$

$$\overset{e}{\leq} \theta_3\left(f_2(\tilde{\mathbf{x}}_2) - f_2(\mathbf{y}_2^k) - \langle \nabla f_2(\mathbf{y}_2^k), \tilde{\mathbf{x}}_2 - \mathbf{y}_2^k \rangle \right)$$

$$+ \mathbb{E}_{i_k}\left(\frac{(1+\theta_3)^2 L_2}{2\theta_3 b}\left\| \mathbf{x}_2^{k+1} - \mathbf{y}_2^k \right\|^2 \right), \tag{5.116}$$

where in the equality $\overset{a}{=}$ we use the fact that

$$\mathbb{E}_{i_k}\left(\nabla f_2(\mathbf{y}_2^k) - \tilde{\nabla} f_2(\mathbf{y}_2^k) \right) = \mathbf{0},$$

and $\mathbf{x}_2^k$, $\mathbf{y}_2^k$, $\tilde{\mathbf{x}}_2$, and $\mathbf{u}_2$ are independent of $i_{k,s}$ (they are known), so

$$\mathbb{E}_{i_k}\langle \nabla f_2(\mathbf{y}_2^k) - \tilde{\nabla} f_2(\mathbf{y}_2^k), \mathbf{y}_2^k \rangle = 0,$$

$$\mathbb{E}_{i_k}\langle \nabla f_2(\mathbf{y}_2^k) - \tilde{\nabla} f_2(\mathbf{y}_2^k), \tilde{\mathbf{x}}_2 \rangle = 0,$$

$$\mathbb{E}_{i_k}\langle \nabla f_2(\mathbf{y}_2^k) - \tilde{\nabla} f_2(\mathbf{y}_2^k), \mathbf{u}_2 \rangle = 0;$$

the equalities $\overset{b}{=}$ and $\overset{c}{=}$ hold similarly; the inequality $\overset{d}{\leq}$ uses the Cauchy–Schwartz inequality; and $\overset{e}{\leq}$ uses (5.104). Taking expectation on (5.115) and adding (5.116), we obtain

$$(1 + \theta_3)\mathbb{E}_{i_k} F_2(\mathbf{x}_2^{k+1})$$

$$\leq -\mathbb{E}_{i_k} \left\langle \mathbf{A}_2^T \bar{\lambda}(\mathbf{x}_1^{k+1}, \mathbf{y}_2^k) + \left(\alpha L_2 + \frac{\beta \|\mathbf{A}_2^T \mathbf{A}_2\|}{\theta_1} \right) \left(\mathbf{x}_2^{k+1} - \mathbf{y}_2^k \right), \right.$$

$$\left. \mathbf{z}^{k+1} - \mathbf{u}_2 + \theta_3(\mathbf{x}_2^{k+1} - \mathbf{y}_2^k) \right\rangle$$

$$+ F_2(\mathbf{u}_2) + \theta_3 F(\tilde{\mathbf{x}}_2) + \mathbb{E}_{i_k} \left(\frac{(1 + \theta_3)\left(1 + \frac{1+\theta_3}{b\theta_3}\right) L_2}{2} \left\| \mathbf{x}_2^{k+1} - \mathbf{y}_2^k \right\|^2 \right)$$

$$\overset{a}{=} -\mathbb{E}_{i_k} \left\langle \mathbf{A}_2^T \bar{\lambda}(\mathbf{x}_1^{k+1}, \mathbf{y}_2^k) + \left(\alpha L_2 + \frac{\beta \|\mathbf{A}_2^T \mathbf{A}_2\|}{\theta_1} \right) \left(\mathbf{x}_2^{k+1} - \mathbf{y}_2^k \right), \right.$$

$$\left. (1 + \theta_3)\mathbf{x}_2^{k+1} - \theta_3 \tilde{\mathbf{x}}_2 - \mathbf{u}_2 \right\rangle$$

$$+ F_2(\mathbf{u}_2) + \theta_3 F(\tilde{\mathbf{x}}_2) + \mathbb{E}_{i_k} \left(\frac{(1 + \theta_3)\left(1 + \frac{1}{b\theta_2}\right) L_2}{2} \left\| \mathbf{x}_2^{k+1} - \mathbf{y}_2^k \right\|^2 \right),$$

where in equality $\overset{a}{=}$ we use (5.111) and set θ_3 satisfying $\theta_2 = \frac{\theta_3}{1+\theta_3}$. Setting $\mathbf{u}_2$ to be $\mathbf{x}_2^k$ and $\mathbf{x}_2^*$, respectively, then multiplying the two inequalities by $1 - \theta_1(1 + \theta_3)$ and $\theta_1(1 + \theta_3)$, respectively, and adding them, we obtain

$$(1 + \theta_3)\mathbb{E}_{i_k} F_2(\mathbf{x}_2^{k+1})$$

$$\leq -\mathbb{E}_{i_k} \left\langle \mathbf{A}_2^T \bar{\lambda}(\mathbf{x}_1^{k+1}, \mathbf{y}_2^k) + \left(\alpha L_2 + \frac{\beta \|\mathbf{A}_2^T \mathbf{A}_2\|}{\theta_1} \right) \left(\mathbf{x}_2^{k+1} - \mathbf{y}_2^k \right), \right.$$

$$\left. (1 + \theta_3)\mathbf{x}_2^{k+1} - \theta_3 \tilde{\mathbf{x}}_2 \right\rangle$$

$$- \mathbb{E}_{i_k} \left\langle \mathbf{A}_2^T \bar{\lambda}(\mathbf{x}_1^{k+1}, \mathbf{y}_2^k) + \left(\alpha L_2 + \frac{\beta \|\mathbf{A}_2^T \mathbf{A}_2\|}{\theta_1} \right) \left(\mathbf{x}_2^{k+1} - \mathbf{y}_2^k \right), \right.$$

$$\left. - [1 - \theta_1(1 + \theta_3)] \mathbf{x}_2^k \right\rangle$$

$$- \mathbb{E}_{i_k} \left\langle \mathbf{A}_2^T \bar{\lambda}(\mathbf{x}_1^{k+1}, \mathbf{y}_2^k) + \left(\alpha L_2 + \frac{\beta \|\mathbf{A}_2^T \mathbf{A}_2\|}{\theta_1} \right) \left(\mathbf{x}_2^{k+1} - \mathbf{y}_2^k \right), \right.$$

$$-\theta_1(1+\theta_3)\mathbf{x}_2^* \Big\rangle$$

$$+\left[1 - \theta_1(1+\theta_3)\right] F_2(\mathbf{x}_2^k) + \theta_1(1+\theta_3) F_2(\mathbf{x}_2^*) + \theta_3 F(\tilde{\mathbf{x}}_2)$$

$$+\mathbb{E}_{i_k}\left(\frac{(1+\theta_3)\left(1+\frac{1}{b\theta_2}\right) L_2}{2}\left\|\mathbf{x}_2^{k+1} - \mathbf{y}_2^k\right\|^2\right). \tag{5.117}$$

Dividing (5.117) by $(1+\theta_3)$, we obtain

$$\mathbb{E}_{i_k} F_2(\mathbf{x}_2^{k+1})$$

$$\leq -\mathbb{E}_{i_k}\left\langle \mathbf{A}_2^T \bar{\lambda}(\mathbf{x}_1^{k+1}, \mathbf{y}_2^k) + \left(\alpha L_2 + \frac{\beta \left\|\mathbf{A}_2^T \mathbf{A}_2\right\|}{\theta_1}\right)\left(\mathbf{x}_2^{k+1} - \mathbf{y}_2^k\right), \mathbf{x}_2^{k+1} - \theta_2 \tilde{\mathbf{x}}_2 \right\rangle$$

$$-\mathbb{E}_{i_k}\left\langle \mathbf{A}_2^T \bar{\lambda}(\mathbf{x}_1^{k+1}, \mathbf{y}_2^k) + \left(\alpha L_2 + \frac{\beta \left\|\mathbf{A}_2^T \mathbf{A}_2\right\|}{\theta_1}\right)\left(\mathbf{x}_2^{k+1} - \mathbf{y}_2^k\right),\right.$$

$$\left.-(1 - \theta_2 - \theta_1)\mathbf{x}_2^k - \theta_1 \mathbf{x}_2^* \right\rangle$$

$$+(1 - \theta_2 - \theta_1) F_2(\mathbf{x}_2^k) + \theta_1 F_2(\mathbf{x}_2^*) + \theta_2 F_2(\tilde{\mathbf{x}}_2)$$

$$+\mathbb{E}_{i_k}\left(\frac{\left(1+\frac{1}{b\theta_2}\right) L_2}{2}\left\|\mathbf{x}_2^{k+1} - \mathbf{y}_2^k\right\|^2\right), \tag{5.118}$$

where we use $\theta_2 = \frac{\theta_3}{1+\theta_3}$ and so $\frac{1-\theta_1(1+\theta_3)}{1+\theta_3} = 1 - \theta_2 - \theta_1$. We obtain (5.107).

Step 3: Setting

$$\hat{\lambda}^k = \tilde{\lambda}^k + \frac{\beta(1-\theta_1)}{\theta_1}(\mathbf{A}_1 \mathbf{x}_1^k + \mathbf{A}_2 \mathbf{x}_2^k - \mathbf{b}), \tag{5.119}$$

we prove that it has the following properties:

$$\hat{\lambda}^{k+1} = \bar{\lambda}(\mathbf{x}_1^{k+1}, \mathbf{x}_2^{k+1}), \tag{5.120}$$

$$\hat{\lambda}^{k+1} - \hat{\lambda}^k = \frac{\beta}{\theta_1}\mathbf{A}_1\left[\mathbf{x}_1^{k+1} - (1 - \theta_1 - \theta_2)\mathbf{x}_1^k - \theta_2 \tilde{\mathbf{x}}_1 - \theta_1 \mathbf{x}_1^*\right]$$

$$+\frac{\beta}{\theta_1}\mathbf{A}_2\left[\mathbf{x}_2^{k+1} - (1 - \theta_1 - \theta_2)\mathbf{x}_2^k - \theta_2 \tilde{\mathbf{x}}_2 - \theta_1 \mathbf{x}_2^*\right], \tag{5.121}$$

$$\hat{\lambda}_s^0 = \hat{\lambda}_{s-1}^m, \quad s \geq 1. \tag{5.122}$$

Indeed, for Algorithm 5.6 we have

$$\boldsymbol{\lambda}^k = \tilde{\boldsymbol{\lambda}}^k + \frac{\beta\theta_2}{\theta_1}\left(\mathbf{A}_1\mathbf{x}_1^k + \mathbf{A}_2\mathbf{x}_2^k - \tilde{\mathbf{b}}\right) \tag{5.123}$$

and

$$\tilde{\boldsymbol{\lambda}}^{k+1} = \boldsymbol{\lambda}^k + \beta\left(\mathbf{A}_1\mathbf{x}_1^{k+1} + \mathbf{A}_2\mathbf{x}_2^{k+1} - \mathbf{b}\right). \tag{5.124}$$

With (5.119) we have

$$
\begin{aligned}
\hat{\boldsymbol{\lambda}}^{k+1} &= \tilde{\boldsymbol{\lambda}}^{k+1} + \beta\left(\frac{1}{\theta_1} - 1\right)(\mathbf{A}_1\mathbf{x}_1^{k+1} + \mathbf{A}_2\mathbf{x}_2^{k+1} - \mathbf{b}) \\
&\overset{a}{=} \boldsymbol{\lambda}^k + \frac{\beta}{\theta_1}(\mathbf{A}_1\mathbf{x}_1^{k+1} + \mathbf{A}_2\mathbf{x}_2^{k+1} - \mathbf{b}) \\
&\overset{b}{=} \tilde{\boldsymbol{\lambda}}^k + \frac{\beta}{\theta_1}\left\{\mathbf{A}_1\mathbf{x}_1^{k+1} + \mathbf{A}_2\mathbf{x}_2^{k+1} - \mathbf{b} + \theta_2\left[\mathbf{A}_1(\mathbf{x}_2^k - \tilde{\mathbf{x}}_1) + \mathbf{A}_2(\mathbf{x}_2^k - \tilde{\mathbf{x}}_2)\right]\right\},
\end{aligned}
\tag{5.125}
$$

where in equality $\overset{a}{=}$ we use (5.124) and the equality $\overset{b}{=}$ is obtained by (5.123) and $\tilde{\mathbf{b}} = \mathbf{A}_1\tilde{\mathbf{x}}_1 + \mathbf{A}_2\tilde{\mathbf{x}}_2$ (see Algorithm 5.7). Together with (5.119) we obtain

$$
\begin{aligned}
\hat{\boldsymbol{\lambda}}^{k+1} - \hat{\boldsymbol{\lambda}}^k &= \frac{\beta}{\theta_1}\mathbf{A}_1\left[\mathbf{x}_1^{k+1} - (1 - \theta_1)\mathbf{x}_1^k - \theta_1\mathbf{x}_1^* + \theta_2(\mathbf{x}_1^k - \tilde{\mathbf{x}}_1)\right] \\
&\quad + \frac{\beta}{\theta_1}\mathbf{A}_2\left[\mathbf{x}_2^{k+1} - (1 - \theta_1)\mathbf{x}_2^k - \theta_1\mathbf{x}_2^* + \theta_2(\mathbf{x}_2^k - \tilde{\mathbf{x}}_2)\right],
\end{aligned}
$$

where we use the fact that $\mathbf{A}_1\mathbf{x}_1^* + \mathbf{A}_2\mathbf{x}_2^* = \mathbf{b}$. So (5.121) is proven.

Since (5.125) equals $\bar{\boldsymbol{\lambda}}(\mathbf{x}_1^{k+1}, \mathbf{x}_2^{k+1})$, we obtain (5.120). Now we prove $\hat{\boldsymbol{\lambda}}_{s-1}^m = \hat{\boldsymbol{\lambda}}_s^0$ when $s \geq 1$.

$$
\begin{aligned}
\hat{\boldsymbol{\lambda}}_s^0 &\overset{a}{=} \tilde{\boldsymbol{\lambda}}_s^0 + \frac{\beta(1 - \theta_{1,s})}{\theta_{1,s}}\left(\mathbf{A}_1\mathbf{x}_{s,1}^m + \mathbf{A}_2\mathbf{x}_{s,2}^m - \mathbf{b}\right) \\
&\overset{b}{=} \tilde{\boldsymbol{\lambda}}_s^0 + \beta\left(\frac{1}{\theta_{1,s-1}} + \tau - 1\right)\left(\mathbf{A}_1\mathbf{x}_{s,1}^m + \mathbf{A}_2\mathbf{x}_{s,2}^m - \mathbf{b}\right) \\
&\overset{c}{=} \boldsymbol{\lambda}_{s-1}^{m-1} - \beta(\tau - 1)\left(\mathbf{A}_1\mathbf{x}_{s,1}^m + \mathbf{A}_2\mathbf{x}_{s,2}^m - \mathbf{b}\right) \\
&\quad + \beta\left(\frac{1}{\theta_{1,s-1}} + \tau - 1\right)\left(\mathbf{A}_1\mathbf{x}_{s,1}^m + \mathbf{A}_2\mathbf{x}_{s,2}^m - \mathbf{b}\right) \\
&= \boldsymbol{\lambda}_{s-1}^{m-1} + \frac{\beta}{\theta_{1,s-1}}\left(\mathbf{A}_1\mathbf{x}_{s,1}^m + \mathbf{A}_2\mathbf{x}_{s,2}^m - \mathbf{b}\right)
\end{aligned}
$$

$$\stackrel{d}{=} \tilde{\lambda}^m_{s-1} - \left(\beta - \frac{\beta}{\theta_{1,s-1}}\right)\left(\mathbf{A}_1\mathbf{x}^m_{s,1} + \mathbf{A}_2\mathbf{x}^m_{s,2} - \mathbf{b}\right)$$

$$= \hat{\lambda}^m_{s-1},\tag{5.126}$$

where $\stackrel{a}{=}$ uses (5.119), $\stackrel{b}{=}$ uses the fact that $\frac{1}{\theta_{1,s}} = \frac{1}{\theta_{1,s-1}} + \tau$, $\stackrel{c}{=}$ uses $\tilde{\lambda}^0_{s+1} = \lambda^{m-1}_s +$
$\beta(1-\tau)(\mathbf{A}_1\mathbf{x}^m_{s,1} + \mathbf{A}_2\mathbf{x}^m_{s,2} - \mathbf{b})$ in Algorithm 5.7, and $\stackrel{d}{=}$ uses (5.124).

Step 4: We now are ready to prove (5.102). Define $L(\mathbf{x}_1, \mathbf{x}_2, \lambda) = F_1(\mathbf{x}_1) - F_1(\mathbf{x}^*_1) + F_2(\mathbf{x}_2) - F_2(\mathbf{x}^*_2) + \langle \lambda, \mathbf{A}_1\mathbf{x}_1 + \mathbf{A}_2\mathbf{x}_2 - \mathbf{b}\rangle$. We have

$$L(\mathbf{x}^{k+1}_1, \mathbf{x}^{k+1}_2, \lambda^*) - \theta_2 L(\tilde{\mathbf{x}}_1, \tilde{\mathbf{x}}_2, \lambda^*) - (1-\theta_1-\theta_2)L(\mathbf{x}^k_1, \mathbf{x}^k_2, \lambda^*)$$
$$= F_1(\mathbf{x}^{k+1}_1) - (1-\theta_2-\theta_1)F_1(\mathbf{x}^k_1) - \theta_1 F_1(\mathbf{x}^*_1) - \theta_2 F_1(\tilde{\mathbf{x}}_1)$$
$$+ F_2(\mathbf{x}^{k+1}_2) - (1-\theta_2-\theta_1)F_2(\mathbf{x}^k_2) - \theta_1 F_2(\mathbf{x}^*_2) - \theta_2 F_2(\tilde{\mathbf{x}}_2)$$
$$+ \left\langle \lambda^*, \mathbf{A}_1\left[\mathbf{x}^{k+1}_1 - (1-\theta_1-\theta_2)\mathbf{x}^k_1 - \theta_2\tilde{\mathbf{x}}_1 - \theta_1\mathbf{x}^*_1\right]\right\rangle$$
$$+ \left\langle \lambda^*, \mathbf{A}_2\left[\mathbf{x}^{k+1}_2 - (1-\theta_1-\theta_2)\mathbf{x}^k_2 - \theta_2\tilde{\mathbf{x}}_2 - \theta_1\mathbf{x}^*_2\right]\right\rangle.$$

Plugging (5.106) and (5.118) into the above, we have

$$\mathbb{E}_{i_k}L(\mathbf{x}^{k+1}_1, \mathbf{x}^{k+1}_2, \lambda^*) - \theta_2 L(\tilde{\mathbf{x}}_1, \tilde{\mathbf{x}}_2, \lambda^*) - (1-\theta_2-\theta_1)L(\mathbf{x}^k_1, \mathbf{x}^k_2, \lambda^*)$$
$$\leq \mathbb{E}_{i_k}\left\langle \lambda^* - \bar{\lambda}(\mathbf{x}^{k+1}_1, \mathbf{y}^k_2), \mathbf{A}_1\left[\mathbf{x}^{k+1}_1 - (1-\theta_1-\theta_2)\mathbf{x}^k_1 - \theta_2\tilde{\mathbf{x}}_1 - \theta_1\mathbf{x}^*_1\right]\right\rangle$$
$$+ \mathbb{E}_{i_k}\left\langle \lambda^* - \bar{\lambda}(\mathbf{x}^{k+1}_1, \mathbf{y}^k_2), \mathbf{A}_2\left[\mathbf{x}^{k+1}_2 - (1-\theta_1-\theta_2)\mathbf{x}^k_2 - \theta_2\tilde{\mathbf{x}}_2 - \theta_1\mathbf{x}^*_2\right]\right\rangle$$
$$- \mathbb{E}_{i_k}\left\langle \mathbf{x}^{k+1}_1 - \mathbf{y}^k_1, \mathbf{x}^{k+1}_1 - (1-\theta_1-\theta_2)\mathbf{x}^k_1 - \theta_2\tilde{\mathbf{x}}_1 - \theta_1\mathbf{x}^*_1\right\rangle_{\mathbf{G}_1}$$
$$- \mathbb{E}_{i_k}\left\langle \mathbf{x}^{k+1}_2 - \mathbf{y}^k_2, \mathbf{x}^{k+1}_2 - (1-\theta_1-\theta_2)\mathbf{x}^k_2 - \theta_2\tilde{\mathbf{x}}_2 - \theta_1\mathbf{x}^*_2\right\rangle_{\left(\alpha L_2 + \frac{\beta\|\mathbf{A}^T_2\mathbf{A}_2\|}{\theta_1}\right)\mathbf{I}}$$
$$+ \frac{L_1}{2}\mathbb{E}_{i_k}\left\|\mathbf{x}^{k+1}_1 - \mathbf{y}^k_1\right\|^2 + \mathbb{E}_{i_k}\left(\frac{\left(1 + \frac{1}{b\theta_2}\right)L_2}{2}\left\|\mathbf{x}^{k+1}_2 - \mathbf{y}^k_2\right\|^2\right)$$
$$\stackrel{a}{=} \mathbb{E}_{i_k}\left\langle \lambda^* - \bar{\lambda}(\mathbf{x}^{k+1}_1, \mathbf{x}^{k+1}_2), \mathbf{A}_1\left[\mathbf{x}^{k+1}_1 - (1-\theta_1-\theta_2)\mathbf{x}^k_1 - \theta_2\tilde{\mathbf{x}}_1 - \theta_1\mathbf{x}^*_1\right]\right\rangle$$
$$+ \mathbb{E}_{i_k}\left\langle \lambda^* - \bar{\lambda}(\mathbf{x}^{k+1}_1, \mathbf{x}^{k+1}_2), \mathbf{A}_2\left[\mathbf{x}^{k+1}_2 - (1-\theta_1-\theta_2)\mathbf{x}^k_2 - \theta_2\tilde{\mathbf{x}}_2 - \theta_1\mathbf{x}^*_2\right]\right\rangle$$
$$- \mathbb{E}_{i_k}\left\langle \mathbf{x}^{k+1}_1 - \mathbf{y}^k_1, \mathbf{x}^{k+1}_1 - (1-\theta_1-\theta_2)\mathbf{x}^k_1 - \theta_2\tilde{\mathbf{x}}_1 - \theta_1\mathbf{x}^*_1\right\rangle_{\mathbf{G}_1}$$
$$- \mathbb{E}_{i_k}\left\langle \mathbf{x}^{k+1}_2 - \mathbf{y}^k_2, \mathbf{x}^{k+1}_2 - (1-\theta_1-\theta_2)\mathbf{x}^k_2\right.$$

$$
\begin{aligned}
&\left. -\theta_2 \tilde{\mathbf{x}}_2 - \theta_1 \mathbf{x}_2^* \right)_{\left(\alpha L_2 + \frac{\beta \|\mathbf{A}_2^T \mathbf{A}_2\|}{\theta_1} \right) \mathbf{I} - \frac{\beta \mathbf{A}_2^T \mathbf{A}_2}{\theta_1}} \\
&+ \frac{L_1}{2} \mathbb{E}_{i_k} \left\| \mathbf{x}_1^{k+1} - \mathbf{y}_1^k \right\|^2 + \mathbb{E}_{i_k} \left(\frac{\left(1 + \frac{1}{b\theta_2} \right) L_2}{2} \left\| \mathbf{x}_2^{k+1} - \mathbf{y}_2^k \right\|^2 \right) \\
&+ \frac{\beta}{\theta_1} \mathbb{E}_{i_k} \left\langle \mathbf{A}_2 \mathbf{x}_2^{k+1} - \mathbf{A}_2 \mathbf{y}_2^k, \mathbf{A}_1 \left[\mathbf{x}_1^{k+1} - (1 - \theta_1 - \theta_2) \mathbf{x}_1^k - \theta_2 \tilde{\mathbf{x}}_1 - \theta_1 \mathbf{x}_1^* \right] \right\rangle,
\end{aligned}
$$
$$(5.127)$$

where in the equality $\stackrel{a}{=}$ we change the term $\bar{\lambda}(\mathbf{x}_1^{k+1}, \mathbf{y}_2^k)$ to $\bar{\lambda}(\mathbf{x}_1^{k+1}, \mathbf{x}_2^{k+1}) - \frac{\beta}{\theta_1} \mathbf{A}_2 (\mathbf{x}_2^{k+1} - \mathbf{y}_2^k)$. For the first two terms in the right-hand side of (5.127), we have

$$
\begin{aligned}
&\left\langle \boldsymbol{\lambda}^* - \bar{\lambda}(\mathbf{x}_1^{k+1}, \mathbf{x}_2^{k+1}), \mathbf{A}_1 \left[\mathbf{x}_1^{k+1} - (1 - \theta_1 - \theta_2) \mathbf{x}_1^k - \theta_2 \tilde{\mathbf{x}}_1 - \theta_1 \mathbf{x}_1^* \right] \right\rangle \\
&+ \left\langle \boldsymbol{\lambda}^* - \bar{\lambda}(\mathbf{x}_1^{k+1}, \mathbf{x}_2^{k+1}), \mathbf{A}_2 \left[\mathbf{x}_2^{k+1} - (1 - \theta_1 - \theta_2) \mathbf{x}_1^k - \theta_2 \tilde{\mathbf{x}}_2 - \theta_1 \mathbf{x}_2^* \right] \right\rangle \\
&\stackrel{a}{=} \frac{\theta_1}{\beta} \langle \boldsymbol{\lambda}^* - \hat{\boldsymbol{\lambda}}^{k+1}, \hat{\boldsymbol{\lambda}}^{k+1} - \hat{\boldsymbol{\lambda}}^k \rangle \\
&\stackrel{b}{=} \frac{\theta_1}{2\beta} \left(\left\| \hat{\boldsymbol{\lambda}}^k - \boldsymbol{\lambda}^* \right\|^2 - \left\| \hat{\boldsymbol{\lambda}}^{k+1} - \boldsymbol{\lambda}^* \right\|^2 - \left\| \hat{\boldsymbol{\lambda}}^{k+1} - \hat{\boldsymbol{\lambda}}^k \right\|^2 \right),
\end{aligned}
$$
$$(5.128)$$

where $\stackrel{a}{=}$ uses (5.120) and (5.121) and $\stackrel{b}{=}$ uses (A.2).

Substituting (5.128) into (5.127), we obtain

$$
\begin{aligned}
&\mathbb{E}_{i_k} L(\mathbf{x}_1^{k+1}, \mathbf{x}_2^{k+1}, \boldsymbol{\lambda}^*) - \theta_2 L(\tilde{\mathbf{x}}_1, \tilde{\mathbf{x}}_2, \boldsymbol{\lambda}^*) - (1 - \theta_2 - \theta_1) L(\mathbf{x}_1^k, \mathbf{x}_2^k, \boldsymbol{\lambda}^*) \\
&\leq \frac{\theta_1}{2\beta} \left(\left\| \hat{\boldsymbol{\lambda}}^k - \boldsymbol{\lambda}^* \right\|^2 - \mathbb{E}_{i_k} \left\| \hat{\boldsymbol{\lambda}}^{k+1} - \boldsymbol{\lambda}^* \right\|^2 - \mathbb{E}_{i_k} \left\| \hat{\boldsymbol{\lambda}}^{k+1} - \hat{\boldsymbol{\lambda}}^k \right\|^2 \right) \\
&\quad - \mathbb{E}_{i_k} \left\langle \mathbf{x}_1^{k+1} - \mathbf{y}_1^k, \mathbf{x}_1^{k+1} - (1 - \theta_1 - \theta_2) \mathbf{x}_1^k - \theta_2 \tilde{\mathbf{x}}_1 - \theta_1 \mathbf{x}_1^* \right\rangle_{\mathbf{G}_1} \\
&\quad - \mathbb{E}_{i_k} \left\langle \mathbf{x}_2^{k+1} - \mathbf{y}_2^k, \mathbf{x}_2^{k+1} - (1 - \theta_1 - \theta_2) \mathbf{x}_2^k \right. \\
&\quad \left. -\theta_2 \tilde{\mathbf{x}}_2 - \theta_1 \mathbf{x}_2^* \right\rangle_{\left(\alpha L_2 + \frac{\beta \|\mathbf{A}_2^T \mathbf{A}_2\|}{\theta_1} \right) \mathbf{I} - \frac{\beta \mathbf{A}_2^T \mathbf{A}_2}{\theta_1}} \\
&\quad + \frac{L_1}{2} \mathbb{E}_{i_k} \left\| \mathbf{x}_1^{k+1} - \mathbf{y}_1^k \right\|^2 + \mathbb{E}_{i_k} \left(\frac{\left(1 + \frac{1}{b\theta_2} \right) L_2}{2} \left\| \mathbf{x}_2^{k+1} - \mathbf{y}_2^k \right\|^2 \right) \\
&\quad + \frac{\beta}{\theta_1} \mathbb{E}_{i_k} \left\langle \mathbf{A}_2 \mathbf{x}_2^{k+1} - \mathbf{A}_2 \mathbf{y}_2^k, \mathbf{A}_1 \left[\mathbf{x}_1^{k+1} - (1 - \theta_1 - \theta_2) \mathbf{x}_1^k - \theta_2 \tilde{\mathbf{x}}_1 - \theta_1 \mathbf{x}_1^* \right] \right\rangle.
\end{aligned}
$$
$$(5.129)$$

Then applying identity (A.1) to the second and the third terms in the right-hand side of (5.129) and rearranging, we have

$$
\mathbb{E}_{i_k} L(\mathbf{x}_1^{k+1}, \mathbf{x}_2^{k+1}, \boldsymbol{\lambda}^*) - \theta_2 L(\tilde{\mathbf{x}}_1, \tilde{\mathbf{x}}_2, \boldsymbol{\lambda}^*) - (1 - \theta_2 - \theta_1) L(\mathbf{x}_1^k, \mathbf{x}_2^k, \boldsymbol{\lambda}^*)
$$

$$
\leq \frac{\theta_1}{2\beta} \left(\left\| \hat{\boldsymbol{\lambda}}^k - \boldsymbol{\lambda}^* \right\|^2 - \mathbb{E}_{i_k} \left\| \hat{\boldsymbol{\lambda}}^{k+1} - \boldsymbol{\lambda}^* \right\|^2 - \mathbb{E}_{i_k} \left\| \hat{\boldsymbol{\lambda}}^{k+1} - \hat{\boldsymbol{\lambda}}^k \right\|^2 \right)
$$

$$
+ \frac{1}{2} \left\| \mathbf{y}_1^k - (1 - \theta_1 - \theta_2)\mathbf{x}_1^k - \theta_2\tilde{\mathbf{x}}_1 - \theta_1\mathbf{x}_1^* \right\|_{\mathbf{G}_1}^2
$$

$$
- \frac{1}{2} \mathbb{E}_{i_k} \left\| \mathbf{x}_1^{k+1} - (1 - \theta_1 - \theta_2)\mathbf{x}_1^k - \theta_2\tilde{\mathbf{x}}_1 - \theta_1\mathbf{x}_1^* \right\|_{\mathbf{G}_1}^2
$$

$$
+ \frac{1}{2} \left\| \mathbf{y}_2^k - (1 - \theta_1 - \theta_2)\mathbf{x}_2^k - \theta_2\tilde{\mathbf{x}}_2 - \theta_1\mathbf{x}_2^* \right\|_{\left(\alpha L_2 + \frac{\beta \|\mathbf{A}_2^T\mathbf{A}_2\|}{\theta_1}\right)\mathbf{I} - \frac{\beta\mathbf{A}_2^T\mathbf{A}_2}{\theta_1}}^2
$$

$$
- \frac{1}{2} \mathbb{E}_{i_k} \left\| \mathbf{x}_2^{k+1} - (1 - \theta_1 - \theta_2)\mathbf{x}_2^k - \theta_2\tilde{\mathbf{x}}_2 - \theta_1\mathbf{x}_2^* \right\|_{\left(\alpha L_2 + \frac{\beta \|\mathbf{A}_2^T\mathbf{A}_2\|}{\theta_1}\right)\mathbf{I} - \frac{\beta\mathbf{A}_2^T\mathbf{A}_2}{\theta_1}}^2
$$

$$
- \frac{1}{2} \mathbb{E}_{i_k} \left\| \mathbf{x}_1^{k+1} - \mathbf{y}_1^k \right\|_{\frac{\beta\|\mathbf{A}_1^T\mathbf{A}_1\|}{\theta_1}\mathbf{I} - \frac{\beta\mathbf{A}_1^T\mathbf{A}_1}{\theta_1}}^2 - \frac{1}{2} \mathbb{E}_{i_k} \left\| \mathbf{x}_2^{k+1} - \mathbf{y}_2^k \right\|_{\frac{\beta\|\mathbf{A}_2^T\mathbf{A}_2\|}{\theta_1}\mathbf{I} - \frac{\beta\mathbf{A}_2^T\mathbf{A}_2}{\theta_1}}^2
$$

$$
+ \frac{\beta}{\theta_1} \mathbb{E}_{i_k} \left\langle \mathbf{A}_2\mathbf{x}_2^{k+1} - \mathbf{A}_2\mathbf{y}_2^k, \mathbf{A}_1 \left[\mathbf{x}_1^{k+1} - (1 - \theta_1 - \theta_2)\mathbf{x}_1^k - \theta_2\tilde{\mathbf{x}}_1 - \theta_1\mathbf{x}_1^* \right] \right\rangle.
$$

$$(5.130)$$

For the last term in the right-hand side of (5.130), we have

$$
\frac{\beta}{\theta_1} \left\langle \mathbf{A}_2\mathbf{x}_2^{k+1} - \mathbf{A}_2\mathbf{y}_2^k, \mathbf{A}_1 \left[\mathbf{x}_1^{k+1} - (1 - \theta_1 - \theta_2)\mathbf{x}_1^k - \theta_2\tilde{\mathbf{x}}_1 - \theta_1\mathbf{x}_1^* \right] \right\rangle
$$

$$
\overset{a}{=} \frac{\beta}{\theta_1} \left\langle \mathbf{A}_2\mathbf{x}_2^{k+1} - \mathbf{A}_2\mathbf{v} - (\mathbf{A}_2\mathbf{y}_2^k - \mathbf{A}_2\mathbf{v}), \right.
$$

$$
\left. \mathbf{A}_1 \left[\mathbf{x}_1^{k+1} - (1 - \theta_1 - \theta_2)\mathbf{x}_1^k - \theta_2\tilde{\mathbf{x}}_1 - \theta_1\mathbf{x}_1^* \right] - \mathbf{0} \right\rangle
$$

$$
\overset{b}{=} \frac{\beta}{2\theta_1} \left\| \mathbf{A}_2\mathbf{x}_2^{k+1} - \mathbf{A}_2\mathbf{v} + \mathbf{A}_1 \left[\mathbf{x}_1^{k+1} - (1 - \theta_1 - \theta_2)\mathbf{x}_1^k - \theta_2\tilde{\mathbf{x}}_1 - \theta_1\mathbf{x}_1^* \right] \right\|^2
$$

$$
- \frac{\beta}{2\theta_1} \left\| \mathbf{A}_2\mathbf{x}_2^{k+1} - \mathbf{A}_2\mathbf{v} \right\|^2 + \frac{\beta}{2\theta_1} \left\| \mathbf{A}_2\mathbf{y}_2^k - \mathbf{A}_2\mathbf{v} \right\|^2
$$

$$
- \frac{\beta}{2\theta_1} \left\| \mathbf{A}_2\mathbf{y}_2^k - \mathbf{A}_2\mathbf{v} + \mathbf{A}_1 \left[\mathbf{x}_1^{k+1} - (1 - \theta_1 - \theta_2)\mathbf{x}_1^k - \theta_2\tilde{\mathbf{x}}_1 - \theta_1\mathbf{x}_1^* \right] \right\|^2
$$

$$
\overset{c}{=} \frac{\theta_1}{2\beta} \left\| \hat{\boldsymbol{\lambda}}^{k+1} - \hat{\boldsymbol{\lambda}}^k \right\|^2 - \frac{\beta}{2\theta_1} \left\| \mathbf{A}_2 \mathbf{x}_2^{k+1} - \mathbf{A}_2 \mathbf{v} \right\|^2 + \frac{\beta}{2\theta_1} \left\| \mathbf{A}_2 \mathbf{y}_2^k - \mathbf{A}_2 \mathbf{v} \right\|^2
$$
$$
- \frac{\beta}{2\theta_1} \left\| \mathbf{A}_2 \mathbf{y}_2^k - \mathbf{A}_2 \mathbf{v} + \mathbf{A}_1 \left[\mathbf{x}_1^{k+1} - (1 - \theta_1 - \theta_2)\mathbf{x}_1^k - \theta_2 \tilde{\mathbf{x}}_1 - \theta_1 \mathbf{x}_1^* \right] \right\|^2,
$$
$$
\tag{5.131}
$$

where in $\overset{a}{=}$ we set $\mathbf{v} = (1 - \theta_1 - \theta_2)\mathbf{x}_2^k + \theta_2 \tilde{\mathbf{x}}_2 + \theta_1 \mathbf{x}_2^*$, $\overset{b}{=}$ uses (A.3), and $\overset{c}{=}$ uses (5.121). Substituting (5.131) into (5.130), we have

$$
\mathbb{E}_{i_k} L(\mathbf{x}_1^{k+1}, \mathbf{x}_2^{k+1}, \boldsymbol{\lambda}^*) - \theta_2 L(\tilde{\mathbf{x}}_1, \tilde{\mathbf{x}}_2, \boldsymbol{\lambda}^*) - (1 - \theta_2 - \theta_1) L(\mathbf{x}_1^k, \mathbf{x}_2^k, \boldsymbol{\lambda}^*)
$$
$$
\leq \frac{\theta_1}{2\beta} \left(\left\| \hat{\boldsymbol{\lambda}}^k - \boldsymbol{\lambda}^* \right\|^2 - \mathbb{E}_{i_k} \left\| \hat{\boldsymbol{\lambda}}^{k+1} - \boldsymbol{\lambda}^* \right\|^2 \right)
$$
$$
+ \frac{1}{2} \left\| \mathbf{y}_1^k - (1 - \theta_1 - \theta_2)\mathbf{x}_1^k - \theta_2 \tilde{\mathbf{x}}_1 - \theta_1 \mathbf{x}_1^* \right\|_{\mathbf{G}_1}^2
$$
$$
- \frac{1}{2} \mathbb{E}_{i_k} \left\| \mathbf{x}_1^{k+1} - (1 - \theta_1 - \theta_2)\mathbf{x}_1^k - \theta_2 \tilde{\mathbf{x}}_1 - \theta_1 \mathbf{x}_1^* \right\|_{\mathbf{G}_1}^2
$$
$$
+ \frac{1}{2} \left\| \mathbf{y}_2^k - (1 - \theta_1 - \theta_2)\mathbf{x}_2^k - \theta_2 \tilde{\mathbf{x}}_2 - \theta_1 \mathbf{x}_2^* \right\|_{\left(\alpha L_2 + \frac{\beta \| \mathbf{A}_2^T \mathbf{A}_2 \|}{\theta_1} \right)\mathbf{I}}^2
$$
$$
- \frac{1}{2} \mathbb{E}_{i_k} \left\| \mathbf{x}_2^{k+1} - (1 - \theta_1 - \theta_2)\mathbf{x}_2^k - \theta_2 \tilde{\mathbf{x}}_2 - \theta_1 \mathbf{x}_2^* \right\|_{\left(\alpha L_2 + \frac{\beta \| \mathbf{A}_2^T \mathbf{A}_2 \|}{\theta_1} \right)\mathbf{I}}^2
$$
$$
- \frac{1}{2} \mathbb{E}_{i_k} \left\| \mathbf{x}_1^{k+1} - \mathbf{y}_1^k \right\|_{\frac{\beta \| \mathbf{A}_1^T \mathbf{A}_1 \|}{\theta_1}\mathbf{I} - \frac{\beta \mathbf{A}_1^T \mathbf{A}_1}{\theta_1}}^2 - \frac{1}{2} \mathbb{E}_{i_k} \left\| \mathbf{x}_2^{k+1} - \mathbf{y}_2^k \right\|_{\frac{\beta \| \mathbf{A}_2^T \mathbf{A}_2 \|}{\theta_1}\mathbf{I} - \frac{\beta \mathbf{A}_2^T \mathbf{A}_2}{\theta_1}}^2
$$
$$
- \frac{\beta}{2\theta_1} \mathbb{E}_{i_k} \left\| \mathbf{A}_2 \mathbf{y}_2^k - \mathbf{A}_2 \mathbf{v} + \mathbf{A}_1 \left[\mathbf{x}_1^{k+1} - (1 - \theta_1 - \theta_2)\mathbf{x}_1^k - \theta_2 \tilde{\mathbf{x}}_1 - \theta_1 \mathbf{x}_1^* \right] \right\|^2.
$$
$$
\tag{5.132}
$$

Since the last three terms in the right-hand side of (5.132) are nonpositive, we obtain (5.102).   □

Theorem 5.10 *If the conditions in Lemma 5.7 hold, then we have*

$$
\mathbb{E} \left(\frac{1}{2\beta} \left\| \frac{\beta(m-1)(\theta_2 + \theta_{1,S}) + \beta}{\theta_{1,S}} \left(\mathbf{A}\hat{\mathbf{x}}_S - \mathbf{b} \right) \right. \right.
$$
$$
\left. \left. - \frac{\beta(m-1)\theta_2}{\theta_{1,0}} \left(\mathbf{A}\mathbf{x}_0^0 - \mathbf{b} \right) + \tilde{\boldsymbol{\lambda}}_0^0 - \boldsymbol{\lambda}^* \right\|^2 \right)
$$
$$
+ \mathbb{E} \left(\frac{(m-1)(\theta_2 + \theta_{1,S}) + 1}{\theta_{1,S}} \left(F(\hat{\mathbf{x}}_S) - F(\mathbf{x}^*) + \langle \boldsymbol{\lambda}^*, \mathbf{A}\hat{\mathbf{x}}_S - \mathbf{b} \rangle \right) \right)
$$
$$
\leq C_3 \left(F(\mathbf{x}_0^0) - F(\mathbf{x}^*) + \langle \boldsymbol{\lambda}^*, \mathbf{A}\mathbf{x}_0^0 - \mathbf{b} \rangle \right)
$$

$$
+ \frac{1}{2\beta} \left\| \tilde{\lambda}_0^0 + \frac{\beta(1-\theta_{1,0})}{\theta_{1,0}} (\mathbf{A}\mathbf{x}_0^0 - \mathbf{b}) - \lambda^* \right\|^2
$$

$$
+ \frac{1}{2} \left\| \mathbf{x}_{0,1}^0 - \mathbf{x}_1^* \right\|_{(\theta_{1,0}L_1+\beta\|\mathbf{A}_1^T\mathbf{A}_1\|)\mathbf{I}-\beta\mathbf{A}_1^T\mathbf{A}_1}^2
$$

$$
+ \frac{1}{2} \left\| \mathbf{x}_{0,2}^0 - \mathbf{x}_2^* \right\|_{\left(\left(1+\frac{1}{b\theta_2}\right)\theta_{1,0}L_2+\beta\|\mathbf{A}_2^T\mathbf{A}_2\|\right)\mathbf{I}}^2, \tag{5.133}
$$

where $C_3 = \frac{1-\theta_{1,0}+(m-1)\theta_2}{\theta_{1,0}}$.

Proof Taking expectation over the first k iterations for (5.102) and dividing θ_1 on both sides of it, we obtain

$$
\frac{1}{\theta_1} \mathbb{E} L(\mathbf{x}_1^{k+1}, \mathbf{x}_2^{k+1}, \lambda^*) - \frac{\theta_2}{\theta_1} L(\tilde{\mathbf{x}}_1, \tilde{\mathbf{x}}_2, \lambda^*) - \frac{1-\theta_2-\theta_1}{\theta_1} L(\mathbf{x}_1^k, \mathbf{x}_2^k, \lambda^*)
$$

$$
\leq \frac{1}{2\beta} \left(\left\| \hat{\lambda}^k - \lambda^* \right\|^2 - \mathbb{E} \left\| \hat{\lambda}^{k+1} - \lambda^* \right\|^2 \right)
$$

$$
+ \frac{\theta_1}{2} \left\| \frac{1}{\theta_1} \left[\mathbf{y}_1^k - (1-\theta_1-\theta_2)\mathbf{x}_1^k - \theta_2\tilde{\mathbf{x}}_1 \right] - \mathbf{x}_1^* \right\|_{\left(L_1+\frac{\beta\|\mathbf{A}_1^T\mathbf{A}_1\|}{\theta_1}\right)\mathbf{I}-\frac{\beta\mathbf{A}_1^T\mathbf{A}_1}{\theta_1}}^2
$$

$$
- \frac{\theta_1}{2} \mathbb{E} \left\| \frac{1}{\theta_1} \left[\mathbf{x}_1^{k+1} - (1-\theta_1-\theta_2)\mathbf{x}_1^k - \theta_2\tilde{\mathbf{x}}_1 \right] - \mathbf{x}_1^* \right\|_{\left(L_1+\frac{\beta\|\mathbf{A}_1^T\mathbf{A}_1\|}{\theta_1}\right)\mathbf{I}-\frac{\beta\mathbf{A}_1^T\mathbf{A}_1}{\theta_1}}^2
$$

$$
+ \frac{\theta_1}{2} \left\| \frac{1}{\theta_1} \left[\mathbf{y}_2^k - (1-\theta_1-\theta_2)\mathbf{x}_2^k - \theta_2\tilde{\mathbf{x}}_2 \right] - \mathbf{x}_2^* \right\|_{\left(\alpha L_2+\frac{\beta\|\mathbf{A}_2^T\mathbf{A}_2\|}{\theta_1}\right)\mathbf{I}}^2
$$

$$
- \frac{\theta_1}{2} \mathbb{E} \left\| \frac{1}{\theta_1} \left[\mathbf{x}_2^{k+1} - (1-\theta_1-\theta_2)\mathbf{x}_2^k - \theta_2\tilde{\mathbf{x}}_2 \right] - \mathbf{x}_2^* \right\|_{\left(\alpha L_2+\frac{\beta\|\mathbf{A}_2^T\mathbf{A}_2\|}{\theta_1}\right)\mathbf{I}}^2, \tag{5.134}
$$

where the expectation is taken under the condition that the randomness under the first s epochs is fixed. Since

$$
\mathbf{y}^k = \mathbf{x}^k + (1-\theta_1-\theta_2)(\mathbf{x}^k - \mathbf{x}^{k-1}), \quad k \geq 1,
$$

we obtain

$$
\frac{1}{\theta_1} \mathbb{E} L(\mathbf{x}_1^{k+1}, \mathbf{x}_2^{k+1}, \lambda^*) - \frac{\theta_2}{\theta_1} L(\tilde{\mathbf{x}}_1, \tilde{\mathbf{x}}_2, \lambda^*) - \frac{1-\theta_2-\theta_1}{\theta_1} L(\mathbf{x}_1^k, \mathbf{x}_2^k, \lambda^*)
$$

$$
\leq \frac{1}{2\beta} \left(\left\| \hat{\lambda}^k - \lambda^* \right\|^2 - \mathbb{E} \left\| \hat{\lambda}^{k+1} - \lambda^* \right\|^2 \right)
$$

$$+ \frac{\theta_1}{2} \left\| \frac{1}{\theta_1} \left[\mathbf{x}_1^k - (1 - \theta_1 - \theta_2) \mathbf{x}_1^{k-1} - \theta_2 \tilde{\mathbf{x}}_1 \right] - \mathbf{x}_1^* \right\|^2_{\left(L_1 + \frac{\beta \|\mathbf{A}_1^T \mathbf{A}_1\|}{\theta_1}\right)\mathbf{I} - \frac{\beta \mathbf{A}_1^T \mathbf{A}_1}{\theta_1}}$$

$$- \frac{\theta_1}{2} \mathbb{E} \left\| \frac{1}{\theta_1} \left[\mathbf{x}_1^{k+1} - (1 - \theta_1 - \theta_2) \mathbf{x}_1^{k} - \theta_2 \tilde{\mathbf{x}}_1 \right] - \mathbf{x}_1^* \right\|^2_{\left(L_1 + \frac{\beta \|\mathbf{A}_1^T \mathbf{A}_1\|}{\theta_1}\right)\mathbf{I} - \frac{\beta \mathbf{A}_1^T \mathbf{A}_1}{\theta_1}}$$

$$+ \frac{\theta_1}{2} \left\| \frac{1}{\theta_1} \left[\mathbf{x}_2^k - (1 - \theta_1 - \theta_2) \mathbf{x}_2^{k-1} - \theta_2 \tilde{\mathbf{x}}_2 \right] - \mathbf{x}_2^* \right\|^2_{\left(\alpha L_2 + \frac{\beta \|\mathbf{A}_2^T \mathbf{A}_2\|}{\theta_1}\right)\mathbf{I}}$$

$$- \frac{\theta_1}{2} \mathbb{E} \left\| \frac{1}{\theta_1} \left[\mathbf{x}_2^{k+1} - (1 - \theta_1 - \theta_2) \mathbf{x}_2^{k} - \theta_2 \tilde{\mathbf{x}}_2 \right] - \mathbf{x}_2^* \right\|^2_{\left(\alpha L_2 + \frac{\beta \|\mathbf{A}_2^T \mathbf{A}_2\|}{\theta_1}\right)\mathbf{I}},$$

$$k \geq 1. \tag{5.135}$$

Adding back the subscript s, taking expectation on the first s epochs, and then summing (5.134) with k from 0 to $m - 1$ (for $k \geq 1$, using (5.135)), we have

$$\frac{1}{\theta_{1,s}} \mathbb{E} \left(L(\mathbf{x}_s^m, \boldsymbol{\lambda}^*) - L(\mathbf{x}^*, \boldsymbol{\lambda}^*) \right) + \frac{\theta_2 + \theta_{1,s}}{\theta_{1,s}} \sum_{k=1}^{m-1} \mathbb{E} \left(L(\mathbf{x}_s^k, \boldsymbol{\lambda}^*) - L(\mathbf{x}^*, \boldsymbol{\lambda}^*) \right)$$

$$\leq \frac{1 - \theta_{1,s} - \theta_2}{\theta_{1,s}} \mathbb{E} \left(L(\mathbf{x}_s^0, \boldsymbol{\lambda}^*) - L(\mathbf{x}^*, \boldsymbol{\lambda}^*) \right) + \frac{m \theta_2}{\theta_{1,s}} \mathbb{E} \left(L(\tilde{\mathbf{x}}_s, \boldsymbol{\lambda}^*) - L(\mathbf{x}^*, \boldsymbol{\lambda}^*) \right)$$

$$+ \frac{1}{2} \mathbb{E} \left\| \frac{1}{\theta_{1,s}} \left[\mathbf{y}_{s,1}^0 - \theta_2 \tilde{\mathbf{x}}_{s,1} - (1 - \theta_{1,s} - \theta_2) \mathbf{x}_{s,1}^0 \right] \right.$$

$$\left. - \mathbf{x}_1^* \right\|^2_{(\theta_{1,s} L_1 + \beta \|\mathbf{A}_1^T \mathbf{A}_1\|) \mathbf{I} - \beta \mathbf{A}_1^T \mathbf{A}_1}$$

$$- \frac{1}{2} \mathbb{E} \left\| \frac{1}{\theta_{1,s}} \left[\mathbf{x}_{s,1}^m - \theta_2 \tilde{\mathbf{x}}_{s,1} - (1 - \theta_{1,s} - \theta_2) \mathbf{x}_{s,1}^{m-1} \right] \right.$$

$$\left. - \mathbf{x}_1^* \right\|^2_{(\theta_{1,s} L_1 + \beta \|\mathbf{A}_1^T \mathbf{A}_1\|) \mathbf{I} - \beta \mathbf{A}_1^T \mathbf{A}_1}$$

$$+ \frac{1}{2} \mathbb{E} \left\| \frac{1}{\theta_{1,s}} \left[\mathbf{y}_{s,2}^0 - \theta_2 \tilde{\mathbf{x}}_{s,2} - (1 - \theta_{1,s} - \theta_2) \mathbf{x}_{s,2}^0 \right] - \mathbf{x}_2^* \right\|^2_{(\alpha \theta_{1,s} L_2 + \beta \|\mathbf{A}_2^T \mathbf{A}_2\|) \mathbf{I}}$$

$$- \frac{1}{2} \mathbb{E} \left\| \frac{1}{\theta_{1,s}} \left[\mathbf{x}_{s,2}^m - \theta_2 \tilde{\mathbf{x}}_{s,2} - (1 - \theta_{1,s} - \theta_2) \mathbf{x}_{s,2}^{m-1} \right] - \mathbf{x}_2^* \right\|^2_{(\alpha \theta_{1,s} L_2 + \beta \|\mathbf{A}_2^T \mathbf{A}_2\|) \mathbf{I}}$$

$$+ \frac{1}{2\beta} \left(\mathbb{E} \left\| \hat{\boldsymbol{\lambda}}_s^0 - \boldsymbol{\lambda}^* \right\|^2 - \mathbb{E} \left\| \hat{\boldsymbol{\lambda}}_s^m - \boldsymbol{\lambda}^* \right\|^2 \right), \quad s \geq 0, \tag{5.136}$$

where we use $L(\mathbf{x}_s^k, \boldsymbol{\lambda}^*)$ and $L(\tilde{\mathbf{x}}_s, \boldsymbol{\lambda}^*)$ to denote $L(\mathbf{x}_{s,1}^k, \mathbf{x}_{s,2}^k, \boldsymbol{\lambda}^*)$ and $L(\tilde{\mathbf{x}}_{s,1}, \tilde{\mathbf{x}}_{s,2}, \boldsymbol{\lambda}^*)$, respectively. Since $L(\mathbf{x}, \boldsymbol{\lambda}^*)$ is convex for $\mathbf{x}$, we have

$$
mL(\tilde{\mathbf{x}}_s, \boldsymbol{\lambda}^*)
$$

$$
= mL\left(\frac{1}{m}\left[\left(1 - \frac{(\tau-1)\theta_{1,s}}{\theta_2}\right)\mathbf{x}_{s-1}^m + \left(1 + \frac{(\tau-1)\theta_{1,s}}{(m-1)\theta_2}\right)\sum_{k=1}^{m-1}\mathbf{x}_{s-1}^k\right], \boldsymbol{\lambda}^*\right)
$$

$$
\leq \left[1 - \frac{(\tau-1)\theta_{1,s}}{\theta_2}\right]L(\mathbf{x}_{s-1}^m, \boldsymbol{\lambda}^*) + \left[1 + \frac{(\tau-1)\theta_{1,s}}{(m-1)\theta_2}\right]\sum_{k=1}^{m-1}L(\mathbf{x}_{s-1}^k, \boldsymbol{\lambda}^*).
$$

$$(5.137)$$

Substituting (5.137) into (5.136), and using $\mathbf{x}_{s-1}^m = \mathbf{x}_s^0$, we have

$$
\frac{1}{\theta_{1,s}}\mathbb{E}\left(L(\mathbf{x}_s^m, \boldsymbol{\lambda}^*) - L(\mathbf{x}^*, \boldsymbol{\lambda}^*)\right) + \frac{\theta_2 + \theta_{1,s}}{\theta_{1,s}}\sum_{k=1}^{m-1}\mathbb{E}\left(L(\mathbf{x}_s^k, \boldsymbol{\lambda}^*) - L(\mathbf{x}^*, \boldsymbol{\lambda}^*)\right)
$$

$$
\leq \frac{1 - \tau\theta_{1,s}}{\theta_{1,s}}\mathbb{E}\left(L(\mathbf{x}_{s-1}^m, \boldsymbol{\lambda}^*) - L(\mathbf{x}^*, \boldsymbol{\lambda}^*)\right)
$$

$$
+ \frac{\theta_2 + \frac{\tau-1}{m-1}\theta_{1,s}}{\theta_{1,s}}\sum_{k=1}^{m-1}\mathbb{E}\left(L(\mathbf{x}_{s-1}^k, \boldsymbol{\lambda}^*) - L(\mathbf{x}^*, \boldsymbol{\lambda}^*)\right)
$$

$$
+ \frac{1}{2}\mathbb{E}\left\|\frac{1}{\theta_{1,s}}\left[\mathbf{y}_{s,1}^0 - \theta_2\tilde{\mathbf{x}}_{s,1} - (1 - \theta_{1,s} - \theta_2)\mathbf{x}_{s,1}^0\right]\right.
$$

$$
\left. -\mathbf{x}_1^*\right\|^2_{(\theta_{1,s}L_1 + \beta\|\mathbf{A}_1^T\mathbf{A}_1\|)\mathbf{I} - \beta\mathbf{A}_1^T\mathbf{A}_1}
$$

$$
- \frac{1}{2}\mathbb{E}\left\|\frac{1}{\theta_{1,s}}\left[\mathbf{x}_{s,1}^m - \theta_2\tilde{\mathbf{x}}_{s,1} - (1 - \theta_{1,s} - \theta_2)\mathbf{x}_{s,1}^{m-1}\right]\right.
$$

$$
\left. -\mathbf{x}_1^*\right\|^2_{(\theta_{1,s}L_1 + \beta\|\mathbf{A}_1^T\mathbf{A}_1\|)\mathbf{I} - \beta\mathbf{A}_1^T\mathbf{A}_1}
$$

$$
+ \frac{1}{2}\mathbb{E}\left\|\frac{1}{\theta_{1,s}}\left[\mathbf{y}_{s,2}^0 - \theta_2\tilde{\mathbf{x}}_{s,2} - (1 - \theta_{1,s} - \theta_2)\mathbf{x}_{s,2}^0\right]\right.
$$

$$
\left. -\mathbf{x}_2^*\right\|^2_{(\alpha\theta_{1,s}L_2 + \beta\|\mathbf{A}_2^T\mathbf{A}_2\|)\mathbf{I}}
$$

$$
- \frac{1}{2}\mathbb{E}\left\|\frac{1}{\theta_{1,s}}\left[\mathbf{x}_{s,2}^m - \theta_2\tilde{\mathbf{x}}_{s,2} - (1 - \theta_{1,s} - \theta_2)\mathbf{x}_{s,2}^{m-1}\right]\right.
$$

$$\left. -\mathbf{x}_2^* \right\|_{(\alpha\theta_{1,s}L_2+\beta\|\mathbf{A}_2^T\mathbf{A}_2\|)\mathbf{I}}^2$$

$$+\frac{1}{2\beta}\left(\mathbb{E}\left\|\hat{\boldsymbol{\lambda}}_s^0-\boldsymbol{\lambda}^*\right\|^2-\mathbb{E}\left\|\hat{\boldsymbol{\lambda}}_s^m-\boldsymbol{\lambda}^*\right\|^2\right),\quad s\geq 0. \tag{5.138}$$

Then from the setting of $\theta_{1,s}=\frac{1}{2+\tau s}$ and $\theta_2=\frac{m-\tau}{\tau(m-1)}$, we have

$$\frac{1}{\theta_{1,s}}=\frac{1-\tau\theta_{1,s+1}}{\theta_{1,s+1}},\quad s\geq 0, \tag{5.139}$$

and

$$\frac{\theta_2+\theta_{1,s}}{\theta_{1,s}}=\frac{\theta_2}{\theta_{1,s+1}}-\tau\theta_2+1=\frac{\theta_2+\frac{\tau-1}{m-1}\theta_{1,s+1}}{\theta_{1,s+1}},\quad s\geq 0. \tag{5.140}$$

Substituting (5.139) into the first term and (5.140) into the second term in the right-hand side of (5.138), we obtain

$$\frac{1}{\theta_{1,s}}\mathbb{E}\left(L(\mathbf{x}_s^m,\boldsymbol{\lambda}^*)-L(\mathbf{x}^*,\boldsymbol{\lambda}^*)\right)+\frac{\theta_2+\theta_{1,s}}{\theta_{1,s}}\sum_{k=1}^{m-1}\mathbb{E}\left(L(\mathbf{x}_s^k,\boldsymbol{\lambda}^*)-L(\mathbf{x}^*,\boldsymbol{\lambda}^*)\right)$$

$$\leq\frac{1}{\theta_{1,s-1}}\mathbb{E}\left(L(\mathbf{x}_{s-1}^m,\boldsymbol{\lambda}^*)-L(\mathbf{x}^*,\boldsymbol{\lambda}^*)\right)$$

$$+\frac{\theta_2+\theta_{1,s-1}}{\theta_{1,s-1}}\sum_{k=1}^{m-1}\mathbb{E}\left(L(\mathbf{x}_{s-1}^k,\boldsymbol{\lambda}^*)-L(\mathbf{x}^*,\boldsymbol{\lambda}^*)\right)$$

$$+\frac{1}{2}\mathbb{E}\left\|\frac{1}{\theta_{1,s}}\left[\mathbf{y}_{s,1}^0-\theta_2\tilde{\mathbf{x}}_{s,1}-(1-\theta_{1,s}-\theta_2)\mathbf{x}_{s,1}^0\right]\right.$$

$$\left.-\mathbf{x}_1^*\right\|_{(\theta_{1,s}L_1+\beta\|\mathbf{A}_1^T\mathbf{A}_1\|)\mathbf{I}-\beta\mathbf{A}_1^T\mathbf{A}_1}^2$$

$$-\frac{1}{2}\mathbb{E}\left\|\frac{1}{\theta_{1,s}}\left[\mathbf{x}_{s,1}^m-\theta_2\tilde{\mathbf{x}}_{s,1}-(1-\theta_{1,s}-\theta_2)\mathbf{x}_{s,1}^{m-1}\right]\right.$$

$$\left.-\mathbf{x}_1^*\right\|_{(\theta_{1,s}L_1+\beta\|\mathbf{A}_1^T\mathbf{A}_1\|)\mathbf{I}-\beta\mathbf{A}_1^T\mathbf{A}_1}^2$$

$$+\frac{1}{2}\mathbb{E}\left\|\frac{1}{\theta_{1,s}}\left[\mathbf{y}_{s,2}^0-\theta_2\tilde{\mathbf{x}}_{s,2}-(1-\theta_{1,s}-\theta_2)\mathbf{x}_{s,2}^0\right]-\mathbf{x}_2^*\right\|_{(\alpha\theta_{1,s}L_2+\beta\|\mathbf{A}_2^T\mathbf{A}_2\|)\mathbf{I}}^2$$

$$-\frac{1}{2}\mathbb{E}\left\|\frac{1}{\theta_{1,s}}\left[\mathbf{x}_{s,2}^m - \theta_2\tilde{\mathbf{x}}_{s,2} - (1 - \theta_{1,s} - \theta_2)\mathbf{x}_{s,2}^{m-1}\right] - \mathbf{x}_2^*\right\|_{(\alpha\theta_{1,s}L_2+\beta\|\mathbf{A}_2^T\mathbf{A}_2\|)\mathbf{I}}^2$$

$$+\frac{1}{2\beta}\left(\mathbb{E}\left\|\hat{\boldsymbol{\lambda}}_s^0 - \boldsymbol{\lambda}^*\right\|^2 - \mathbb{E}\left\|\hat{\boldsymbol{\lambda}}_s^m - \boldsymbol{\lambda}^*\right\|^2\right), \quad s \geq 0. \tag{5.141}$$

When $k = 0$, for

$$\mathbf{y}_{s+1}^0 = (1 - \theta_2)\mathbf{x}_s^m + \theta_2\tilde{\mathbf{x}}_{s+1}$$

$$+\frac{\theta_{1,s+1}}{\theta_{1,s}}\left[(1 - \theta_{1,s})\mathbf{x}_s^m - (1 - \theta_{1,s} - \theta_2)\mathbf{x}_s^{m-1} - \theta_2\tilde{\mathbf{x}}_s\right],$$

we obtain

$$\frac{1}{\theta_{1,s}}\left[\mathbf{x}_s^m - \theta_2\tilde{\mathbf{x}}_s - (1 - \theta_{1,s} - \theta_2)\mathbf{x}_s^{m-1}\right]$$

$$= \frac{1}{\theta_{1,s+1}}\left[\mathbf{y}_{s+1}^0 - \theta_2\tilde{\mathbf{x}}_{s+1} - (1 - \theta_{1,s+1} - \theta_2)\mathbf{x}_{s+1}^0\right]. \tag{5.142}$$

Substituting (5.142) into the third and the fifth terms in the right-hand side of (5.141) and substituting (5.122) into the last term in the right-hand side of (5.141), we obtain

$$\frac{1}{\theta_{1,s}}\mathbb{E}\left(L(\mathbf{x}_s^m, \boldsymbol{\lambda}^*) - L(\mathbf{x}^*, \boldsymbol{\lambda}^*)\right) + \frac{\theta_2 + \theta_{1,s}}{\theta_{1,s}}\sum_{k=1}^{m-1}\mathbb{E}\left(L(\mathbf{x}_s^k, \boldsymbol{\lambda}^*) - L(\mathbf{x}^*, \boldsymbol{\lambda}^*)\right)$$

$$\leq \frac{1}{\theta_{1,s-1}}\mathbb{E}\left(L(\mathbf{x}_{s-1}^m, \boldsymbol{\lambda}^*) - L(\mathbf{x}^*, \boldsymbol{\lambda}^*)\right)$$

$$+\frac{\theta_2 + \theta_{1,s-1}}{\theta_{1,s-1}}\sum_{k=1}^{m-1}\mathbb{E}\left(L(\mathbf{x}_{s-1}^k, \boldsymbol{\lambda}^*) - L(\mathbf{x}^*, \boldsymbol{\lambda}^*)\right)$$

$$+\frac{1}{2}\mathbb{E}\left\|\frac{1}{\theta_{1,s-1}}\left[\mathbf{x}_{s-1,1}^m - \theta_2\tilde{\mathbf{x}}_{s-1,1} - (1 - \theta_{1,s-1} - \theta_2)\mathbf{x}_{s-1,1}^{m-1}\right]\right.$$

$$\left.-\mathbf{x}_1^*\right\|_{(\theta_{1,s}L_1+\beta\|\mathbf{A}_1^T\mathbf{A}_1\|)\mathbf{I}-\beta\mathbf{A}_1^T\mathbf{A}_1}^2$$

$$-\frac{1}{2}\mathbb{E}\left\|\frac{1}{\theta_{1,s}}\left[\mathbf{x}_{s,1}^m - \theta_2\tilde{\mathbf{x}}_{s,1} - (1 - \theta_{1,s} - \theta_2)\mathbf{x}_{s,1}^{m-1}\right]\right.$$

$$\left.-\mathbf{x}_1^*\right\|_{(\theta_{1,s}L_1+\beta\|\mathbf{A}_1^T\mathbf{A}_1\|)\mathbf{I}-\beta\mathbf{A}_1^T\mathbf{A}_1}^2$$

$$+ \frac{1}{2} \mathbb{E} \left\| \frac{1}{\theta_{1,s-1}} \left[\mathbf{x}_{s-1,2}^m - \theta_2 \tilde{\mathbf{x}}_{s-1,2} - (1 - \theta_{1,s-1} - \theta_2) \mathbf{x}_{s-1,2}^{m-1} \right] \right.$$

$$\left. - \mathbf{x}_2^* \right\|_{(\alpha \theta_{1,s} L_2 + \beta \|\mathbf{A}_2^T \mathbf{A}_2\|) \mathbf{I}}^2$$

$$- \frac{1}{2} \mathbb{E} \left\| \frac{1}{\theta_{1,s}} \left[\mathbf{x}_{s,2}^m - \theta_2 \tilde{\mathbf{x}}_{s,2} - (1 - \theta_{1,s} - \theta_2) \mathbf{x}_{s,2}^{m-1} \right] \right.$$

$$\left. - \mathbf{x}_2^* \right\|_{(\alpha \theta_{1,s} L_2 + \beta \|\mathbf{A}_2^T \mathbf{A}_2\|) \mathbf{I}}^2$$

$$+ \frac{1}{2\beta} \left(\mathbb{E} \left\| \hat{\boldsymbol{\lambda}}_{s-1}^m - \boldsymbol{\lambda}^* \right\|^2 - \mathbb{E} \left\| \hat{\boldsymbol{\lambda}}_s^m - \boldsymbol{\lambda}^* \right\|^2 \right), \quad s \geq 1.$$

For $\theta_{1,s-1} \geq \theta_{1,s}$, we have $\|\mathbf{x}\|_{\theta_{1,s-1}L}^2 \geq \|\mathbf{x}\|_{\theta_{1,s}L}^2$, thus

$$\frac{1}{\theta_{1,s}} \mathbb{E} \left(L(\mathbf{x}_s^m, \boldsymbol{\lambda}^*) - L(\mathbf{x}^*, \boldsymbol{\lambda}^*) \right) + \frac{\theta_2 + \theta_{1,s}}{\theta_{1,s}} \sum_{k=1}^{m-1} \mathbb{E} \left(L(\mathbf{x}_s^k, \boldsymbol{\lambda}^*) - L(\mathbf{x}^*, \boldsymbol{\lambda}^*) \right)$$

$$\leq \frac{1}{\theta_{1,s-1}} \mathbb{E} \left(L(\mathbf{x}_{s-1}^m, \boldsymbol{\lambda}^*) - L(\mathbf{x}^*, \boldsymbol{\lambda}^*) \right)$$

$$+ \frac{\theta_2 + \theta_{1,s-1}}{\theta_{1,s-1}} \sum_{k=1}^{m-1} \mathbb{E} \left(L(\mathbf{x}_{s-1}^k, \boldsymbol{\lambda}^*) - L(\mathbf{x}^*, \boldsymbol{\lambda}^*) \right)$$

$$+ \frac{1}{2} \mathbb{E} \left\| \frac{1}{\theta_{1,s-1}} \left[\mathbf{x}_{s-1,1}^m - \theta_2 \tilde{\mathbf{x}}_{s-1,1} - (1 - \theta_{1,s-1} - \theta_2) \mathbf{x}_{s-1,1}^{m-1} \right] \right.$$

$$\left. - \mathbf{x}_1^* \right\|_{(\theta_{1,s-1} L_1 + \beta \|\mathbf{A}_1^T \mathbf{A}_1\|) \mathbf{I} - \beta \mathbf{A}_1^T \mathbf{A}_1}^2$$

$$- \frac{1}{2} \mathbb{E} \left\| \frac{1}{\theta_{1,s}} \left[\mathbf{x}_{s,1}^m - \theta_2 \tilde{\mathbf{x}}_{s,1} - (1 - \theta_{1,s} - \theta_2) \mathbf{x}_{s,1}^{m-1} \right] \right.$$

$$\left. - \mathbf{x}_1^* \right\|_{(\theta_{1,s} L_1 + \beta \|\mathbf{A}_1^T \mathbf{A}_1\|) \mathbf{I} - \beta \mathbf{A}_1^T \mathbf{A}_1}^2$$

$$+ \frac{1}{2} \mathbb{E} \left\| \frac{1}{\theta_{1,s-1}} \left[\mathbf{x}_{s-1,2}^m - \theta_2 \tilde{\mathbf{x}}_{s-1,2} - (1 - \theta_{1,s-1} - \theta_2) \mathbf{x}_{s-1,2}^{m-1} \right] \right.$$

$$\left. - \mathbf{x}_2^* \right\|_{(\alpha \theta_{1,s-1} L_2 + \beta \|\mathbf{A}_2^T \mathbf{A}_2\|) \mathbf{I}}^2$$

$$-\frac{1}{2}\mathbb{E}\left\|\frac{1}{\theta_{1,s}}\left[\mathbf{x}_{s,2}^m - \theta_2\tilde{\mathbf{x}}_{s,2} - (1 - \theta_{1,s} - \theta_2)\mathbf{x}_{s,2}^{m-1}\right] - \mathbf{x}_2^*\right\|_{(\alpha\theta_{1,s}L_2+\beta\|\mathbf{A}_2^T\mathbf{A}_2\|)\mathbf{I}}^2$$

$$+\frac{1}{2\beta}\left(\mathbb{E}\left\|\hat{\boldsymbol{\lambda}}_{s-1}^m - \boldsymbol{\lambda}^*\right\|^2 - \mathbb{E}\left\|\hat{\boldsymbol{\lambda}}_s^m - \boldsymbol{\lambda}^*\right\|^2\right), \quad s \geq 1. \tag{5.143}$$

When $s = 0$, via (5.136) and using $\mathbf{y}_{0,1}^0 = \tilde{\mathbf{x}}_{0,1} = \mathbf{x}_{0,1}^0$, $\mathbf{y}_{0,2}^0 = \tilde{\mathbf{x}}_{0,2} = \mathbf{x}_{0,2}^0$, and $\theta_{1,0} \geq \theta_{1,1}$, we obtain

$$\frac{1}{\theta_{1,0}}\mathbb{E}\left(L(\mathbf{x}_0^m, \boldsymbol{\lambda}^*) - L(\mathbf{x}^*, \boldsymbol{\lambda}^*)\right) + \frac{\theta_2 + \theta_{1,0}}{\theta_{1,0}}\sum_{k=1}^{m-1}\mathbb{E}\left(L(\mathbf{x}_0^k, \boldsymbol{\lambda}^*) - L(\mathbf{x}^*, \boldsymbol{\lambda}^*)\right)$$

$$\leq \frac{1 - \theta_{1,0} + (m-1)\theta_2}{\theta_{1,0}}\left(L(\mathbf{x}_0^0, \boldsymbol{\lambda}^*) - L(\mathbf{x}^*, \boldsymbol{\lambda}^*)\right)$$

$$+\frac{1}{2}\left\|\mathbf{x}_{0,1}^0 - \mathbf{x}_1^*\right\|_{(\theta_{1,0}L_1+\beta\|\mathbf{A}_1^T\mathbf{A}_1\|)\mathbf{I}-\beta\mathbf{A}_1^T\mathbf{A}_1}^2$$

$$-\frac{1}{2}\mathbb{E}\left\|\frac{1}{\theta_{1,0}}\left[\mathbf{x}_{0,1}^m - \theta_2\tilde{\mathbf{x}}_{0,1} - (1 - \theta_{1,0} - \theta_2)\mathbf{x}_{0,1}^{m-1}\right]\right.$$

$$\left.-\mathbf{x}_1^*\right\|_{(\theta_{1,1}L_1+\beta\|\mathbf{A}_1^T\mathbf{A}_1\|)\mathbf{I}-\beta\mathbf{A}_1^T\mathbf{A}_1}^2$$

$$+\frac{1}{2}\left\|\mathbf{x}_{0,2}^0 - \mathbf{x}_1^*\right\|_{(\alpha\theta_{1,0}L_2+\beta\|\mathbf{A}_2^T\mathbf{A}_2\|)\mathbf{I}}^2$$

$$-\frac{1}{2}\mathbb{E}\left\|\frac{1}{\theta_{1,0}}\left[\mathbf{x}_{0,2}^m - \theta_2\tilde{\mathbf{x}}_{0,2} - (1 - \theta_{1,0} - \theta_2)\mathbf{x}_{0,2}^{m-1}\right]\right.$$

$$\left.-\mathbf{x}_2^*\right\|_{(\alpha\theta_{1,1}L_2+\beta\|\mathbf{A}_2^T\mathbf{A}_2\|)\mathbf{I}}^2$$

$$+\frac{1}{2\beta}\left(\left\|\hat{\boldsymbol{\lambda}}_0^0 - \boldsymbol{\lambda}^*\right\|^2 - \mathbb{E}\left\|\hat{\boldsymbol{\lambda}}_0^m - \boldsymbol{\lambda}^*\right\|^2\right), \tag{5.144}$$

where we use $\theta_{1,0} \geq \theta_{1,1}$ in the fourth and the sixth lines.

Summing (5.143) with s from 1 to $S-1$ and adding (5.144), we have

$$\frac{1}{\theta_{1,S}}\mathbb{E}\left(L(\mathbf{x}_S^m, \boldsymbol{\lambda}^*) - L(\mathbf{x}^*, \boldsymbol{\lambda}^*)\right) + \frac{\theta_{1,S} + \theta_2}{\theta_{1,S}}\sum_{k=1}^{m-1}\mathbb{E}\left(L(\mathbf{x}_S^k, \boldsymbol{\lambda}^*) - L(\mathbf{x}^*, \boldsymbol{\lambda}^*)\right)$$

$$\leq \frac{1 - \theta_{1,0} + (m-1)\theta_2}{\theta_{1,0}}\left(L(\mathbf{x}_0^0, \boldsymbol{\lambda}^*) - L(\mathbf{x}^*, \boldsymbol{\lambda}^*)\right)$$

$$+\frac{1}{2}\left\|\mathbf{x}_{0,1}^0 - \mathbf{x}_1^*\right\|_{(\theta_{1,0}L_1+\beta\|\mathbf{A}_1^T\mathbf{A}_1\|)\mathbf{I}-\beta\mathbf{A}_1^T\mathbf{A}_1}^2 + \frac{1}{2}\left\|\mathbf{x}_{0,2}^0 - \mathbf{x}_2^*\right\|_{(\alpha\theta_{1,0}L_2+\beta\|\mathbf{A}_2^T\mathbf{A}_2\|)\mathbf{I}}^2$$

$$+ \frac{1}{2\beta} \left(\left\| \hat{\boldsymbol{\lambda}}_0^0 - \boldsymbol{\lambda}^* \right\|^2 - \mathbb{E} \left\| \hat{\boldsymbol{\lambda}}_S^m - \boldsymbol{\lambda}^* \right\|^2 \right)$$

$$- \frac{1}{2} \mathbb{E} \left\| \frac{1}{\theta_{1,S}} \left[\mathbf{x}_{S,1}^m - \theta_2 \tilde{\mathbf{x}}_{S,1} - (1 - \theta_{1,s} - \theta_2) \mathbf{x}_{S,1}^{m-1} \right] \right.$$

$$\left. - \mathbf{x}_1^* \right\|^2_{\left(\theta_{1,S} L_1 + \beta \| \mathbf{A}_1^T \mathbf{A}_1 \| \right) \mathbf{I} - \beta \mathbf{A}_1^T \mathbf{A}_1}$$

$$- \frac{1}{2} \mathbb{E} \left\| \frac{1}{\theta_{1,S}} \left[\mathbf{x}_{S,2}^m - \theta_2 \tilde{\mathbf{x}}_{S,2} - (1 - \theta_{1,S} - \theta_2) \mathbf{x}_{S,2}^{m-1} \right] - \mathbf{x}_2^* \right\|^2_{\left(\alpha \theta_{1,S} L_2 + \beta \| \mathbf{A}_2^T \mathbf{A}_2 \| \right) \mathbf{I}}$$

$$\leq \frac{1 - \theta_{1,0} + (m-1)\theta_2}{\theta_{1,0}} \left(L(\mathbf{x}_0^0, \boldsymbol{\lambda}^*) - L(\mathbf{x}^*, \boldsymbol{\lambda}^*) \right)$$

$$+ \frac{1}{2} \left\| \mathbf{x}_{0,1}^0 - \mathbf{x}_1^* \right\|^2_{\left(\theta_{1,0} L_1 + \beta \| \mathbf{A}_1^T \mathbf{A}_1 \| \right) \mathbf{I} - \beta \mathbf{A}_1^T \mathbf{A}_1} + \frac{1}{2} \left\| \mathbf{x}_{0,2}^0 - \mathbf{x}_2^* \right\|^2_{\left(\alpha \theta_{1,0} L_2 + \beta \| \mathbf{A}_2^T \mathbf{A}_2 \| \right) \mathbf{I}}$$

$$+ \frac{1}{2\beta} \left(\left\| \hat{\boldsymbol{\lambda}}_0^0 - \boldsymbol{\lambda}^* \right\|^2 - \mathbb{E} \left\| \hat{\boldsymbol{\lambda}}_S^m - \boldsymbol{\lambda}^* \right\|^2 \right). \tag{5.145}$$

Now we analyze $\| \hat{\boldsymbol{\lambda}}_S^m - \boldsymbol{\lambda}^* \|^2$. From (5.126), for $s \geq 1$ we have

$$\hat{\boldsymbol{\lambda}}_s^m - \hat{\boldsymbol{\lambda}}_{s-1}^m = \hat{\boldsymbol{\lambda}}_s^m - \hat{\boldsymbol{\lambda}}_s^0 = \sum_{k=1}^{m} \left(\hat{\boldsymbol{\lambda}}_s^k - \hat{\boldsymbol{\lambda}}_s^{k-1} \right)$$

$$\overset{a}{=} \beta \sum_{k=1}^{m} \left[\frac{1}{\theta_{1,s}} \left(\mathbf{A} \mathbf{x}_s^k - \mathbf{b} \right) - \frac{1 - \theta_{1,s} - \theta_2}{\theta_{1,s}} \left(\mathbf{A} \mathbf{x}_s^{k-1} - \mathbf{b} \right) - \frac{\theta_2}{\theta_{1,s}} \left(\mathbf{A} \tilde{\mathbf{x}}_s - \mathbf{b} \right) \right]$$

$$= \frac{\beta}{\theta_{1,s}} \left(\mathbf{A} \mathbf{x}_s^m - \mathbf{b} \right) + \frac{\beta(\theta_2 + \theta_{1,s})}{\theta_{1,s}} \sum_{k=1}^{m-1} \left(\mathbf{A} \mathbf{x}_s^k - \mathbf{b} \right)$$

$$- \frac{\beta(1 - \theta_{1,s} - \theta_2)}{\theta_{1,s}} \left(\mathbf{A} \mathbf{x}_{s-1}^m - \mathbf{b} \right) - \frac{m \beta \theta_2}{\theta_{1,s}} \left(\mathbf{A} \tilde{\mathbf{x}}_s - \mathbf{b} \right)$$

$$\overset{b}{=} \frac{\beta}{\theta_{1,s}} \left(\mathbf{A} \mathbf{x}_s^m - \mathbf{b} \right) + \frac{\beta(\theta_2 + \theta_{1,s})}{\theta_{1,s}} \sum_{k=1}^{m-1} \left(\mathbf{A} \mathbf{x}_s^k - \mathbf{b} \right)$$

$$- \beta \left[\frac{1 - \theta_{1,s} - (\tau - 1)\theta_{1,s}}{\theta_{1,s}} \left(\mathbf{A} \mathbf{x}_{s-1}^m - \mathbf{b} \right) \right.$$

$$\left. + \frac{\theta_2 + \frac{\tau-1}{m-1}\theta_{1,s}}{\theta_{1,s}} \sum_{k=1}^{m-1} \left(\mathbf{A} \mathbf{x}_{s-1}^k - \mathbf{b} \right) \right]$$

$$\overset{c}{=} \frac{\beta}{\theta_{1,s}} \left(\mathbf{A}\mathbf{x}_s^m - \mathbf{b} \right) + \frac{\beta(\theta_2 + \theta_{1,s})}{\theta_{1,s}} \sum_{k=1}^{m-1} \left(\mathbf{A}\mathbf{x}_s^k - \mathbf{b} \right)$$

$$- \frac{\beta}{\theta_{1,s-1}} \left(\mathbf{A}\mathbf{x}_{s-1}^m - \mathbf{b} \right) - \frac{\beta(\theta_2 + \theta_{1,s-1})}{\theta_{1,s-1}} \sum_{k=1}^{m-1} \left(\mathbf{A}\mathbf{x}_{s-1}^k - \mathbf{b} \right), \tag{5.146}$$

where $\overset{a}{=}$ uses (5.121), $\overset{b}{=}$ uses the definition of $\tilde{\mathbf{x}}_s$, and $\overset{c}{=}$ uses (5.139) and (5.140). When $s = 0$, we can obtain

$$\hat{\boldsymbol{\lambda}}_0^m - \hat{\boldsymbol{\lambda}}_0^0 = \sum_{k=1}^m \left(\hat{\boldsymbol{\lambda}}_0^k - \hat{\boldsymbol{\lambda}}_0^{k-1} \right)$$

$$= \sum_{k=1}^m \left[\frac{\beta}{\theta_{1,0}} \left(\mathbf{A}\mathbf{x}_0^k - \mathbf{b} \right) - \frac{\beta(1 - \theta_{1,0} - \theta_2)}{\theta_{1,0}} \left(\mathbf{A}\mathbf{x}_0^{k-1} - \mathbf{b} \right) \right.$$

$$\left. - \frac{\theta_2 \beta}{\theta_{1,0}} \left(\mathbf{A}\mathbf{x}_0^0 - \mathbf{b} \right) \right]$$

$$= \frac{\beta}{\theta_{1,0}} \left(\mathbf{A}\mathbf{x}_0^m - \mathbf{b} \right) + \frac{\beta(\theta_2 + \theta_{1,0})}{\theta_{1,0}} \sum_{k=1}^{m-1} \left(\mathbf{A}\mathbf{x}_0^k - \mathbf{b} \right)$$

$$- \frac{\beta[1 - \theta_{1,0} + (m-1)\theta_2]}{\theta_{1,0}} \left(\mathbf{A}\mathbf{x}_0^0 - \mathbf{b} \right). \tag{5.147}$$

Summing (5.146) with s from 1 to $S - 1$ and adding (5.147), we have

$$\hat{\boldsymbol{\lambda}}_S^m - \boldsymbol{\lambda}^* = \hat{\boldsymbol{\lambda}}_S^m - \hat{\boldsymbol{\lambda}}_0^0 + \hat{\boldsymbol{\lambda}}_0^0 - \boldsymbol{\lambda}^*$$

$$= \frac{\beta}{\theta_{1,S}} \left(\mathbf{A}\mathbf{x}_S^m - \mathbf{b} \right) + \frac{\beta(\theta_2 + \theta_{1,S})}{\theta_{1,S}} \sum_{k=1}^{m-1} \left(\mathbf{A}\mathbf{x}_S^k - \mathbf{b} \right)$$

$$- \frac{\beta \left[1 - \theta_{1,0} + (m-1)\theta_2 \right]}{\theta_{1,0}} \left(\mathbf{A}\mathbf{x}_0^0 - \mathbf{b} \right)$$

$$+ \tilde{\boldsymbol{\lambda}}_0^0 + \frac{\beta(1 - \theta_{1,0})}{\theta_{1,0}} \left(\mathbf{A}\mathbf{x}_0^0 - \mathbf{b} \right) - \boldsymbol{\lambda}^*$$

$$\overset{a}{=} \frac{(m-1)(\theta_2 + \theta_{1,S})\beta + \beta}{\theta_{1,S}} \left(\mathbf{A}\hat{\mathbf{x}}_S - \mathbf{b} \right)$$

$$+ \tilde{\boldsymbol{\lambda}}_0^0 - \frac{\beta(m-1)\theta_2}{\theta_{1,0}} \left(\mathbf{A}\mathbf{x}_0^0 - \mathbf{b} \right) - \boldsymbol{\lambda}^*, \tag{5.148}$$

where the equality $\stackrel{a}{=}$ uses the definition of $\hat{\mathbf{x}}_S$. Substituting (5.148) into (5.145) and using that $L(\mathbf{x}, \boldsymbol{\lambda})$ is convex in $\mathbf{x}$, we have

$$
\mathbb{E}\left(\frac{1}{2\beta}\left\|\frac{\beta(m-1)(\theta_2 + \theta_{1,S}) + \beta}{\theta_{1,S}}\left(\mathbf{A}\hat{\mathbf{x}}_S - \mathbf{b}\right)\right.\right.
$$
$$
\left.\left.-\frac{\beta(m-1)\theta_2}{\theta_{1,0}}\left(\mathbf{A}\mathbf{x}_0^0 - \mathbf{b}\right) + \tilde{\boldsymbol{\lambda}}_0^0 - \boldsymbol{\lambda}^*\right\|^2\right)
$$
$$
+\mathbb{E}\left(\frac{(m-1)(\theta_2 + \theta_{1,S}) + 1}{\theta_{1,S}}\left(L(\hat{\mathbf{x}}_S, \boldsymbol{\lambda}^*) - L(\mathbf{x}^*, \boldsymbol{\lambda}^*)\right)\right)
$$
$$
\leq \frac{1 - \theta_{1,0} + (m-1)\theta_2}{\theta_{1,0}}\left(L(\mathbf{x}_0^0, \boldsymbol{\lambda}^*) - L(\mathbf{x}^*, \boldsymbol{\lambda}^*)\right) + \frac{1}{2\beta}\left\|\hat{\boldsymbol{\lambda}}_0^0 - \boldsymbol{\lambda}^*\right\|^2
$$
$$
+\frac{1}{2}\left\|\mathbf{x}_{0,1}^0 - \mathbf{x}_1^*\right\|_{(\theta_{1,0}L_1 + \beta\|\mathbf{A}_1^T\mathbf{A}_1\|)\mathbf{I} - \beta\mathbf{A}_1^T\mathbf{A}_1}^2
$$
$$
+\frac{1}{2}\left\|\mathbf{x}_{0,2}^0 - \mathbf{x}_2^*\right\|_{(\alpha\theta_{1,0}L_2 + \beta\|\mathbf{A}_2^T\mathbf{A}_2\|)\mathbf{I}}^2 .
$$

By the definitions of $L(\mathbf{x}, \boldsymbol{\lambda})$ and $\hat{\boldsymbol{\lambda}}_0^0$ we can obtain (5.133).   □

With Theorem 5.10 in hand, we can check that the convergence rate of Acc-SADMM to solve problem (5.98) is exactly $O(1/(mS))$. However, for non-accelerated methods, the convergence rate for ADMM in the non-ergodic sense is $O(1/\sqrt{mS})$ [5].

5.5   The Infinite Case

In the previous section, the momentum technique is used to achieve faster convergence rate. We introduce the other benefit: the momentum technique can increase the mini-batch size. Unlike the previous sections, here we focus on the infinite case which differs from the finite case, since the true gradient for the infinite case is unavailable. We show that by using the momentum technique, the mini-batch size can be largely increased. We consider the problem:

$$
\min_{\mathbf{x}} f(\mathbf{x}) \equiv \mathbb{E}[F(\mathbf{x}; \xi)], \tag{5.149}
$$

where $f(\mathbf{x})$ is μ-strongly convex and L-smooth and its stochastic gradient has finite variance bound σ, i.e.,

$$
\mathbb{E}\|\nabla F(\mathbf{x}; \xi) - \nabla f(\mathbf{x})\|^2 \leq \sigma^2.
$$

Algorithm 5.8 Stochastic accelerated gradient descent (SAGD)

Input $\eta = \frac{1}{2L}$, $\theta = \sqrt{\mu/(2L)}$, $\mathbf{x}^0$, and mini-batch size b.

for $k = 0$ to K **do**

$\mathbf{y}^k = \frac{2}{1+\theta}\mathbf{x}^k - \frac{1-\theta}{1+\theta}\mathbf{x}^{k-1}$,

Randomly sample b functions whose index set is denoted as I^k,

Set $\tilde{\nabla} f(\mathbf{y}^k) = \frac{1}{b} \sum_{i \in I^k} \nabla F(\mathbf{y}^k; i)$,

$\mathbf{x}^{k+1} = \mathbf{y}^k - \eta \tilde{\nabla} f(\mathbf{y}^k)$.

end for k

Output $\mathbf{x}^{K+1}$.

We apply mini-batch AGD to solve (5.149). The algorithm is shown in Algorithm 5.8.

Theorem 5.11 *For Algorithm 5.8, set the mini-batch size* $b = \frac{\sigma^2}{2L\theta\epsilon}$, *where* $\theta = \sqrt{\mu/(2L)}$. *After running* $K = \log_{1-\sqrt{\frac{\mu}{2L}}} \left(f(\mathbf{x}^0) - f(\mathbf{x}^*) + L\|\mathbf{x}^0 - \mathbf{x}^*\|^2 \right) - \log_{1-\sqrt{\frac{\mu}{2L}}}(\epsilon) \sim O\left(\sqrt{\frac{L}{\mu}} \log(L\|\mathbf{x}^0 - \mathbf{x}^*\|^2/\epsilon) \right)$ *iterations, we have*

$$\mathbb{E} f(\mathbf{x}^{K+1}) - f(\mathbf{x}^*) + L\mathbb{E} \left\| \mathbf{x}^{K+1} - (1-\theta)\mathbf{x}^K - \theta\mathbf{x}^* \right\|^2 \leq 2\epsilon.$$

Proof For $f(\mathbf{x})$ is L-smooth, we have

$$f(\mathbf{x}^{k+1}) \leq f(\mathbf{y}^k) + \langle \nabla f(\mathbf{y}^k), \mathbf{x}^{k+1} - \mathbf{y}^k \rangle + \frac{L}{2} \left\| \mathbf{x}^{k+1} - \mathbf{y}^k \right\|^2. \quad (5.150)$$

For $f(\mathbf{y})$ is μ-strongly convex, we have

$$f(\mathbf{y}^k) \leq f(\mathbf{x}^*) + \left\langle \nabla f(\mathbf{y}^k), \mathbf{y}^k - \mathbf{x}^* \right\rangle - \frac{\mu}{2} \left\| \mathbf{y}^k - \mathbf{x}^* \right\|^2, \quad (5.151)$$

and

$$f(\mathbf{y}^k) \leq f(\mathbf{x}^k) + \left\langle \nabla f(\mathbf{y}^k), \mathbf{y}^k - \mathbf{x}^k \right\rangle. \quad (5.152)$$

Multiplying (5.151) by θ and (5.152) by $1-\theta$ and summing the results with (5.150), we have

$$f(\mathbf{x}^{k+1}) \leq \theta f(\mathbf{x}^*) + (1-\theta)f(\mathbf{x}^k) + \left\langle \nabla f(\mathbf{y}^k), \mathbf{x}^{k+1} - (1-\theta)\mathbf{x}^k - \theta\mathbf{x}^* \right\rangle$$

$$+ \frac{L}{2} \left\| \mathbf{x}^{k+1} - \mathbf{y}^k \right\|^2 - \frac{\mu\theta}{2} \left\| \mathbf{y}^k - \mathbf{x}^* \right\|^2$$

$$= \theta f(\mathbf{x}^*) + (1-\theta)f(\mathbf{x}^k)$$

$$+ \left\langle \nabla f(\mathbf{y}^k) - \tilde{\nabla} f(\mathbf{y}^k), \mathbf{x}^{k+1} - (1-\theta)\mathbf{x}^k - \theta\mathbf{x}^* \right\rangle$$

$$+ \frac{L}{2} \left\| \mathbf{x}^{k+1} - \mathbf{y}^k \right\|^2 - \frac{\mu\theta}{2} \left\| \mathbf{y}^k - \mathbf{x}^* \right\|^2$$
$$+ \left\langle \tilde{\nabla} f(\mathbf{y}^k), \mathbf{x}^{k+1} - (1-\theta)\mathbf{x}^k - \theta\mathbf{x}^* \right\rangle,$$

where we will set $\theta = \sqrt{\mu/(2L)}$. By taking the expectation on the random number in $\mathcal{I}^k$, we have

$$\mathbb{E}_k \left\langle \nabla f(\mathbf{y}^k) - \tilde{\nabla} f(\mathbf{y}^k), \mathbf{x}^{k+1} - (1-\theta)\mathbf{x}^k - \theta\mathbf{x}^* \right\rangle$$
$$\overset{a}{=} \mathbb{E}_k \left\langle \nabla f(\mathbf{y}^k) - \tilde{\nabla} f(\mathbf{y}^k), \mathbf{x}^{k+1} - \mathbf{y}^k \right\rangle$$
$$\leq \frac{1}{2L} \mathbb{E}_k \left\| \nabla f(\mathbf{y}^k) - \tilde{\nabla} f(\mathbf{y}^k) \right\|^2 + \frac{L}{2} \mathbb{E}_k \left\| \mathbf{x}^{k+1} - \mathbf{y}^k \right\|^2,$$

where in $\overset{a}{=}$ we use $\mathbb{E}_k \left(\nabla f(\mathbf{y}^k) - \tilde{\nabla} f(\mathbf{y}^k) \right) = \mathbf{0}$. Thus we have

$$\mathbb{E}_k f(\mathbf{x}^{k+1})$$
$$\leq \theta f(\mathbf{x}^*) + (1-\theta) f(\mathbf{x}^k) - 2L\mathbb{E}_k \left\langle \mathbf{x}^{k+1} - \mathbf{y}^k, \mathbf{x}^{k+1} - (1-\theta)\mathbf{x}^k - \theta\mathbf{x}^* \right\rangle$$
$$+ \frac{L}{2} \mathbb{E}_k \left\| \mathbf{x}^{k+1} - \mathbf{y}^k \right\|^2 - \frac{\mu\theta}{2} \left\| \mathbf{y}^k - \mathbf{x}^* \right\|^2 + \frac{L}{2} \mathbb{E}_k \left\| \mathbf{x}^{k+1} - \mathbf{y}^k \right\|^2$$
$$+ \frac{1}{2L} \mathbb{E}_k \left\| \nabla f(\mathbf{y}^k) - \tilde{\nabla} f(\mathbf{y}^k) \right\|^2$$
$$\leq \theta f(\mathbf{x}^*) + (1-\theta) f(\mathbf{x}^k) + \frac{1}{2L} \mathbb{E}_k \left\| \nabla f(\mathbf{y}^k) - \tilde{\nabla} f(\mathbf{y}^k) \right\|^2$$
$$+ L \left\| \mathbf{y}^k - (1-\theta)\mathbf{x}^k - \theta\mathbf{x}^* \right\|^2 - L\mathbb{E}_k \left\| \mathbf{x}^{k+1} - (1-\theta)\mathbf{x}^k - \theta\mathbf{x}^* \right\|^2$$
$$- \frac{\mu\theta}{2} \left\| \mathbf{y}^k - \mathbf{x}^* \right\|^2.$$

Then using

$$L \left\| \mathbf{y}^k - (1-\theta)\mathbf{x}^k - \theta\mathbf{x}^* \right\|^2$$
$$= \theta^2 L \left\| \frac{1}{\theta} \mathbf{y}^k - \theta\mathbf{x}^* - \frac{1-\theta}{\theta} \mathbf{x}^k - (1-\theta)\mathbf{x}^* \right\|^2$$
$$= \theta^2 L \left\| \theta(\mathbf{y}^k - \mathbf{x}^*) + \left(\frac{1}{\theta} - \theta \right) \mathbf{y}^k - \frac{1-\theta}{\theta} \mathbf{x}^k - (1-\theta)\mathbf{x}^* \right\|^2$$
$$= \theta^2 L \left\| \theta(\mathbf{y}^k - \mathbf{x}^*) + (1-\theta) \left[\left(1 + \frac{1}{\theta} \right) \mathbf{y}^k - \frac{1}{\theta} \mathbf{x}^k - \mathbf{x}^* \right] \right\|^2$$

$$\overset{a}{\leq} \theta^3 L \left\| \mathbf{y}^k - \mathbf{x}^* \right\|^2 + \theta^2(1-\theta)L \left\| \left(1 + \frac{1}{\theta}\right)\mathbf{y}^k - \frac{1}{\theta}\mathbf{x}^k - \mathbf{x}^* \right\|^2$$

$$\overset{b}{=} \frac{\theta\mu}{2} \left\| \mathbf{y}^k - \mathbf{x}^* \right\|^2 + \theta^2(1-\theta)L \left\| \left(1 + \frac{1}{\theta}\right)\mathbf{y}^k - \frac{1}{\theta}\mathbf{x}^k - \mathbf{x}^* \right\|^2,$$

where in $\overset{a}{\leq}$ we use the convexity of $\|\cdot\|^2$ and in $\overset{b}{=}$ we use $\theta = \sqrt{\mu/(2L)}$. Then from the update rule, we have

$$\frac{1}{\theta}\left[\mathbf{x}^k - (1-\theta)\mathbf{x}^{k-1}\right] = \left(1 + \frac{1}{\theta}\right)\mathbf{y}^k - \frac{1}{\theta}\mathbf{x}^k.$$

So we obtain

$$\mathbb{E}_k f(\mathbf{x}^{k+1}) - f(\mathbf{x}^*) + L\mathbb{E}_k \left\| \mathbf{x}^{k+1} - (1-\theta)\mathbf{x}^k - \theta\mathbf{x}^* \right\|^2$$

$$\leq (1-\theta)\left(f(\mathbf{x}^k) - f(\mathbf{x}^*) + L \left\| \mathbf{x}^k - (1-\theta)\mathbf{x}^{k-1} - \theta\mathbf{x}^* \right\|^2 \right)$$

$$+ \frac{1}{2L}\mathbb{E}_k \left\| \nabla f(\mathbf{y}^k) - \tilde{\nabla} f(\mathbf{y}^k) \right\|^2.$$

Setting $K = \log_{1-\theta}\left(f(\mathbf{x}^0) - f(\mathbf{x}^*) + L\|\mathbf{x}^0 - \mathbf{x}^*\|^2\right) + \log_{1-\theta}(\epsilon^{-1}) \sim O\left(\sqrt{\frac{L}{\mu}}\log(1/\epsilon)\right)$ and taking the first K iterations into account, we have

$$\mathbb{E} f(\mathbf{x}^{K+1}) - f(\mathbf{x}^*) + L\mathbb{E} \left\| \mathbf{x}^{K+1} - (1-\theta)\mathbf{x}^K - \theta\mathbf{x}^* \right\|^2$$

$$\leq \epsilon + \frac{1}{2L}\sum_{i=0}^{K}(1-\theta)^{K-i}\mathbb{E} \left\| \nabla f(\mathbf{y}^i) - \tilde{\nabla} f(\mathbf{y}^i) \right\|^2.$$

By the setting of $b = \frac{\sigma^2}{2L\theta\epsilon}$ mini-batch size, we have

$$\mathbb{E} \left\| \nabla f(\mathbf{y}^k) - \tilde{\nabla} f(\mathbf{y}^k) \right\|^2 \leq 2L\theta\epsilon, \quad k \geq 0.$$

So we have

$$\mathbb{E} f(\mathbf{x}^{K+1}) - f(\mathbf{x}^*) + L\mathbb{E} \left\| \mathbf{x}^{K+1} - (1-\theta)\mathbf{x}^K - \theta\mathbf{x}^* \right\|^2 \leq 2\epsilon.$$

$\square$

Theorem 5.11 indicates that the total stochastic gradient calls is $\tilde{O}\left(\frac{\sigma^2}{\mu\epsilon}\right)$. However, the algorithm converges in $\tilde{O}\left(\sqrt{\frac{L}{\mu}}\right)$ steps, which is the fastest rate even in deterministic algorithms. So the momentum technique enlarges the mini-batch size by $\Omega\left(\sqrt{\frac{L}{\mu}}\right)$ times.

References

1. Z. Allen-Zhu, Katyusha: the first truly accelerated stochastic gradient method, in *Proceedings of the 49th Annual ACM SIGACT Symposium on Theory of Computing*, Montreal, (2017), pp. 1200–1206
2. Z. Allen-Zhu, E. Hazan, Optimal black-box reductions between optimization objectives, in *Advances in Neural Information Processing Systems*, Barcelona, vol. 29 (2016), pp. 1614–1622
3. Z. Allen-Zhu, Y. Li, Neon2: finding local minima via first-order oracles, in *Advances in Neural Information Processing Systems*, Montreal, vol. 31 (2018), pp. 3716–3726
4. Z. Allen-Zhu, Y. Yuan, Improved SVRG for non-strongly-convex or sum-of-non-convex objectives, in *Proceedings of the 33th International Conference on Machine Learning*, New York, (2016), pp. 1080–1089
5. D. Davis, W. Yin, Convergence rate analysis of several splitting schemes, in *Splitting Methods in Communication, Imaging, Science, and Engineering* (Springer, New York, 2016), pp. 115–163
6. A. Defazio, F. Bach, S. Lacoste-Julien, SAGA: a fast incremental gradient method with support for non-strongly convex composite objectives, in *Advances in Neural Information Processing Systems*, Montreal, vol. 27 (2014), pp. 1646–1654
7. C. Fang, C.J. Li, Z. Lin, T. Zhang, SPIDER: near-optimal non-convex optimization via stochastic path-integrated differential estimator, in *Advances in Neural Information Processing Systems*, Montreal, vol. 31 (2018), pp. 689–699
8. O. Fercoq, P. Richtárik, Accelerated, parallel, and proximal coordinate descent. SIAM J. Optim. **25**(4), 1997–2023 (2015)
9. D. Garber, E. Hazan, C. Jin, S.M. Kakade, C. Musco, P. Netrapalli, A. Sidford, Faster eigenvector computation via shift-and-invert preconditioning, in *Proceedings of the 33th International Conference on Machine Learning*, New York, (2016), pp. 2626–2634
10. R. Johnson, T. Zhang, Accelerating stochastic gradient descent using predictive variance reduction, in *Advances in Neural Information Processing Systems*, Lake Tahoe, vol. 26 (2013), pp. 315–323
11. Q. Lin, Z. Lu, L. Xiao, An accelerated proximal coordinate gradient method, in *Advances in Neural Information Processing Systems*, Montreal, vol. 27 (2014), pp. 3059–3067
12. H. Lin, J. Mairal, Z. Harchaoui, A universal catalyst for first-order optimization, in *Advances in Neural Information Processing Systems*, Montreal, vol. 28 (2015), pp. 3384–3392
13. J. Mairal, Optimization with first-order surrogate functions, in *Proceedings of the 30th International Conference on Machine Learning*, Atlanta, (2013), pp. 783–791
14. Y. Nesterov, A method for unconstrained convex minimization problem with the rate of convergence $O(1/k^2)$. Sov. Math. Dokl. **27**(2), 372–376 (1983)
15. Y. Nesterov, *Introductory Lectures on Convex Optimization: A Basic Course* (Springer, New York, 2004)
16. Y. Nesterov, Efficiency of coordinate descent methods on huge-scale optimization problems. SIAM J. Optim. **22**(2), 341–362 (2012)

17. M. Schmidt, N. Le Roux, F. Bach, Minimizing finite sums with the stochastic average gradient. Math. Program. **162**(1–2), 83–112 (2017)
18. S. Shalev-Shwartz, T. Zhang, Stochastic dual coordinate ascent methods for regularized loss minimization. J. Mach. Learn. Res. **14**, 567–599 (2013)
19. S. Shalev-Shwartz, T. Zhang, Accelerated proximal stochastic dual coordinate ascent for regularized loss minimization, in *Proceedings of the 31th International Conference on Machine Learning*, Beijing, (2014), pp. 64–72

Chapter 6
Accelerated Parallel Algorithms

Along with the popularity of multi-core computers and the crucial demands for handling large-scale data in machine learning, designing parallel algorithms has attracted lots of interests in recent years. In this section, we introduce how to apply the momentum technique to accelerate parallel algorithms.

6.1 Accelerated Asynchronous Algorithms

Parallel algorithms can be implemented in two fashions: asynchronous updates and synchronous updates. For asynchronous update, the threads (machines) are allowed to work independently and concurrently and none of them needs to wait for the others to finish computing. Therefore, a large overhead is reduced when we compare asynchronous algorithms with synchronous ones, especially when the computation costs for each thread are different, or a large load imbalance exists.

For synchronous algorithms, the updates are essentially identical to the serial one with variants only on implementation. However, for asynchronous algorithms the states of the parameters for computing the gradient are different. This is because when one thread is computing the gradient, other threads might have updated the parameters. So the gradient might be computed on an old state. We consider a simple algorithm, asynchronous gradient descent, in which all the threads concurrently load the parameters and compute the gradients, and update on the latest parameters. If we assign a global counter k to indicate each update from any thread, the iteration can be formulated as

$$\mathbf{x}^{k+1} = \mathbf{x}^k - \gamma \nabla f(\mathbf{x}^{j(k)}),$$

© Springer Nature Singapore Pte Ltd. 2020
Z. Lin et al., *Accelerated Optimization for Machine Learning*,
https://doi.org/10.1007/978-981-15-2910-8_6

where γ is the step size and $\mathbf{x}^{j(k)}$ is the state of $\mathbf{x}$ at the reading time. Typically, $\mathbf{x}^{j(k)}$ can be any of $\{\mathbf{x}^1, \cdots, \mathbf{x}^k\}$ when the parameters are updated with locks. In other words, for asynchronous algorithms the gradient might be delayed.

Up to now, lots of plain asynchronous algorithms have been designed. Niu et al. [12] and Agarwal et al. [1] proposed Asynchronous Stochastic Gradient Descent (ASGD). Some variance reduction (VR) based asynchronous algorithms were also designed later, such as [4, 13] in the primal space and [8] in the dual. Mania et al. [10] introduced a unified approach to analyze asynchronous stochastic algorithms by viewing them as serial methods operating on noisy inputs. Under the proposed framework of perturbed iterate analysis, lots of asynchronous algorithms, e.g., ASGD, asynchronous SVRG (ASVRG), and asynchronous SCD, can be studied in a unified scheme. Notably, linear convergence rates of asynchronous SCD and asynchronous sparse SVRG are achievable. However, to the best of our knowledge, the framework cannot incorporate the momentum based acceleration algorithms.

In the following, we illustrate how to apply the momentum technique to asynchronous algorithms. The challenges lying in designing asynchronous algorithms are twofold:

- In serial accelerated schemes, the extrapolation points are subtly and strictly connected with $\mathbf{x}^k$ and $\mathbf{x}^{k-1}$, i.e., $\mathbf{y}^k = \mathbf{x}^k + \frac{\theta_k(1-\theta_{k-1})}{\theta_k}\left(\mathbf{x}^k - \mathbf{x}^{k-1}\right)$. However, such information might not be available for asynchronous algorithms because there are unknown delays in updating the parameters.
- Since $\mathbf{x}^{k+1}$ is updated based on $\mathbf{y}^k$, i.e., $\mathbf{x}^{k+1} = \mathbf{y}^k - \frac{1}{L}\nabla f(\mathbf{y}^k)$, $\mathbf{x}^{k+1}$ is related to the old past updates (to generate $\mathbf{y}^k$). This is different from unaccelerated algorithms, e.g., in gradient descent, $\mathbf{x}^{k+1} = \mathbf{x}^k - \frac{1}{L}\nabla f(\mathbf{x}^k)$, in which $\mathbf{x}^{k+1}$ only depends on $\mathbf{x}^k$.

The main technique to tackle the above difficulties is by the "momentum compensation" [5]. We will show that doing only one original step of momentum prevents us from bounding the distance between delayed gradient and the latest one. Instead, by "momentum compensation" our algorithm is able to achieve a faster rate by order. Later, we will demonstrate that the technique can be applied to modern stochastic algorithms.

6.1.1 Asynchronous Accelerated Gradient Descent

Because the gradients are delayed for asynchronous algorithms, we make an assumption on a bounded delay for analysis:

Assumption 6.1 *All the updates before the $(k - \tau - 1)$-th iteration are completed before the "read" step of the k-th iteration.*

In most asynchronous parallelism, there are typically two schemes:

- Atom (consistent read) scheme: The parameter $\mathbf{x}$ is updated as an atom. When $\mathbf{x}$ is read or updated in the central node, it will be locked. So $\mathbf{x}^{j(k)} \in \{\mathbf{x}^0, \mathbf{x}^1, \cdots, \mathbf{x}^k\}$.
- Wild (inconsistent read) scheme: To further reduce the system overhead, there is no lock in implementation. All the threads may perform modifications on $\mathbf{x}$ at the same time [12]. Obviously, analysis becomes more complicated in this situation.

In this book, we only consider the atom scheme. If Assumption 6.1 holds, we have

$$j(k) \in \{k - \tau, k - \tau + 1, \cdots, k\}.$$

We consider the following objective function:

$$\min_{\mathbf{x}} f(\mathbf{x}) + h(\mathbf{x}),$$

where $f(\mathbf{x})$ is L-smooth and both $f(\mathbf{x})$ and $h(\mathbf{x})$ are convex. Recall the serial accelerated gradient descent (AGD) [11] described in Sect. 2.2.1 (called Accelerated Proximal Gradient Method when $h(\mathbf{x}) \neq 0$). If we directly implement AGD [11] asynchronously, we can only obtain the gradient $\nabla f(\mathbf{y}^{j(k)})$ due to the delay. Now we measure the distance between $\mathbf{y}^{j(k)}$ and $\mathbf{y}^k$. We have

$$\mathbf{y}^k = \mathbf{x}^k + \frac{\theta_k(1 - \theta_{k-1})}{\theta_{k-1}}(\mathbf{x}^k - \mathbf{x}^{k-1}). \tag{6.1}$$

Setting $a_k = \frac{\theta_k(1-\theta_{k-1})}{\theta_{k-1}}$, we have $a_k \leq 1$, because θ_k is monotonically non-increasing and $0 \leq \theta_k \leq 1$ in our setting. We have

$$
\begin{aligned}
\mathbf{y}^k &= \mathbf{x}^k + a_k(\mathbf{x}^k - \mathbf{y}^{k-1}) + a_k(\mathbf{y}^{k-1} - \mathbf{x}^{k-1}) \\
&= \mathbf{y}^{k-1} + (a_k + 1)(\mathbf{x}^k - \mathbf{y}^{k-1}) + a_k a_{k-1}(\mathbf{x}^{k-1} - \mathbf{x}^{k-2}), \quad k \geq 2.
\end{aligned}
\tag{6.2}
$$

For $\mathbf{x}^{k-1} - \mathbf{x}^{k-2}$, where $k \geq j(k) + 2 \geq 2$, we have

$$
\begin{aligned}
&\mathbf{x}^{k-1} - \mathbf{x}^{k-2} \\
&= \mathbf{x}^{k-1} - \mathbf{y}^{k-2} + \mathbf{y}^{k-2} - \mathbf{x}^{k-2} \\
&= \mathbf{x}^{k-1} - \mathbf{y}^{k-2} + a_{k-2}(\mathbf{x}^{k-2} - \mathbf{x}^{k-3}) \\
&= \mathbf{x}^{k-1} - \mathbf{y}^{k-2} + a_{k-2}(\mathbf{x}^{k-2} - \mathbf{y}^{k-3}) + a_{k-2}a_{k-3}(\mathbf{x}^{k-3} - \mathbf{x}^{k-4})
\end{aligned}
$$

$$= \mathbf{x}^{k-1} - \mathbf{y}^{k-2} + \sum_{i=j(k)+1}^{k-2} \left[\left(\prod_{l=i}^{k-2} a_l \right) (\mathbf{x}^i - \mathbf{y}^{i-1}) \right]$$

$$+ \left(\prod_{l=j(k)}^{k-2} a_l \right) (\mathbf{x}^{j(k)} - \mathbf{x}^{j(k)-1}). \tag{6.3}$$

Set $b(l, k) = \prod_{i=l}^{k} a_i$, where $l \le k$. Substituting (6.3) into (6.2), we have

$$\mathbf{y}^k = \mathbf{y}^{k-1} + (b(k, k) + 1)(\mathbf{x}^k - \mathbf{y}^{k-1}) + b(k-1, k)(\mathbf{x}^{k-1} - \mathbf{y}^{k-2})$$

$$+ \sum_{i=j(k)+1}^{k-2} \left(b(i, k)(\mathbf{x}^i - \mathbf{y}^{i-1}) \right) + b(j(k), k)(\mathbf{x}^{j(k)} - \mathbf{x}^{j(k)-1})$$

$$= \mathbf{y}^{k-1} + (\mathbf{x}^k - \mathbf{y}^{k-1}) + \sum_{i=j(k)+1}^{k} \left(b(i, k)(\mathbf{x}^i - \mathbf{y}^{i-1}) \right)$$

$$+ b(j(k), k)(\mathbf{x}^{j(k)} - \mathbf{x}^{j(k)-1}). \tag{6.4}$$

By checking, when $k = j(k)$ and $k = j(k) + 1$, (6.4) is also right. Summing (6.4) with $k = j(k) + 1$ to k, we have

$$\mathbf{y}^k = \mathbf{y}^{j(k)} + \sum_{i=j(k)+1}^{k} (\mathbf{x}^i - \mathbf{y}^{i-1}) + \sum_{l=j(k)+1}^{k} \sum_{i=j(k)+1}^{l} b(i, l)(\mathbf{x}^i - \mathbf{y}^{i-1})$$

$$+ \left(\sum_{i=j(k)+1}^{k} b(j(k), i) \right) (\mathbf{x}^{j(k)} - \mathbf{x}^{j(k)-1})$$

$$\overset{a}{=} \mathbf{y}^{j(k)} + \sum_{i=j(k)+1}^{k} (\mathbf{x}^i - \mathbf{y}^{i-1}) + \sum_{i=j(k)+1}^{k} \left(\sum_{l=i}^{k} b(i, l) \right) (\mathbf{x}^i - \mathbf{y}^{i-1})$$

$$+ \left(\sum_{i=j(k)+1}^{k} b(j(k), i) \right) (\mathbf{x}^{j(k)} - \mathbf{x}^{j(k)-1}), \tag{6.5}$$

where $\overset{a}{=}$ is obtained by reordering the summations on i and l. Notice that $\mathbf{x}^{j(k)} - \mathbf{x}^{j(k)-1}$ is related to all the past updates before $j(k)$. If we directly implement AGD asynchronously like most asynchronous algorithms, then $j(k) < k$ (due to delay), so $\sum_{i=j(k)+1}^{k} b(j(k), i) > 0$. Because $\mathbf{x}^{j(k)} - \mathbf{x}^{j(k)-1}$ is hard to bound, it causes difficulty in obtaining the accelerated convergence rate.

Algorithm 6.1 Asynchronous accelerated gradient descent (AAGD)

Input θ_k, step size γ, $\mathbf{x}^0 = \mathbf{0}$, $\mathbf{z}^0 = \mathbf{0}$, $a_k = \frac{\theta_k(1-\theta_{k-1})}{\theta_{k-1}}$, and $b(l,k) = \prod_{i=l}^{k} a_i$.

for $k = 0$ to K **do**

1   $\mathbf{w}^{j(k)} = \mathbf{x}^{j(k)} + \left(\sum_{i=j(k)}^{k} b(j(k), i) \right) (\mathbf{x}^{j(k)} - \mathbf{x}^{j(k)-1})$,

2   $\boldsymbol{\delta}^k = \arg\min_{\boldsymbol{\delta}} \left(h(\mathbf{z}^k + \boldsymbol{\delta}) + \langle \nabla f(\mathbf{w}^{j(k)}), \boldsymbol{\delta} \rangle + \frac{\theta_k}{2\gamma} \|\boldsymbol{\delta}\|^2 \right)$,

3   $\mathbf{z}^{k+1} = \mathbf{z}^k + \boldsymbol{\delta}^k$,

4   $\mathbf{x}^{k+1} = \theta_k \mathbf{z}^{k+1} + (1 - \theta_k)\mathbf{x}^k$.

end for k

Output $\mathbf{x}^{K+1}$.

Instead, one can compensate the momentum term and introduce a new extrapolation point $\mathbf{w}^{j(k)}$, such that

$$\mathbf{w}^{j(k)} = \mathbf{x}^{j(k)} + \left(\sum_{i=j(k)}^{k} b(j(k), i) \right) (\mathbf{x}^{j(k)} - \mathbf{x}^{j(k)-1}).$$

One can find that these are actually several steps of momentum. Then the difference between $\mathbf{y}^k$ and $\mathbf{w}^{j(k)}$ can be directly bounded by the norm of several latest updates, namely $\left\| \sum_{i=j(k)+1}^{k} \left(1 + \sum_{l=i}^{k} b(i,l)\right) (\mathbf{x}^i - \mathbf{y}^{i-1}) \right\|^2$ (see (6.23)). So we are able to obtain the accelerated rate. The algorithm is shown in Algorithm 6.1.

We have the following theorem.

Theorem 6.1 *Under Assumption 6.1, for Algorithm 6.1, when $h(\mathbf{x})$ is generally convex, if the step size γ satisfies $2\gamma L + 3\gamma L(\tau^2 + 3\tau)^2 \leq 1$, $\theta_k = \frac{2}{k+2}$, and the first τ iterations are updated in serial,[1] we have*

$$F(\mathbf{x}^{K+1}) - F(\mathbf{x}^*) \leq \frac{\theta_K^2}{2\gamma} \|\mathbf{x}^0 - \mathbf{x}^*\|^2. \tag{6.6}$$

When $h(\mathbf{x})$ is μ-strongly convex ($\mu \leq L$), the step size γ satisfies $\frac{5}{2}\gamma L + \gamma L(\tau^2 + 3\tau)^2 \leq 1$, and $\theta_k = \frac{-\gamma\mu + \sqrt{\gamma\mu^2 + 4\gamma\mu}}{2}$, denoted as θ instead, we have

$$F(\mathbf{x}^{K+1}) - F(\mathbf{x}^*) \leq (1-\theta)^{K+1} \left[F(\mathbf{x}^0) - F(\mathbf{x}^*) + \left(\frac{\theta^2}{2\gamma} + \frac{\mu\theta}{2} \right) \|\mathbf{x}^0 - \mathbf{x}^*\|^2 \right]. \tag{6.7}$$

Proof We introduce

$$\mathbf{y}^k = (1 - \theta_k)\mathbf{x}^k + \theta_k \mathbf{z}^k. \tag{6.8}$$

---

[1] This assumption is used only for simplifying the proof.

Because $\mathbf{x}^k = \theta_{k-1}\mathbf{z}^k + (1 - \theta_{k-1})\mathbf{x}^{k-1}$, we can eliminate $\mathbf{z}^k$ and obtain $\mathbf{y}^k = \mathbf{x}^k + \frac{\theta_k(1-\theta_{k-1})}{\theta_{k-1}}(\mathbf{x}^k - \mathbf{x}^{k-1})$, which is (6.1). We can further have

$$\mathbf{x}^{k+1} - \mathbf{y}^k = \theta_k(\mathbf{z}^{k+1} - \mathbf{z}^k). \tag{6.9}$$

From (6.5) and $\mathbf{y}^{j(k)} = \mathbf{x}^{j(k)} + a_{j(k)}\left(\mathbf{x}^{j(k)} - \mathbf{x}^{j(k)-1}\right)$, we have

$$\mathbf{y}^k = \mathbf{x}^{j(k)} + \sum_{i=j(k)+1}^{k}(\mathbf{x}^i - \mathbf{y}^{i-1}) + \sum_{i=j(k)+1}^{k}\left(\sum_{l=i}^{k}b(i,l)\right)(\mathbf{x}^i - \mathbf{y}^{i-1})$$

$$+ \left(\sum_{i=j(k)}^{k}b(j(k),i)\right)(\mathbf{x}^{j(k)} - \mathbf{x}^{j(k)-1}). \tag{6.10}$$

Through the optimality of $\mathbf{z}^{k+1}$ in Step 2 of Algorithm 6.1, we have

$$\theta_k(\mathbf{z}^{k+1} - \mathbf{z}^k) + \gamma\nabla f(\mathbf{w}^{j(k)}) + \gamma\boldsymbol{\xi}^k = \mathbf{0}, \tag{6.11}$$

where $\boldsymbol{\xi}^k \in \partial h(\mathbf{z}^{k+1})$ and

$$(\mathbf{x}^{k+1} - \mathbf{y}^k) + \gamma\nabla f(\mathbf{w}^{j(k)}) + \gamma\boldsymbol{\xi}^k = \mathbf{0}, \tag{6.12}$$

thanks to (6.9).

For f is L-smooth, we obtain

$$f(\mathbf{x}^{k+1}) \leq f(\mathbf{y}^k) + \langle\nabla f(\mathbf{y}^k), \mathbf{x}^{k+1} - \mathbf{y}^k\rangle + \frac{L}{2}\left\|\mathbf{x}^{k+1} - \mathbf{y}^k\right\|^2$$

$$\overset{a}{=} f(\mathbf{y}^k) - \gamma\langle\nabla f(\mathbf{y}^k), \nabla f(\mathbf{w}^{j(k)}) + \boldsymbol{\xi}^k\rangle + \frac{L}{2}\left\|\mathbf{x}^{k+1} - \mathbf{y}^k\right\|^2$$

$$\overset{b}{=} f(\mathbf{y}^k) - \gamma\langle\nabla f(\mathbf{w}^{j(k)}) + \boldsymbol{\xi}^k, \nabla f(\mathbf{w}^{j(k)}) + \boldsymbol{\xi}^k\rangle + \frac{L}{2}\left\|\mathbf{x}^{k+1} - \mathbf{y}^k\right\|^2$$

$$+ \gamma\langle\boldsymbol{\xi}^k, \nabla f(\mathbf{w}^{j(k)}) + \boldsymbol{\xi}^k\rangle + \gamma\langle\nabla f(\mathbf{w}^{j(k)}) - \nabla f(\mathbf{y}^k), \nabla f(\mathbf{w}^{j(k)}) + \boldsymbol{\xi}^k\rangle$$

$$\overset{c}{=} f(\mathbf{y}^k) - \gamma\left(1 - \frac{\gamma L}{2}\right)\left\|\frac{1}{\gamma}\left(\mathbf{x}^{k+1} - \mathbf{y}^k\right)\right\|^2 - \langle\boldsymbol{\xi}^k, \mathbf{x}^{k+1} - \mathbf{y}^k\rangle$$

$$- \langle\nabla f(\mathbf{w}^{j(k)}) - \nabla f(\mathbf{y}^k), \mathbf{x}^{k+1} - \mathbf{y}^k\rangle, \tag{6.13}$$

where in $\overset{a}{=}$ we use (6.12), in $\overset{b}{=}$ we use $-\nabla f(\mathbf{y}^k) = -\left(\nabla f(\mathbf{w}^{j(k)}) + \boldsymbol{\xi}^k\right) + \boldsymbol{\xi}^k + \left(\nabla f(\mathbf{w}^{j(k)}) - \nabla f(\mathbf{y}^k)\right)$, and in $\overset{c}{=}$ we reuse (6.12).

For the last term of (6.13), applying the Cauchy–Schwartz inequality, we have

$$\left\langle \nabla f(\mathbf{w}^{j(k)}) - \nabla f(\mathbf{y}^k), \mathbf{x}^{k+1} - \mathbf{y}^k \right\rangle$$

$$\leq \frac{\gamma}{2C_1} \left\| \nabla f(\mathbf{w}^{j(k)}) - \nabla f(\mathbf{y}^k) \right\|^2 + \frac{\gamma C_1}{2} \left\| \frac{1}{\gamma} \left(\mathbf{x}^{k+1} - \mathbf{y}^k \right) \right\|^2$$

$$\leq \frac{\gamma L^2}{2C_1} \left\| \mathbf{w}^{j(k)} - \mathbf{y}^k \right\|^2 + \frac{\gamma C_1}{2} \left\| \frac{1}{\gamma} \left(\mathbf{x}^{k+1} - \mathbf{y}^k \right) \right\|^2. \tag{6.14}$$

Substituting (6.14) into (6.13), we obtain

$$f(\mathbf{x}^{k+1}) \leq f(\mathbf{y}^k) - \gamma \left(1 - \frac{\gamma L}{2} \right) \left\| \frac{1}{\gamma} \left(\mathbf{x}^{k+1} - \mathbf{y}^k \right) \right\|^2 - \langle \boldsymbol{\xi}^k, \mathbf{x}^{k+1} - \mathbf{y}^k \rangle$$

$$+ \frac{\gamma L^2}{2C_1} \left\| \mathbf{w}^{j(k)} - \mathbf{y}^k \right\|^2 + \frac{\gamma C_1}{2} \left\| \frac{1}{\gamma} \left(\mathbf{x}^{k+1} - \mathbf{y}^k \right) \right\|^2. \tag{6.15}$$

Considering $\|\mathbf{z}^{k+1} - \mathbf{x}^*\|^2$, we have

$$\frac{1}{2\gamma} \left\| \theta_k \mathbf{z}^{k+1} - \theta_k \mathbf{x}^* \right\|^2$$

$$= \frac{1}{2\gamma} \left\| \theta_k \mathbf{z}^k - \theta_k \mathbf{x}^* + \theta_k \mathbf{z}^{k+1} - \theta_k \mathbf{z}^k \right\|^2$$

$$= \frac{1}{2\gamma} \left\| \theta_k \mathbf{z}^k - \theta_k \mathbf{x}^* \right\|^2 + \frac{1}{2\gamma} \left\| \theta_k \mathbf{z}^{k+1} - \theta_k \mathbf{z}^k \right\|^2$$

$$+ \frac{1}{\gamma} \left\langle \theta_k \left(\mathbf{z}^{k+1} - \mathbf{z}^k \right), \theta_k \mathbf{z}^k - \theta_k \mathbf{x}^* \right\rangle$$

$$\overset{a}{=} \frac{1}{2\gamma} \left\| \theta_k \mathbf{z}^k - \theta_k \mathbf{x}^* \right\|^2 + \frac{1}{2\gamma} \left\| \mathbf{x}^{k+1} - \mathbf{y}^k \right\|^2 - \langle \nabla f(\mathbf{w}^{j(k)}) + \boldsymbol{\xi}^k, \theta_k \mathbf{z}^k - \theta_k \mathbf{x}^* \rangle, \tag{6.16}$$

where in $\overset{a}{=}$ we use (6.11). For the last term, we have

$$-\langle \nabla f(\mathbf{w}^{j(k)}), \theta_k \mathbf{z}^k - \theta_k \mathbf{x}^* \rangle$$

$$\overset{a}{=} -\langle \nabla f(\mathbf{w}^{j(k)}), \mathbf{y}^k - (1 - \theta_k)\mathbf{x}^k - \theta_k \mathbf{x}^* \rangle$$

$$\overset{b}{=} -\langle \nabla f(\mathbf{w}^{j(k)}), \mathbf{w}^{j(k)} - (1 - \theta_k)\mathbf{x}^k - \theta_k \mathbf{x}^* \rangle - \langle \nabla f(\mathbf{w}^{j(k)}), \mathbf{y}^k - \mathbf{w}^{j(k)} \rangle$$

$$\overset{c}{\leq} (1 - \theta_k)f(\mathbf{x}^k) + \theta_k f(\mathbf{x}^*) - f(\mathbf{w}^{j(k)}) - \langle \nabla f(\mathbf{w}^{j(k)}), \mathbf{y}^k - \mathbf{w}^{j(k)} \rangle$$

$$\overset{d}{\leq} (1 - \theta_k)f(\mathbf{x}^k) + \theta_k f(\mathbf{x}^*) - f(\mathbf{y}^k) + \langle \nabla f(\mathbf{y}^k) - \nabla f(\mathbf{w}^{j(k)}), \mathbf{y}^k - \mathbf{w}^{j(k)} \rangle, \tag{6.17}$$

where $\overset{a}{=}$ uses (6.8), in $\overset{b}{=}$ we insert $\mathbf{w}^{j(k)}$, in $\overset{c}{\leq}$ we use the convexity of f, namely applying

$$f(\mathbf{w}^{j(k)}) + \langle \nabla f(\mathbf{w}^{j(k)}), \mathbf{a} - \mathbf{w}^{j(k)} \rangle \leq f(\mathbf{a})$$

on $\mathbf{a} = \mathbf{x}^*$ and $\mathbf{a} = \mathbf{x}^k$, respectively, and in $\overset{d}{\leq}$ we use

$$-f(\mathbf{w}^{j(k)}) \leq -f(\mathbf{y}^k) + \langle \nabla f(\mathbf{y}^k), \mathbf{y}^k - \mathbf{w}^{j(k)} \rangle.$$

Substituting (6.17) into (6.16), we have

$$
\frac{1}{2\gamma} \left\| \theta_k \mathbf{z}^{k+1} - \theta_k \mathbf{x}^* \right\|^2
$$
$$
\leq \frac{1}{2\gamma} \left\| \theta_k \mathbf{z}^k - \theta_k \mathbf{x}^* \right\|^2 + \frac{1}{2\gamma} \left\| \mathbf{x}^{k+1} - \mathbf{y}^k \right\|^2 - \langle \boldsymbol{\xi}^k, \theta_k \mathbf{z}^k - \theta_k \mathbf{x}^* \rangle
$$
$$
+ (1 - \theta_k) f(\mathbf{x}^k) + \theta_k f(\mathbf{x}^*) - f(\mathbf{y}^k) + \left\langle \nabla f(\mathbf{y}^k) - \nabla f(\mathbf{w}^{j(k)}), \mathbf{y}^k - \mathbf{w}^{j(k)} \right\rangle.
$$
$$\tag{6.18}$$

Adding (6.15) and (6.18), we have

$$
f(\mathbf{x}^{k+1}) \leq (1 - \theta_k) f(\mathbf{x}^k) + \theta_k f(\mathbf{x}^*) - \gamma \left(\frac{1}{2} - \frac{\gamma L}{2} \right) \left\| \frac{1}{\gamma} \left(\mathbf{x}^{k+1} - \mathbf{y}^k \right) \right\|^2
$$
$$
- \langle \boldsymbol{\xi}^k, \mathbf{x}^{k+1} - \mathbf{y}^k \rangle + \frac{\gamma L^2}{2 C_1} \left\| \mathbf{w}^{j(k)} - \mathbf{y}^k \right\|^2 + \frac{\gamma C_1}{2} \left\| \frac{1}{\gamma} \left(\mathbf{x}^{k+1} - \mathbf{y}^k \right) \right\|^2
$$
$$
- \left\langle \boldsymbol{\xi}^k, \theta_k \mathbf{z}^k - \theta_k \mathbf{x}^* \right\rangle + \left\langle \nabla f(\mathbf{y}^k) - \nabla f(\mathbf{w}^{j(k)}), \mathbf{y}^k - \mathbf{w}^{j(k)} \right\rangle
$$
$$
+ \frac{1}{2\gamma} \left\| \theta_k \mathbf{z}^k - \theta_k \mathbf{x}^* \right\|^2 - \frac{1}{2\gamma} \left\| \theta_k \mathbf{z}^{k+1} - \theta_k \mathbf{x}^* \right\|^2
$$
$$
\overset{a}{\leq} (1 - \theta_k) f(\mathbf{x}^k) + \theta_k f(\mathbf{x}^*) - \gamma \left(\frac{1}{2} - \frac{\gamma L}{2} \right) \left\| \frac{1}{\gamma} \left(\mathbf{x}^{k+1} - \mathbf{y}^k \right) \right\|^2
$$
$$
- \langle \boldsymbol{\xi}^k, \mathbf{x}^{k+1} - \mathbf{y}^k \rangle + \left(\frac{\gamma L^2}{2 C_1} + L \right) \left\| \mathbf{w}^{j(k)} - \mathbf{y}^k \right\|^2
$$
$$
+ \frac{\gamma C_1}{2} \left\| \frac{1}{\gamma} \left(\mathbf{x}^{k+1} - \mathbf{y}^k \right) \right\|^2 - \langle \boldsymbol{\xi}^k, \theta_k \mathbf{z}^k - \theta_k \mathbf{x}^* \rangle + \frac{1}{2\gamma} \left\| \theta_k \mathbf{z}^k - \theta_k \mathbf{x}^* \right\|^2
$$
$$
- \frac{1}{2\gamma} \left\| \theta_k \mathbf{z}^{k+1} - \theta_k \mathbf{x}^* \right\|^2,
$$
$$\tag{6.19}$$

where in $\overset{a}{\leq}$ we use $\langle \nabla f(\mathbf{y}^k) - \nabla f(\mathbf{w}^{j(k)}), \mathbf{y}^k - \mathbf{w}^{j(k)} \rangle \leq L \|\mathbf{y}^k - \mathbf{w}^{j(k)}\|^2$, by the Cauchy–Schwartz inequality and the L-smoothness of $f(\cdot)$. Since $\boldsymbol{\xi}^k \in \partial h(\mathbf{z}^{k+1})$, we also have

$$-\langle \boldsymbol{\xi}^k, \mathbf{x}^{k+1} - \mathbf{y}^k \rangle - \langle \boldsymbol{\xi}^k, \theta_k \mathbf{z}^k - \theta_k \mathbf{x}^* \rangle$$
$$= \theta_k \langle \boldsymbol{\xi}^k, \mathbf{x}^* - \mathbf{z}^{k+1} \rangle$$
$$\leq \theta_k h(\mathbf{x}^*) - \theta_k h(\mathbf{z}^{k+1}) - \frac{\mu \theta_k}{2} \left\| \mathbf{z}^{k+1} - \mathbf{x}^* \right\|^2. \tag{6.20}$$

For the convexity of $h(\mathbf{z}^{k+1})$, we have

$$\theta_k h(\mathbf{z}^{k+1}) + (1 - \theta_k) h(\mathbf{x}^k) \geq h(\mathbf{x}^{k+1}). \tag{6.21}$$

Substituting (6.20) into (6.19) and using (6.21), we have

$$F(\mathbf{x}^{k+1}) \leq (1 - \theta_k) F(\mathbf{x}^k) + \theta_k F(\mathbf{x}^*) - \gamma \left(\frac{1}{2} - \frac{\gamma L}{2} \right) \left\| \frac{1}{\gamma} \left(\mathbf{x}^{k+1} - \mathbf{y}^k \right) \right\|^2$$
$$+ \left(\frac{\gamma L^2}{2C_1} + L \right) \left\| \mathbf{w}^{j(k)} - \mathbf{y}^k \right\|^2 + \frac{\gamma C_1}{2} \left\| \frac{1}{\gamma} \left(\mathbf{x}^{k+1} - \mathbf{y}^k \right) \right\|^2$$
$$+ \frac{1}{2\gamma} \left\| \theta_k \mathbf{z}^k - \theta_k \mathbf{x}^* \right\|^2 - \left(\frac{1}{2\gamma} + \frac{\mu}{2\theta_k} \right) \left\| \theta_k \mathbf{z}^{k+1} - \theta_k \mathbf{x}^* \right\|^2. \tag{6.22}$$

We first consider the generally convex case. Through (6.10), we have

$$\left\| \mathbf{w}^{j(k)} - \mathbf{y}^k \right\|^2$$
$$= \left\| \sum_{i=j(k)+1}^{k} \left(1 + \sum_{l=i}^{k} b(i,l) \right) (\mathbf{x}^i - \mathbf{y}^{i-1}) \right\|^2$$
$$\overset{a}{\leq} \left[\sum_{i=j(k)+1}^{k} \left(1 + \sum_{l=i}^{k} b(i,l) \right) \right] \sum_{i=j(k)+1}^{k} \left(1 + \sum_{l=i}^{k} b(i,l) \right) \left\| \mathbf{x}^i - \mathbf{y}^{i-1} \right\|^2$$
$$\overset{b}{\leq} \left[\sum_{i=j(k)+1}^{k} \left(1 + \sum_{l=1}^{k-i+1} 1 \right) \right] \sum_{i=j(k)+1}^{k} \left(1 + \sum_{l=1}^{k-i+1} 1 \right) \left\| \mathbf{x}^i - \mathbf{y}^{i-1} \right\|^2$$
$$\overset{c}{\leq} \left[\sum_{ii=1}^{k-j(k)} \left(1 + \sum_{l=1}^{ii} 1 \right) \right] \sum_{ii=1}^{k-j(k)} \left(1 + \sum_{l=1}^{ii} 1 \right) \left\| \mathbf{x}^{k-ii+1} - \mathbf{y}^{k-ii} \right\|^2$$

$$\overset{d}{\leq} \left[\sum_{ii=1}^{\min(\tau,k)} \left(1 + \sum_{l=1}^{ii} 1 \right) \right] \sum_{ii=1}^{\tau} \left(1 + \sum_{l=1}^{ii} 1 \right) \left\| \mathbf{x}^{k-ii+1} - \mathbf{y}^{k-ii} \right\|^2$$

$$\leq \frac{\tau^2 + 3\tau}{2} \sum_{i=1}^{\min(\tau,k)} (i+1) \left\| \mathbf{x}^{k-i+1} - \mathbf{y}^{k-i} \right\|^2, \tag{6.23}$$

where in $\overset{a}{\leq}$ we use the fact that for $c_i \geq 0$, $0 \leq i \leq n$,

$$\| c_1 \mathbf{a}_1 + c_2 \mathbf{a}_2 + \cdots c_n \mathbf{a}_n \|^2$$
$$\leq (c_1 + c_2 + \cdots + c_n)\left(c_1 \| \mathbf{a}_1 \|^2 + c_2 \| \mathbf{a}_2 \|^2 + \cdots c_n \| \mathbf{a}_n \|^2 \right),$$

because function $\phi(\mathbf{x}) = \|\mathbf{x}\|^2$ is convex, in $\overset{b}{\leq}$ we use $b(i,l) \leq 1$, in $\overset{c}{\leq}$ we change variable $ii = k - i + 1$, and in $\overset{d}{\leq}$ we use $k - j(k) \leq \tau$.

As we are more interested in the limiting case, when k is large (e.g., $k \geq 2(\tau - 1)$) we assume that at the first τ steps, we run our algorithm in serial. Dividing θ_k^2 on both sides of (6.23) and summing the results with $k = 0$ to K, we have

$$\sum_{k=0}^{K} \frac{1}{\theta_k^2} \left\| \mathbf{w}^{j(k)} - \mathbf{y}^k \right\|^2 = \sum_{k=\tau}^{K} \frac{1}{\theta_k^2} \left\| \mathbf{w}^{j(k)} - \mathbf{y}^k \right\|^2$$

$$\leq \frac{\tau^2 + 3\tau}{2} \sum_{k=\tau}^{K} \sum_{i=1}^{\min(\tau, k-\tau)} \frac{i+1}{\theta_k^2} \left\| \mathbf{x}^{k-i+1} - \mathbf{y}^{k-i} \right\|^2$$

$$\overset{a}{\leq} \frac{\tau^2 + 3\tau}{2} \sum_{k=\tau}^{K} \sum_{i=1}^{\min(\tau, k-\tau)} \frac{4(i+1)}{\theta_{k-i}^2} \left\| \mathbf{x}^{k-i+1} - \mathbf{y}^{k-i} \right\|^2$$

$$\leq \frac{\tau^2 + 3\tau}{2} \sum_{i=1}^{\tau} \sum_{k=i+\tau}^{K} \frac{4(i+1)}{\theta_{k-i}^2} \left\| \mathbf{x}^{k-i+1} - \mathbf{y}^{k-i} \right\|^2$$

$$= \frac{\tau^2 + 3\tau}{2} \sum_{i=1}^{\tau} [4(i+1)] \sum_{k'=\tau}^{K-i} \frac{1}{\theta_{k'}^2} \left\| \mathbf{x}^{k'+1} - \mathbf{y}^{k'} \right\|^2$$

$$\leq \frac{\tau^2 + 3\tau}{2} \sum_{i=1}^{\tau} [4(i+1)] \sum_{k=0}^{K} \frac{1}{\theta_k^2} \left\| \mathbf{x}^{k+1} - \mathbf{y}^k \right\|^2$$

$$\overset{b}{=} (\tau^2 + 3\tau)^2 \sum_{k=0}^{K} \frac{1}{\theta_k^2} \left\| \mathbf{x}^{k+1} - \mathbf{y}^k \right\|^2, \tag{6.24}$$

where in $\overset{a}{\leq}$ we use the fact that $\frac{1}{\theta_k^2} \leq \frac{4}{\theta_{k-i}^2}$ for $\theta_k = \frac{2}{k+2}$, $k \geq 2(\tau - 1)$, and $i \leq \min(\tau, k - \tau)$, and $\overset{b}{=}$ is because $\sum_{i=1}^{\tau}(i + 1) = \frac{1}{2}(\tau^2 + 3\tau)$.

Dividing θ_k^2 on both sides of (6.22) and using $\mu = 0$, we have

$$\frac{F(\mathbf{x}^{k+1}) - F(\mathbf{x}^*)}{\theta_k^2}$$

$$\leq \frac{(1 - \theta_k)(F(\mathbf{x}^k) - F(\mathbf{x}^*))}{\theta_k^2} - \frac{\gamma}{\theta_k^2}\left(\frac{1}{2} - \frac{\gamma L}{2}\right)\left\|\frac{1}{\gamma}\left(\mathbf{x}^{k+1} - \mathbf{y}^k\right)\right\|^2$$

$$+ \frac{\gamma}{\theta_k^2}\left(\frac{\gamma^2 L^2}{2C_1} + \gamma L\right)\left\|\frac{1}{\gamma}\left(\mathbf{w}^{j(k)} - \mathbf{y}^k\right)\right\|^2 + \frac{\gamma C_1}{2\theta_k^2}\left\|\frac{1}{\gamma}\left(\mathbf{x}^{k+1} - \mathbf{y}^k\right)\right\|^2$$

$$+ \frac{1}{2\gamma}\left\|\mathbf{z}^k - \mathbf{x}^*\right\|^2 - \frac{1}{2\gamma}\left\|\mathbf{z}^{k+1} - \mathbf{x}^*\right\|^2$$

$$\overset{a}{\leq} \frac{F(\mathbf{x}^k) - F(\mathbf{x}^*)}{\theta_{k-1}^2} - \frac{\gamma}{\theta_k^2}\left(\frac{1}{2} - \frac{\gamma L}{2}\right)\left\|\frac{1}{\gamma}\left(\mathbf{x}^{k+1} - \mathbf{y}^k\right)\right\|^2$$

$$+ \frac{\gamma}{\theta_k^2}\left(\frac{\gamma^2 L^2}{2C_1} + L\gamma\right)\left\|\frac{1}{\gamma}\left(\mathbf{w}^{j(k)} - \mathbf{y}^k\right)\right\|^2 + \frac{\gamma C_1}{2\theta_k^2}\left\|\frac{1}{\gamma}\left(\mathbf{x}^{k+1} - \mathbf{y}^k\right)\right\|^2$$

$$+ \frac{1}{2\gamma}\left\|\mathbf{z}^k - \mathbf{x}^*\right\|^2 - \frac{1}{2\gamma}\left\|\mathbf{z}^{k+1} - \mathbf{x}^*\right\|^2, \tag{6.25}$$

where in $\overset{a}{\leq}$ we use $\frac{1 - \theta_k}{\theta_k^2} \leq \frac{1}{\theta_{k-1}^2}$ for $\theta_k = \frac{2}{k+2}$, $k \geq 1$.

Telescoping (6.25) with k from 0 to K and applying (6.24), and when $k = 0$, for $\frac{1 - \theta_0}{\theta_0} = 0$, we have

$$\frac{F(\mathbf{x}^{K+1}) - F(\mathbf{x}^*)}{\theta_k^2}$$

$$\leq -\sum_{k=0}^{K}\frac{\gamma}{\theta_k^2}\left(\frac{1}{2} - \frac{\gamma L}{2}\right)\left\|\frac{1}{\gamma}\left(\mathbf{x}^{k+1} - \mathbf{y}^k\right)\right\|^2$$

$$+ \sum_{k=0}^{K}\frac{\gamma}{\theta_k^2}\left(\frac{\gamma^2 L^2}{2C_1} + \gamma L\right)\left\|\frac{1}{\gamma}\left(\mathbf{w}^{j(k)} - \mathbf{y}^k\right)\right\|^2$$

$$+ \sum_{k=0}^{K}\frac{\gamma C_1}{2\theta_k^2}\left\|\frac{1}{\gamma}\left(\mathbf{x}^{k+1} - \mathbf{y}^k\right)\right\|^2$$

$$+ \frac{1}{2\gamma}\left\|\mathbf{z}^0 - \mathbf{x}^*\right\|^2 - \frac{1}{2\gamma}\left\|\mathbf{z}^{K+1} - \mathbf{x}^*\right\|^2$$

$$\leq \frac{1}{2\gamma} \left\| \mathbf{z}^0 - \mathbf{x}^* \right\|^2 - \frac{1}{2\gamma} \left\| \mathbf{z}^{K+1} - \mathbf{x}^* \right\|^2$$

$$- \left[\frac{1}{2} - \frac{\gamma L}{2} - \frac{C_1}{2} - \left(\frac{\gamma^2 L^2}{2C_1} + \gamma L \right) (\tau^2 + 3\tau)^2 \right]$$

$$\times \sum_{k=0}^{K} \frac{\gamma}{\theta_k^2} \left\| \frac{1}{\gamma} \left(\mathbf{x}^{k+1} - \mathbf{y}^k \right) \right\|^2.$$

Setting $C_1 = \gamma L$, we have that $2\gamma L + 3\gamma L (\tau^2 + 3\tau)^2 \leq 1$ implies

$$\frac{1}{2} - \frac{\gamma L}{2} - \frac{C_1}{2} - \left(\frac{\gamma^2 L^2}{2C_1} + \gamma L \right) (\tau^2 + 3\tau)^2 \geq 0.$$

So

$$\frac{F(\mathbf{x}^{K+1}) - F(\mathbf{x}^*)}{\theta_K^2} + \frac{1}{2\gamma} \left\| \mathbf{z}^{K+1} - \mathbf{x}^* \right\|^2 \leq \frac{1}{2\gamma} \left\| \mathbf{z}^0 - \mathbf{x}^* \right\|^2,$$

and we have (6.6).

Now we consider the strongly convex case. In the following, we set $\theta_k = \theta$ in the previous deductions. Multiplying (6.23) with $(1 - \theta)^{K-k}$ and telescoping the results with k from 0 to K, we have

$$\sum_{k=0}^{K} (1 - \theta)^{K-k} \left\| \mathbf{w}^{j(k)} - \mathbf{y}^k \right\|^2$$

$$\leq \frac{\tau^2 + 3\tau}{2} \sum_{k=0}^{K} \sum_{i=1}^{\min(\tau,k)} (i + 1)(1 - \theta)^{K-k} \left\| \mathbf{x}^{k-i+1} - \mathbf{y}^{k-i} \right\|^2$$

$$= \frac{\tau^2 + 3\tau}{2} \sum_{k=0}^{K} \sum_{i=1}^{\min(\tau,k)} (1 - \theta)^{-i} (i + 1)(1 - \theta)^{K-(k-i)} \left\| \mathbf{x}^{k-i+1} - \mathbf{y}^{k-i} \right\|^2$$

$$\leq \frac{\tau^2 + 3\tau}{2(1 - \theta)^\tau} \sum_{k=0}^{K} \sum_{i=1}^{\min(\tau,k)} (i + 1)(1 - \theta)^{K-(k-i)} \left\| \mathbf{x}^{k-i+1} - \mathbf{y}^{k-i} \right\|^2$$

$$= \frac{\tau^2 + 3\tau}{2(1 - \theta)^\tau} \sum_{i=1}^{\tau} (i + 1) \sum_{k=i}^{K} (1 - \theta)^{K-(k-i)} \left\| \mathbf{x}^{k-i+1} - \mathbf{y}^{k-i} \right\|^2$$

$$= \frac{\tau^2 + 3\tau}{2(1 - \theta)^\tau} \sum_{i=1}^{\tau} (i + 1) \sum_{k'=0}^{K-i} (1 - \theta)^{K-k'} \left\| \mathbf{x}^{k'+1} - \mathbf{y}^{k'} \right\|^2$$

$$\leq \frac{\tau^2 + 3\tau}{2(1-\theta)^\tau} \sum_{i=1}^{\tau} (i+1) \sum_{k'=0}^{K} (1-\theta)^{K-k'} \left\| \mathbf{x}^{k'+1} - \mathbf{y}^{k'} \right\|^2$$

$$\overset{a}{=} \frac{(\tau^2 + 3\tau)^2}{4(1-\theta)^\tau} \sum_{k=0}^{K} (1-\theta)^{K-k} \left\| \mathbf{x}^{k+1} - \mathbf{y}^k \right\|^2, \tag{6.26}$$

where $\overset{a}{=}$ is because $\sum_{i=1}^{\tau}(i+1) = \frac{1}{2}(\tau^2 + 3\tau)$.

By rearranging terms in (6.22), we have

$$F(\mathbf{x}^{k+1}) - F(\mathbf{x}^*) + \left(\frac{\theta^2}{2\gamma} + \frac{\mu\theta}{2} \right) \left\| \mathbf{z}^{k+1} - \mathbf{x}^* \right\|^2$$

$$\leq (1-\theta) \left[F(\mathbf{x}^k) - F(\mathbf{x}^*) + \left(\frac{\theta^2}{2\gamma} + \frac{\mu\theta}{2} \right) \left\| \mathbf{z}^k - \mathbf{x}^* \right\|^2 \right]$$

$$- \gamma \left(\frac{1}{2} - \frac{\gamma L}{2} - \frac{C_1}{2} \right) \left\| \frac{1}{\gamma} \left(\mathbf{x}^{k+1} - \mathbf{y}^k \right) \right\|^2$$

$$+ \gamma \left(\frac{\gamma^2 L^2}{2C_1} + \gamma L \right) \left\| \frac{1}{\gamma} \left(\mathbf{w}^{j(k)} - \mathbf{y}^k \right) \right\|^2, \tag{6.27}$$

where we use $\theta = \frac{-\gamma\mu + \sqrt{\gamma\mu^2 + 4\gamma\mu}}{2}$, which satisfies

$$\left(\frac{\theta^2}{2\gamma} + \frac{\mu\theta}{2} \right) (1-\theta) = \frac{\theta^2}{2\gamma}.$$

Equivalently, θ is the root of $x^2 + \mu\gamma x - \mu\gamma = 0$ and so $\sqrt{\gamma\mu}/2 \leq \theta \leq \sqrt{\gamma\mu}$ when $\gamma\mu \leq 1$. For the assumption on γ, we have

$$9\gamma L\tau^2 \leq \frac{5}{2}\gamma L + \gamma L(\tau^2 + 3\tau)^2 \leq 1. \tag{6.28}$$

We then consider $\frac{1}{(1-\theta)^\tau}$. Without loss of generality, we assume that $\tau \geq 2$. Then we have

$$\frac{1}{(1-\theta)^\tau} \overset{a}{\leq} \frac{1}{(1-\sqrt{\gamma\mu})^\tau} \overset{b}{\leq} \frac{1}{\left(1 - \frac{1}{3\tau}\sqrt{\mu/L}\right)^\tau}$$

$$\overset{c}{\leq} \frac{1}{(1-\frac{1}{3\tau})^\tau} \overset{d}{\leq} \frac{1}{(1-\frac{1}{3})^1} = \frac{3}{2}, \tag{6.29}$$

where in $\overset{a}{\leq}$ we use $\theta \leq \sqrt{\gamma\mu}$, in $\overset{b}{\leq}$ we use (6.28), in $\overset{c}{\leq}$ we use $\frac{\mu}{L} \leq 1$, and in $\overset{d}{\leq}$ we use the fact that function $g(x) = (1 - \frac{1}{3x})^{-x}$ is monotonically decreasing for $x \in [1, \infty)$.

Multiplying (6.27) with θ^{K-k} and summing the result with k from 0 to K, we have

$$F(\mathbf{x}^{K+1}) - F(\mathbf{x}^*) + \left(\frac{\theta^2}{2\gamma} + \frac{\mu\theta}{2}\right) \left\|\mathbf{z}^{K+1} - \mathbf{x}^*\right\|^2$$

$$\leq (1-\theta)^{K+1} \left[F(\mathbf{x}^0) - F(\mathbf{x}^*) + \left(\frac{\theta^2}{2\gamma} + \frac{\mu\theta}{2}\right) \left\|\mathbf{z}^0 - \mathbf{x}^*\right\|^2\right]$$

$$-\gamma\left(\frac{1}{2} - \frac{\gamma L}{2} - \frac{C_1}{2}\right) \sum_{i=0}^{K} (1-\theta)^{K-k} \left\|\mathbf{x}^{k+1} - \mathbf{y}^k\right\|^2$$

$$+\gamma\left(\frac{\gamma^2 L^2}{2C_1} + \gamma L\right) \sum_{k=0}^{K} (1-\theta)^{K-k} \left\|\frac{1}{\gamma}\left(\mathbf{w}^{j(k)} - \mathbf{y}^k\right)\right\|^2$$

$$\overset{a}{\leq} (1-\theta)^{K+1} \left[F(\mathbf{x}^0) - F(\mathbf{x}^*) + \left(\frac{\theta^2}{2\gamma} + \frac{\mu\theta}{2}\right) \left\|\mathbf{z}^0 - \mathbf{x}^*\right\|^2\right]$$

$$-\gamma\left[\frac{1}{2} - \frac{\gamma L}{2} - \frac{C_1}{2} - \left(\frac{\gamma^2 L^2}{2C_1} + \gamma L\right) \frac{(\tau^2 + 3\tau)^2}{4(1-\theta)^\tau}\right]$$

$$\times \sum_{i=0}^{K} (1-\theta)^{K-k} \left\|\mathbf{x}^{k+1} - \mathbf{y}^k\right\|^2$$

$$\overset{b}{\leq} (1-\theta)^{K+1} \left[F(\mathbf{x}^0) - F(\mathbf{x}^*) + \left(\frac{\theta^2}{2\gamma} + \frac{\mu\theta}{2}\right) \left\|\mathbf{z}^0 - \mathbf{x}^*\right\|^2\right]$$

$$-\gamma\left[\frac{1}{2} - \frac{\gamma L}{2} - \frac{C_1}{2} - \left(\frac{\gamma^2 L^2}{2C_1} + \gamma L\right) \frac{3(\tau^2 + 3\tau)^2}{8}\right]$$

$$\times \sum_{i=0}^{K} (1-\theta)^{K-k} \left\|\mathbf{x}^{k+1} - \mathbf{y}^k\right\|^2,$$

where in $\overset{a}{\leq}$ we use (6.26) and in $\overset{b}{\leq}$ we use (6.29). Setting $C_1 = \gamma L$, by the assumption on the choice of γ, we have

$$\frac{1}{2} - \frac{\gamma L}{2} - \frac{C_1}{2} - \left(\frac{\gamma^2 L^2}{2C_1} + \gamma L\right) \frac{3(\tau^2 + 3\tau)^2}{8} \geq 0.$$

So (6.7) is proven.   □

Algorithm 6.2 Asynchronous accelerated stochastic coordinate descent (AASCD)

Input θ_k, step size γ, $\mathbf{x}^0 = \mathbf{0}$, and $\mathbf{z}^0 = \mathbf{0}$.
Define $a_k = \frac{\theta_k(1-\theta_{k-1})}{\theta_{k-1}}$, $b(l,k) = \prod_{i=l}^{k} a_i$.
for $k = 0$ **to** K **do**
1   $\mathbf{w}^{j(k)} = \mathbf{y}^{j(k)} + \sum_{i=j(k)+1}^{k} b(j(k)+1, i)(\mathbf{y}^{j(k)} - \mathbf{x}^{j(k)})$,
2   Randomly choose an index i_k from $[n]$,
3   $\delta_k = \operatorname{argmin}_\delta \left(h_{i_k}(\mathbf{z}_{i_k}^k + \delta) + \nabla_{i_k} f(\mathbf{w}^{j(k)}) \cdot \delta + \frac{\theta_k}{2\gamma} \delta^2 \right)$,
4   $\mathbf{z}_{i_k}^{k+1} = \mathbf{z}_{i_k}^k + \delta_k$ with other coordinates unchanged,
5   $\mathbf{y}^k = (1 - \theta_k)\mathbf{x}^k + \theta_k \mathbf{z}^k$,
6   $\mathbf{x}^{k+1} = (1 - \theta_k)\mathbf{x}^k + n\theta_k \mathbf{z}^{k+1} - (n-1)\theta_k \mathbf{z}^k$.
end for k
Output $\mathbf{x}^{K+1}$.

6.1.2 Asynchronous Accelerated Stochastic Coordinate Descent

To meet the demand of large-scale machine learning, most asynchronous algorithms are designed in a stochastic fashion. Momentum compensation can further be applied to accelerate modern state-of-the-art stochastic asynchronous algorithms, such as asynchronous SCD [8] and ASVRG [13]. We illustrate Asynchronous Accelerated Stochastic Coordinate Descent (AASCD) as an example. The readers can refer to [5] for fusing momentum compensation with variance reduction.

Stochastic coordinate descent algorithms have been described in Sect. 5.1.1. We solve the following problem:

$$\min_{\mathbf{x} \in \mathbb{R}^n} f(\mathbf{x}) + h(\mathbf{x}),$$

where $f(\mathbf{x})$ has L_c-coordinate Lipschitz continuous gradient (see (A.4)), $h(\mathbf{x})$ has coordinate separable structure, i.e., $h(\mathbf{x}) = \sum_{i=1}^{n} h_i(\mathbf{x}_i)$ with $\mathbf{x} = (\mathbf{x}_1^T, \cdots, \mathbf{x}_n^T)^T$, and $f(\mathbf{x})$ and $h_i(\mathbf{x}_i)$ are convex. We illustrate AASCD. Like AAGD, we compute the distance between the delayed extrapolation points and the latest ones, and introduce a new extrapolation term to compensate the "lost" momentum. The algorithm is shown in Algorithm 6.2.

Theorem 6.2 *Assume that $f(\mathbf{x})$ and $h_i(\mathbf{x}_i)$ are convex, $f(\mathbf{x})$ has L_c-coordinate Lipschitz continuous gradient, and $\tau \leq \sqrt{n}$. For Algorithm 6.2, if the step size γ satisfies $2\gamma L_c + \left(2 + \frac{1}{n}\right)\gamma L_c \left[(\tau^2 + \tau)/n + 2\tau\right]^2 \leq 1$ and $\theta_k = \frac{2}{2n+k}$, we have*

$$\frac{\mathbb{E}F(\mathbf{x}^{K+1}) - F(\mathbf{x}^*)}{\theta_K^2} + \frac{n^2}{2\gamma}\mathbb{E}\|\mathbf{z}^{K+1} - \mathbf{x}^*\|^2 \leq \frac{F(\mathbf{x}^0) - F(\mathbf{x}^*)}{\theta_{-1}^2} + \frac{n^2}{2\gamma}\|\mathbf{z}^0 - \mathbf{x}^*\|^2.$$

When $h(\mathbf{x})$ is μ-strongly convex with $\mu \leq L_c$, setting the step size γ to satisfy that
$2\gamma L_c + (\frac{3}{4} + \frac{3}{8n})\gamma L_c \left[(\tau^2 + \tau)/n + 2\tau\right]^2 \leq 1$ *and* $\theta_k = \frac{-\gamma\mu + \sqrt{\gamma^2\mu^2 + 4\gamma\mu}}{2n}$, *denoted as* θ *instead, we have*

$$\mathbb{E}F(\mathbf{x}^{K+1}) - F(\mathbf{x}^*) + \frac{n^2\theta^2 + n\theta\mu\gamma}{2\gamma}\mathbb{E}\|\mathbf{z}^{K+1} - \mathbf{x}^*\|^2$$

$$\leq (1 - \theta)^{K+1}\left(F(\mathbf{x}^0) - F(\mathbf{x}^*) + \frac{n^2\theta^2 + n\theta\mu\gamma}{2\gamma}\|\mathbf{x}^0 - \mathbf{x}^*\|^2\right).$$

Proof From Steps 5 and 6 of Algorithm 6.2, we have

$$\theta_k\mathbf{z}^k = \mathbf{y}^k - (1 - \theta_k)\mathbf{x}^k, \tag{6.30}$$

$$n\theta_k\mathbf{z}^{k+1} = \mathbf{x}^{k+1} - (1 - \theta_k)\mathbf{x}^k + (n - 1)\theta_k\mathbf{z}^k, \tag{6.31}$$

and

$$\mathbf{x}^{k+1} = \mathbf{y}^k + n\theta_k(\mathbf{z}^{k+1} - \mathbf{z}^k). \tag{6.32}$$

Multiplying (6.30) with $(n - 1)$ and adding with (6.31), we have

$$n\theta_k\mathbf{z}^{k+1} = \mathbf{x}^{k+1} - (1 - \theta_k)\mathbf{x}^k + (n - 1)\mathbf{y}^k - (n - 1)(1 - \theta_k)\mathbf{x}^k. \tag{6.33}$$

Eliminating $\mathbf{z}^k$ using (6.33) and (6.30), for $k \geq 1$, we have

$$\frac{1}{\theta_k}[\mathbf{y}^k - (1 - \theta_k)\mathbf{x}^k]$$

$$= \frac{1}{n\theta_{k-1}}[\mathbf{x}^k - (1 - \theta_{k-1})\mathbf{x}^{k-1} + (n - 1)\mathbf{y}^{k-1} - (n - 1)(1 - \theta_{k-1})\mathbf{x}^{k-1}].$$
$$\tag{6.34}$$

Computing out $\mathbf{y}^k$ through (6.34), we have

$$\mathbf{y}^k = \mathbf{x}^k - \theta_k\mathbf{x}^k + \frac{\theta_k}{n\theta_{k-1}}\mathbf{x}^k - \frac{\theta_k(1 - \theta_{k-1})}{n\theta_{k-1}}\mathbf{x}^{k-1}$$

$$- \frac{(n - 1)\theta_k(1 - \theta_{k-1})}{n\theta_{k-1}}\mathbf{x}^{k-1} + \frac{(n - 1)\theta_k}{n\theta_{k-1}}\mathbf{y}^{k-1}$$

$$= \mathbf{x}^k + \frac{\theta_k}{\theta_{k-1}}\left(\frac{1}{n} - \theta_{k-1}\right)(\mathbf{x}^k - \mathbf{y}^{k-1}) + \frac{\theta_k(1 - \theta_{k-1})}{\theta_{k-1}}(\mathbf{y}^{k-1} - \mathbf{x}^{k-1}).$$

We set $a_k = \frac{\theta_k(1-\theta_{k-1})}{\theta_{k-1}}$ and $b(l,k) = \prod_{i=l}^{k} a_i$ when $l \le k$ and $b(l,k) = 0$ when $l > k$. Then by setting $c_k = \frac{\theta_k}{\theta_{k-1}}(\frac{1}{n} - \theta_{k-1})$, we have

$$
\begin{aligned}
\mathbf{y}^k &= \mathbf{x}^k + c_k(\mathbf{x}^k - \mathbf{y}^{k-1}) + a_k(\mathbf{y}^{k-1} - \mathbf{x}^{k-1}) \\
&= \mathbf{y}^{k-1} + (1+c_k)(\mathbf{x}^k - \mathbf{y}^{k-1}) + a_k(\mathbf{y}^{k-1} - \mathbf{x}^{k-1}) \\
&= \mathbf{y}^{k-1} + (1+c_k)(\mathbf{x}^k - \mathbf{y}^{k-1}) + a_k c_{k-1}(\mathbf{x}^{k-1} - \mathbf{y}^{k-2}) + a_k a_{k-1}(\mathbf{y}^{k-2} - \mathbf{x}^{k-2}) \\
&= \mathbf{y}^{k-1} + (1+c_k)(\mathbf{x}^k - \mathbf{y}^{k-1}) + \sum_{i=j(k)+1}^{k-1} b(i+1,k)c_i(\mathbf{x}^i - \mathbf{y}^{i-1}) \\
&\quad + b(j(k)+1,k)(\mathbf{y}^{j(k)} - \mathbf{x}^{j(k)}), \quad k \ge j(k)+1 \ge 1.
\end{aligned}
\tag{6.35}
$$

Summing (6.35) with $k = j(k)+1$ to k, we have

$$
\begin{aligned}
\mathbf{y}^k &= \mathbf{y}^{j(k)} + \sum_{i=j(k)+1}^{k} (1+c_i)(\mathbf{x}^i - \mathbf{y}^{i-1}) \\
&\quad + \sum_{l=j(k)+1}^{k} \sum_{i=j(k)+1}^{l-1} c_i b(i+1,l)(\mathbf{x}^i - \mathbf{y}^{i-1}) \\
&\quad + \left(\sum_{i=j(k)+1}^{k} b(j(k)+1,i) \right)(\mathbf{y}^{j(k)} - \mathbf{x}^{j(k)}) \\
&= \mathbf{y}^{j(k)} + \sum_{i=j(k)+1}^{k} (1+c_i)(\mathbf{x}^i - \mathbf{y}^{i-1}) \\
&\quad + \sum_{i=j(k)+1}^{k-1} \left(\sum_{l=i+1}^{k} c_i b(i+1,l) \right)(\mathbf{x}^i - \mathbf{y}^{i-1}) \\
&\quad + \left(\sum_{i=j(k)+1}^{k} b(j(k)+1,i) \right)(\mathbf{y}^{j(k)} - \mathbf{x}^{j(k)}) \\
&= \mathbf{y}^{j(k)} + \sum_{i=j(k)+1}^{k} (1+c_i)(\mathbf{x}^i - \mathbf{y}^{i-1}) \\
&\quad + \sum_{i=j(k)+1}^{k} c_i \left(\sum_{l=i+1}^{k} b(i+1,l) \right)(\mathbf{x}^i - \mathbf{y}^{i-1}) \\
&\quad + \left(\sum_{i=j(k)+1}^{k} b(j(k)+1,i) \right)(\mathbf{y}^{j(k)} - \mathbf{x}^{j(k)}),
\end{aligned}
\tag{6.36}
$$

where in the last equality, we use $b(k + 1, l) = 0$ for all $l \leq k$. Then from the optimality of $\mathbf{z}_{i_k}^{k+1}$ in Steps 3 and 4, we have

$$n\theta_k(\mathbf{z}_{i_k}^{k+1} - \mathbf{z}_{i_k}^k) + \gamma \nabla_{i_k} f(\mathbf{w}^{j(k)}) + \gamma \boldsymbol{\xi}_{i_k}^k = 0, \tag{6.37}$$

where $\boldsymbol{\xi}_{i_k}^k \in \partial h_{i_k}(\mathbf{z}_{i_k}^{k+1})$. From (6.32),

$$\mathbf{x}_{i_k}^{k+1} - \mathbf{y}_{i_k}^k + \gamma \nabla_{i_k} f(\mathbf{w}^{j(k)}) + \gamma \boldsymbol{\xi}_{i_k}^k = 0. \tag{6.38}$$

Since f has coordinate Lipschitz continuous gradients and $\mathbf{x}^{k+1}$ and $\mathbf{y}^k$ only differ in the i_k-th entry, we have

$$\begin{aligned}
f(\mathbf{x}^{k+1}) &\leq f(\mathbf{y}^k) + \langle \nabla_{i_k} f(\mathbf{y}^k), \mathbf{x}_{i_k}^{k+1} - \mathbf{y}_{i_k}^k \rangle + \frac{L_c}{2}\left(\mathbf{x}_{i_k}^{k+1} - \mathbf{y}_{i_k}^k\right)^2 \\
&\overset{a}{=} f(\mathbf{y}^k) - \gamma \langle \nabla_{i_k} f(\mathbf{y}^k), \nabla_{i_k} f(\mathbf{w}^{j(k)}) + \boldsymbol{\xi}_{i_k}^k \rangle + \frac{L_c}{2}\left(\mathbf{x}_{i_k}^{k+1} - \mathbf{y}_{i_k}^k\right)^2 \\
&\overset{b}{=} f(\mathbf{y}^k) - \gamma \langle \nabla_{i_k} f(\mathbf{w}^{j(k)}) + \boldsymbol{\xi}_{i_k}^k, \nabla_{i_k} f(\mathbf{w}^{j(k)}) + \boldsymbol{\xi}_{i_k}^k \rangle \\
&\quad + \frac{L_c}{2}\left(\mathbf{x}_{i_k}^{k+1} - \mathbf{y}_{i_k}^k\right)^2 + \gamma \langle \boldsymbol{\xi}_{i_k}^k, \nabla_{i_k} f(\mathbf{w}^{j(k)}) + \boldsymbol{\xi}_{i_k}^k \rangle \\
&\quad + \gamma \langle \nabla_{i_k} f(\mathbf{w}^{j(k)}) - \nabla_{i_k} f(\mathbf{y}^k), \nabla_{i_k} f(\mathbf{w}^{j(k)}) + \boldsymbol{\xi}_{i_k}^k \rangle \\
&= f(\mathbf{y}^k) - \gamma\left(1 - \frac{\gamma L_c}{2}\right)\left(\frac{\mathbf{x}_{i_k}^{k+1} - \mathbf{y}_{i_k}^k}{\gamma}\right)^2 - \langle \boldsymbol{\xi}_{i_k}^k, \mathbf{x}_{i_k}^{k+1} - \mathbf{y}_{i_k}^k \rangle \\
&\quad - \langle \nabla_{i_k} f(\mathbf{w}^{j(k)}) - \nabla_{i_k} f(\mathbf{y}^k), \mathbf{x}_{i_k}^{k+1} - \mathbf{y}_{i_k}^k \rangle \\
&\overset{c}{\leq} f(\mathbf{y}^k) - \gamma\left(1 - \frac{\gamma L_c}{2}\right)\left(\frac{\mathbf{x}_{i_k}^{k+1} - \mathbf{y}_{i_k}^k}{\gamma}\right)^2 - \langle \boldsymbol{\xi}_{i_k}^k, \mathbf{x}_{i_k}^{k+1} - \mathbf{y}_{i_k}^k \rangle \\
&\quad + \frac{\gamma L_c^2}{2C_2}\left(\mathbf{w}_{i_k}^{j(k)} - \mathbf{y}_{i_k}^k\right)^2 + \frac{\gamma C_2}{2}\left(\frac{\mathbf{x}_{i_k}^{k+1} - \mathbf{y}_{i_k}^k}{\gamma}\right)^2, \tag{6.39}
\end{aligned}$$

where in $\overset{a}{\leq}$ we use (6.38), in $\overset{b}{\leq}$ we insert $-\gamma \langle \nabla_{i_k} f(\mathbf{w}^{j(k)}) + \boldsymbol{\xi}_{i_k}^k, \nabla_{i_k} f(\mathbf{w}^{j(k)}) + \boldsymbol{\xi}_{i_k}^k \rangle$, and in $\overset{c}{\leq}$ we use the Cauchy–Schwartz inequality and the coordinate Lipschitz continuity of ∇f. $C_2 > 0$ will be chosen later.

We consider $\|\mathbf{z}^k - \mathbf{z}^*\|^2$ and have

$$\frac{n^2}{2\gamma}\left\|\theta_k\mathbf{z}^{k+1} - \theta_k\mathbf{x}^*\right\|^2 = \frac{n^2}{2\gamma}\left\|\theta_k\mathbf{z}^k - \theta_k\mathbf{x}^* + \theta_k\mathbf{z}^{k+1} - \theta_k\mathbf{z}^k\right\|^2$$

$$= \frac{n^2}{2\gamma}\left\|\theta_k\mathbf{z}^k - \theta_k\mathbf{x}^*\right\|^2 + \frac{n^2}{2\gamma}\left(\theta_k z_{i_k}^{k+1} - \theta_k z_{i_k}^k\right)^2$$

$$+ \frac{n^2}{\gamma}\left\langle\theta_k\left(z_{i_k}^{k+1} - z_{i_k}^k\right), \theta_k z_{i_k}^k - \theta_k x_{i_k}^*\right\rangle$$

$$\stackrel{a}{=} \frac{n^2}{2\gamma}\left\|\theta_k\mathbf{z}^k - \theta_k\mathbf{x}^*\right\|^2 + \frac{1}{2\gamma}\left(x_{i_k}^{k+1} - y_{i_k}^k\right)^2$$

$$- n\left\langle\nabla_{i_k}f(\mathbf{w}^{j(k)}) + \xi_{i_k}^k, \theta_k z_{i_k}^k - \theta_k x_{i_k}^*\right\rangle, \tag{6.40}$$

where $\stackrel{a}{=}$ uses (6.32) and (6.37).

So taking expectation on (6.40), we have

$$\frac{n^2}{2\gamma}\mathbb{E}_{i_k}\left\|\theta_k\mathbf{z}^{k+1} - \theta_k\mathbf{x}^*\right\|^2$$

$$= \frac{n^2}{2\gamma}\left\|\theta_k\mathbf{z}^k - \theta_k\mathbf{x}^*\right\|^2 + \frac{1}{2\gamma n}\sum_{i_k=1}^n\left(x_{i_k}^{k+1} - y_{i_k}^k\right)^2$$

$$- \langle\nabla f(\mathbf{w}^{j(k)}), \theta_k\mathbf{z}^k - \theta_k\mathbf{x}^*\rangle - \sum_{i_k=1}^n\langle\xi_{i_k}^k, \theta_k z_{i_k}^k - \theta_k x_{i_k}^*\rangle. \tag{6.41}$$

By the same technique of (6.17), for the last but one term of (6.41), we have

$$-\langle\nabla f(\mathbf{w}^{j(k)}), \theta_k\mathbf{z}^k - \theta_k\mathbf{x}^*\rangle$$

$$= -\langle\nabla f(\mathbf{w}^{j(k)}), \mathbf{y}^k - (1-\theta_k)\mathbf{x}^k - \theta_k\mathbf{x}^*\rangle$$

$$= -\langle\nabla f(\mathbf{w}^{j(k)}), \mathbf{w}^{j(k)} - (1-\theta_k)\mathbf{x}^k - \theta_k\mathbf{x}^*\rangle - \langle\nabla f(\mathbf{w}^{j(k)}), \mathbf{y}^k - \mathbf{w}^{j(k)}\rangle$$

$$\leq (1-\theta_k)f(\mathbf{x}^k) + \theta_k f(\mathbf{x}^*) - f(\mathbf{w}^{j(k)}) - \langle\nabla f(\mathbf{w}^{j(k)}), \mathbf{y}^k - \mathbf{w}^{j(k)}\rangle$$

$$\leq (1-\theta_k)f(\mathbf{x}^k) + \theta_k f(\mathbf{x}^*) - f(\mathbf{y}^k) + \left\langle\nabla f(\mathbf{y}^k) - \nabla f(\mathbf{w}^{j(k)}), \mathbf{y}^k - \mathbf{w}^{j(k)}\right\rangle.$$

$$\tag{6.42}$$

Substituting (6.42) into (6.41), we have

$$\frac{n^2}{2\gamma}\mathbb{E}_{i_k}\left\|\theta_k\mathbf{z}^{k+1}-\theta_k\mathbf{x}^*\right\|^2$$

$$\leq \frac{n^2}{2\gamma}\left\|\theta_k\mathbf{z}^k-\theta_k\mathbf{x}^*\right\|^2+\frac{1}{2\gamma n}\sum_{i_k=1}^{n}\left(\mathbf{x}_{i_k}^{k+1}-\mathbf{y}_{i_k}^k\right)^2-\sum_{i_k=1}^{n}\langle\boldsymbol{\xi}_{i_k}^k,\theta_k\mathbf{z}_{i_k}^k-\theta_k\mathbf{x}_{i_k}^*\rangle.$$

$$+(1-\theta_k)f(\mathbf{x}^k)+\theta_k f(\mathbf{x}^*)-f(\mathbf{y}^k)+\langle\nabla f(\mathbf{y}^k)-\nabla f(\mathbf{w}^{j(k)}),\mathbf{y}^k-\mathbf{w}^{j(k)}\rangle$$

$$\leq \frac{n^2}{2\gamma}\left\|\theta_k\mathbf{z}^k-\theta_k\mathbf{x}^*\right\|^2+\frac{1}{2\gamma n}\sum_{i_k=1}^{n}\left(\mathbf{x}_{i_k}^{k+1}-\mathbf{y}_{i_k}^k\right)^2-\sum_{i_k=1}^{n}\langle\boldsymbol{\xi}_{i_k}^k,\theta_k\mathbf{z}_{i_k}^k-\theta_k\mathbf{x}_{i_k}^*\rangle$$

$$+(1-\theta_k)f(\mathbf{x}^k)+\theta_k f(\mathbf{x}^*)-f(\mathbf{y}^k)+L_c\left\|\mathbf{y}^k-\mathbf{w}^{j(k)}\right\|^2. \tag{6.43}$$

Taking expectation on the random index i_k for (6.39), we have

$$\mathbb{E}_{i_k}f(\mathbf{x}^{k+1})\leq f(\mathbf{y}^k)-\gamma\left(1-\frac{\gamma L_c}{2}-\frac{C_2}{2}\right)\frac{1}{n}\sum_{i_k=1}^{n}\left(\frac{\mathbf{x}_{i_k}^{k+1}-\mathbf{y}_{i_k}^k}{\gamma}\right)^2$$

$$-\frac{1}{n}\sum_{i_k=1}^{n}\langle\boldsymbol{\xi}_{i_k}^k,\mathbf{x}_{i_k}^{k+1}-\mathbf{y}_{i_k}^k\rangle+\frac{\gamma L_c^2}{2nC_2}\left\|\mathbf{w}^{j(k)}-\mathbf{y}^k\right\|^2. \tag{6.44}$$

Adding (6.44) and (6.43) we have

$$\mathbb{E}_{i_k}f(\mathbf{x}^{k+1})$$

$$\leq (1-\theta_k)f(\mathbf{x}^k)+\theta_k f(\mathbf{x}^*)-\frac{\gamma}{n}\left(\frac{1}{2}-\frac{\gamma L_c}{2}-\frac{C_2}{2}\right)\sum_{i_k=1}^{n}\left(\frac{\mathbf{x}_{i_k}^{k+1}-\mathbf{y}_{i_k}^k}{\gamma}\right)^2$$

$$+\left(\frac{\gamma L_c^2}{2nC_2}+L_c\right)\left\|\mathbf{w}^{j(k)}-\mathbf{y}^k\right\|^2$$

$$-\sum_{i_k=1}^{n}\left\langle\boldsymbol{\xi}_{i_k}^k,\theta_k\mathbf{z}_{i_k}^k-\theta_k\mathbf{x}_{i_k}^*+\frac{1}{n}\left(\mathbf{x}_{i_k}^{k+1}-\mathbf{y}_{i_k}^k\right)\right\rangle$$

$$+\frac{n^2}{2\gamma}\left\|\theta_k\mathbf{z}^k-\theta_k\mathbf{x}^*\right\|^2-\frac{n^2}{2\gamma}\mathbb{E}_{i_k}\left\|\theta_k\mathbf{z}^{k+1}-\theta_k\mathbf{x}^*\right\|^2$$

$$\overset{a}{=}(1-\theta_k)f(\mathbf{x}^k)+\theta_k f(\mathbf{x}^*)-\frac{\gamma}{n}\left(\frac{1}{2}-\frac{\gamma L_c}{2}-\frac{C_2}{2}\right)\sum_{i_k=1}^{n}\left(\frac{\mathbf{x}_{i_k}^{k+1}-\mathbf{y}_{i_k}^k}{\gamma}\right)^2$$

$$+ \left(\frac{\gamma L_c^2}{2nC_2} + L_c \right) \left\| \mathbf{w}^{j(k)} - \mathbf{y}^k \right\|^2 - \sum_{i_k=1}^{n} \langle \boldsymbol{\xi}_{i_k}^k, \theta_k \mathbf{z}_{i_k}^{k+1} - \theta_k \mathbf{x}_{i_k}^* \rangle$$

$$+ \frac{n^2}{2\gamma} \left\| \theta_k \mathbf{z}^k - \theta_k \mathbf{x}^* \right\|^2 - \frac{n^2}{2\gamma} \mathbb{E}_{i_k} \left\| \theta_k \mathbf{z}^{k+1} - \theta_k \mathbf{x}^* \right\|^2, \tag{6.45}$$

where in $\overset{a}{=}$ we use (6.32).

For h_{i_k} is μ-strongly convex, we have

$$\theta_k \langle \boldsymbol{\xi}_{i_k}^k, \mathbf{x}_{i_k}^* - \mathbf{z}_{i_k}^{k+1} \rangle \le \theta_k h_{i_k}(\mathbf{x}^*) - \theta_k h_{i_k}(\mathbf{z}_{i_k}^{k+1}) - \frac{\mu \theta_k}{2} \left(\mathbf{z}_{i_k}^{k+1} - \mathbf{x}_{i_k}^* \right)^2. \tag{6.46}$$

Analyzing the expectation, we have

$$\mathbb{E}_{i_k} \left\| \mathbf{z}^{k+1} - \mathbf{x}^* \right\|^2 = \frac{1}{n} \sum_{i_k=1}^{n} \left[\left(\mathbf{z}_{i_k}^{k+1} - \mathbf{x}_{i_k}^* \right)^2 + \sum_{j \neq i_k} \left(\mathbf{z}_j^{k+1} - \mathbf{x}_j^* \right)^2 \right]$$

$$\overset{a}{=} \frac{1}{n} \sum_{i_k=1}^{n} \left[\left(\mathbf{z}_{i_k}^{k+1} - \mathbf{x}_{i_k}^* \right)^2 + \sum_{j \neq i_k} \left(\mathbf{z}_j^k - \mathbf{x}_j^* \right)^2 \right]$$

$$= \frac{1}{n} \sum_{i_k=1}^{n} \left(\mathbf{z}_{i_k}^{k+1} - \mathbf{x}_{i_k}^* \right)^2 + \frac{n-1}{n} \left\| \mathbf{z}^k - \mathbf{x}^* \right\|^2, \tag{6.47}$$

where $\overset{a}{=}$ uses the fact that $\mathbf{z}^{k+1}$ and $\mathbf{z}^k$ only differ at the i_k-th entry. Similar to (6.47), we can find that

$$\mathbb{E}_{i_k} \left\| \mathbf{x}^{k+1} - \mathbf{y}^k \right\|^2 = \frac{1}{n} \sum_{i_k=1}^{n} \left[\left(\mathbf{x}_{i_k}^{k+1} - \mathbf{y}_{i_k}^k \right)^2 + \sum_{j \neq i_k} \left(\mathbf{x}_j^{k+1} - \mathbf{y}_j^k \right)^2 \right]$$

$$\overset{a}{=} \frac{1}{n} \sum_{i_k=1}^{n} \left(\mathbf{x}_{i_k}^{k+1} - \mathbf{y}_{i_k}^k \right)^2, \tag{6.48}$$

where $\overset{a}{=}$ uses (6.32) and that $\mathbf{x}^{k+1}$ and $\mathbf{y}^k$ only differ at the i_k-th entry.

Then plugging (6.46) into (6.45), we have

$$\mathbb{E}_{i_k} f(\mathbf{x}^{k+1}) + \theta_k \sum_{i_k=1}^{n} h_{i_k}(\mathbf{z}_{i_k}^{k+1})$$

$$\le (1 - \theta_k) f(\mathbf{x}^k) + \theta_k F(\mathbf{x}^*) - \frac{\gamma}{n} \left(\frac{1}{2} - \frac{\gamma L}{2} - \frac{C_2}{2} \right) \sum_{i_k=1}^{n} \left(\frac{\mathbf{x}_{i_k}^{k+1} - \mathbf{y}_{i_k}^k}{\gamma} \right)^2$$

$$+ \left(\frac{\gamma L_c^2}{2nC_2} + L_c \right) \left\| \mathbf{w}^{j(k)} - \mathbf{y}^k \right\|^2 - \sum_{i_k=1}^{n} \frac{\mu \theta_k}{2} \left(\mathbf{z}_{i_k}^{k+1} - \mathbf{x}_{i_k}^* \right)^2$$

$$+ \frac{n^2}{2\gamma} \left\| \theta_k \mathbf{z}^k - \theta_k \mathbf{x}^* \right\|^2 - \frac{n^2}{2\gamma} \mathbb{E}_{i_k} \left\| \theta_k \mathbf{z}^{k+1} - \theta_k \mathbf{x}^* \right\|^2$$

$$\overset{a}{=} (1 - \theta_k) f(\mathbf{x}^k) + \theta_k F(\mathbf{x}^*) - \frac{1}{\gamma} \left(\frac{1}{2} - \frac{\gamma L}{2} - \frac{C_2}{2} \right) \mathbb{E}_{i_k} \left\| \mathbf{x}^{k+1} - \mathbf{y}^k \right\|^2$$

$$+ \left(\frac{\gamma L_c^2}{2nC_2} + L_c \right) \left\| \mathbf{w}^{j(k)} - \mathbf{y}^k \right\|^2 + \frac{n^2 \theta_k^2 + (n-1)\theta_k \mu \gamma}{2\gamma} \left\| \mathbf{z}^k - \mathbf{x}^* \right\|^2$$

$$- \frac{n^2 \theta_k^2 + n\theta_k \mu \gamma}{2\gamma} \mathbb{E}_{i_k} \left\| \mathbf{z}^{k+1} - \mathbf{x}^* \right\|^2, \tag{6.49}$$

where in $\overset{a}{=}$ we use (6.47) and (6.48).

On the other hand, because $\mathbf{x}^{k+1} = (1 - \theta_k)\mathbf{x}^k + n\theta_k \mathbf{z}^{k+1} - (n-1)\theta_k \mathbf{z}^k = (1 - \theta_k)\mathbf{x}^k + \theta_k \mathbf{z}^k + n\theta_k(\mathbf{z}^{k+1} - \mathbf{z}^k)$, we can define $\hat{h}_k = \sum_{i=0}^{k} e_{k,i} h(\mathbf{z}^i)$ and obtain the same result of Lemma 5.1 in ASCD (Sect. 5.1.1). Thus

$$\mathbb{E}_{i_k} \hat{h}_{k+1} = (1 - \theta_k)\hat{h}_k + \theta_k \sum_{i_k=1}^{n} h_{i_k}(\mathbf{z}_{i_k}^{k+1}). \tag{6.50}$$

From (6.36), using the same technique of (6.23) we have

$$\left\| \mathbf{w}^{j(k)} - \mathbf{y}^k \right\|^2$$

$$= \left\| \sum_{i=j(k)+1}^{k} \left(1 + c_i + c_i \sum_{l=i+1}^{k} b(i+1, l) \right) (\mathbf{x}^i - \mathbf{y}^{i-1}) \right\|^2$$

$$\overset{a}{\leq} \left[\sum_{i=j(k)+1}^{k} \left(1 + c_i + c_i \sum_{l=i+1}^{k} b(i+1, l) \right) \right]$$

$$\times \sum_{i=j(k)+1}^{k} \left(1 + c_i + c_i \sum_{l=i+1}^{k} b(i+1, l) \right) \left\| \mathbf{x}^i - \mathbf{y}^{i-1} \right\|^2$$

$$\overset{b}{\leq} \left[\sum_{i=j(k)+1}^{k} \left(1 + c_i \sum_{l=1}^{k-i+1} 1 \right) \right] \sum_{i=j(k)+1}^{k} \left(1 + c_i \sum_{l=1}^{k-i+1} 1 \right) \left\| \mathbf{x}^i - \mathbf{y}^{i-1} \right\|^2$$

$$\overset{c}{\leq} \left[\sum_{ii=1}^{k-j(k)} \left(1 + \frac{1}{n} \sum_{l=1}^{ii} 1 \right) \right] \sum_{ii=1}^{k-j(k)} \left(1 + \frac{1}{n} \sum_{l=1}^{ii} 1 \right) \left\| \mathbf{x}^{k-ii+1} - \mathbf{y}^{k-ii} \right\|^2$$

$$\frac{d}{\leq} \left[\sum_{ii=1}^{\min(\tau,k)} \left(1 + \frac{1}{n} \sum_{l=1}^{ii} 1\right) \right] \sum_{ii=1}^{\tau} \left(1 + \frac{1}{n} \sum_{l=1}^{ii} 1\right) \left\| \mathbf{x}^{k-ii+1} - \mathbf{y}^{k-ii} \right\|^2$$

$$\leq \left(\frac{\tau^2 + \tau}{2n} + \tau\right) \sum_{i=1}^{\min(\tau,k)} \left(\frac{i}{n} + 1\right) \left\| \mathbf{x}^{k-i+1} - \mathbf{y}^{k-i} \right\|^2, \tag{6.51}$$

where $\overset{a}{\leq}$ uses the convexity of $\|\cdot\|^2$, in $\overset{b}{\leq}$ we use $b(i,l) \leq 1$, in $\overset{c}{\leq}$ we change variable $ii = k - i + 1$ and use $c_k \leq \frac{1}{n}$, and in $\overset{d}{\leq}$ we use $k - j(k) \leq \tau$.

As in Theorem 6.1, we are more interested in the limiting case, when k is large (e.g., $k \geq 2(\tau - 1)$) we assume that at the first τ steps, we run our algorithm in serial. Dividing θ_k^2 on both sides of (6.51) and summing the results with $k = 0$ to K, we have

$$\sum_{k=0}^{K} \frac{1}{\theta_k^2} \left\| \mathbf{w}^{j(k)} - \mathbf{y}^k \right\|^2$$

$$\overset{a}{\leq} \left(\frac{\tau^2 + \tau}{2n} + \tau\right) \sum_{k=0}^{K} \sum_{i=1}^{\min(\tau,k)} \frac{4(\frac{i}{n} + 1)}{\theta_{k-i}^2} \left\| \mathbf{x}^{k-i+1} - \mathbf{y}^{k-i} \right\|^2$$

$$\leq \left(\frac{\tau^2 + \tau}{2n} + \tau\right) \sum_{i=1}^{\tau} \sum_{k=i+\tau}^{K} \frac{4(\frac{i}{n} + 1)}{\theta_{k-i}^2} \left\| \mathbf{x}^{k-i+1} - \mathbf{y}^{k-i} \right\|^2$$

$$= \left(\frac{\tau^2 + \tau}{2n} + \tau\right) \sum_{i=1}^{\tau} \left[4\left(\frac{i}{n} + 1\right)\right] \sum_{k'=\tau}^{K-i} \frac{1}{\theta_{k'}^2} \left\| \mathbf{x}^{k'+1} - \mathbf{y}^{k'} \right\|^2$$

$$\leq \left(\frac{\tau^2 + \tau}{2n} + \tau\right) \sum_{i=1}^{\tau} \left[4\left(\frac{i}{n} + 1\right)\right] \sum_{k=0}^{K} \frac{1}{\theta_k^2} \left\| \mathbf{x}^{k+1} - \mathbf{y}^k \right\|^2$$

$$= \left(\frac{\tau^2 + \tau}{n} + 2\tau\right)^2 \sum_{k=0}^{K} \frac{1}{\theta_k^2} \left\| \mathbf{x}^{k+1} - \mathbf{y}^k \right\|^2, \tag{6.52}$$

where in $\overset{a}{\leq}$ we use $\frac{1}{\theta_k^2} \leq \frac{4}{\theta_{k-i}^2}$, since $\theta_k = \frac{2}{2n+k}$ and $2i \leq 2\tau \leq 2n + k$.

We first consider the generally convex case. Clearly, we can check that the setting of θ_k satisfies the assumptions of Lemma 5.1 in Sect. 5.1.1. Now we consider (6.49).

By (6.50), replacing $\theta_k \sum_{i_k=1}^{n} h_{i_k}(\mathbf{z}_{i_k}^{k+1})$ with $\mathbb{E}_{i_k}\hat{h}_{k+1} - (1-\theta_k)\hat{h}_k$, dividing both sides of (6.49) by θ_k^2, and using $\mu = 0$, we have

$$
\frac{\mathbb{E}_{i_k} f(\mathbf{x}^{k+1}) + \mathbb{E}_{i_k}\hat{h}_{k+1} - F(\mathbf{x}^*)}{\theta_k^2} + \frac{n^2}{2\gamma}\mathbb{E}_{i_k}\left\|\mathbf{z}^{k+1} - \mathbf{x}^*\right\|^2
$$

$$
\leq \frac{1-\theta_k}{\theta_k^2}\left(f(\mathbf{x}^k) + \hat{h}_k - F(\mathbf{x}^*)\right) - \frac{1}{\gamma\theta_k^2}\left(\frac{1}{2} - \frac{\gamma L}{2} - \frac{C_2}{2}\right)\mathbb{E}_{i_k}\left\|\mathbf{x}^{k+1} - \mathbf{y}^k\right\|^2
$$

$$
+ \frac{1}{\theta_k^2}\left(\frac{\gamma L_c^2}{2nC_2} + L_c\right)\left\|\mathbf{w}^{j(k)} - \mathbf{y}^k\right\|^2 + \frac{n^2}{2\gamma}\left\|\mathbf{z}^k - \mathbf{x}^*\right\|^2
$$

$$
\overset{a}{\leq} \frac{1}{\theta_{k-1}^2}\left(f(\mathbf{x}^k) + \hat{h}_k - F(\mathbf{x}^*)\right) - \frac{1}{\gamma\theta_k^2}\left(1 - \frac{\gamma L}{2} - \frac{C_2}{2}\right)\mathbb{E}_{i_k}\left\|\mathbf{x}^{k+1} - \mathbf{y}^k\right\|^2
$$

$$
+ \frac{1}{\theta_k^2}\left(\frac{\gamma L_c^2}{2nC_2} + L_c\right)\left\|\mathbf{w}^{j(k)} - \mathbf{y}^k\right\|^2 + \frac{n^2}{2\gamma}\left\|\mathbf{z}^k - \mathbf{x}^*\right\|^2, \tag{6.53}
$$

where in $\overset{a}{\leq}$ we use $\frac{1-\theta_k}{\theta_k^2} \leq \frac{1}{\theta_{k-1}^2}$.

Taking expectation on the first k iterations for (6.53) and summing it with k from 0 to K, we have

$$
\frac{\mathbb{E} f(\mathbf{x}^{K+1}) + \mathbb{E}\hat{h}_{K+1} - F(\mathbf{x}^*)}{\theta_K^2} + \frac{n^2}{2\gamma}\mathbb{E}\left\|\mathbf{z}^{K+1} - \mathbf{x}^*\right\|^2
$$

$$
\leq \frac{f(\mathbf{x}^0) + \hat{h}_0 - F(\mathbf{x}^*)}{\theta_{-1}^2} - \left(\frac{1}{2} - \frac{\gamma L}{2} - \frac{C_2}{2}\right)\sum_{k=0}^{K}\frac{1}{\gamma\theta_k^2}\mathbb{E}\left\|\mathbf{x}^{k+1} - \mathbf{y}^k\right\|^2
$$

$$
+ \left(\frac{\gamma L_c^2}{2nC_2} + L_c\right)\sum_{k=0}^{K}\frac{1}{\theta_k^2}\mathbb{E}\left\|\mathbf{w}^{j(k)} - \mathbf{y}^k\right\|^2 + \frac{n^2}{2\gamma}\left\|\mathbf{z}^0 - \mathbf{x}^*\right\|^2
$$

$$
\overset{a}{\leq} \frac{f(\mathbf{x}^0) + \hat{h}_0 - F(\mathbf{x}^*)}{\theta_{-1}^2} + \frac{n^2}{2\gamma}\left\|\mathbf{z}^0 - \mathbf{x}^*\right\|^2
$$

$$
- \left[\frac{1}{2} - \frac{\gamma L_c}{2} - \frac{C_2}{2} - \gamma\left(\frac{\gamma L_c^2}{2nC_2} + L_c\right)\left(\frac{\tau^2 + \tau}{n} + 2\tau\right)^2\right]
$$

$$
\times \sum_{k=0}^{K}\frac{\gamma}{\theta_k^2}\mathbb{E}\left\|\frac{\mathbf{x}^{k+1} - \mathbf{y}^k}{\gamma}\right\|^2,
$$

where $\overset{a}{\leq}$ uses (6.52).

Setting $C_2 = \gamma L_c$, by the assumption

$$2\gamma L + \left(2 + \frac{1}{n}\right)\gamma L_c \left(\frac{\tau^2 + \tau}{n} + 2\tau\right)^2 \leq 1,$$

we have

$$\frac{1}{2} - \frac{\gamma L_c}{2} - \frac{C_2}{2} - \gamma\left(\frac{\gamma L_c^2}{2nC_2} + L_c\right)\left(\frac{\tau^2 + \tau}{n} + 2\tau\right)^2 \geq 0.$$

So

$$\frac{\mathbb{E}F(\mathbf{x}^{K+1}) - F(\mathbf{x}^*)}{\theta_K^2} + \frac{n^2}{2\gamma}\mathbb{E}\left\|\mathbf{z}^{K+1} - \mathbf{x}^*\right\|^2$$

$$\overset{a}{\leq} \frac{\mathbb{E}f(\mathbf{x}^{K+1}) + \mathbb{E}\hat{h}_{K+1} - F(\mathbf{x}^*)}{\theta_K^2} + \frac{n^2}{2\gamma}\mathbb{E}\left\|\mathbf{z}^{K+1} - \mathbf{x}^*\right\|^2$$

$$\leq \frac{f(\mathbf{x}^0) + \hat{h}_0 - F(\mathbf{x}^*)}{\theta_{-1}^2} + \frac{n^2}{\gamma^2}\left\|\mathbf{z}^0 - \mathbf{x}^*\right\|^2$$

$$\overset{b}{=} \frac{F(\mathbf{x}^0) - F(\mathbf{x}^*)}{\theta_{-1}^2} + \frac{n^2}{2\gamma}\left\|\mathbf{z}^0 - \mathbf{x}^*\right\|^2,$$

where in $\overset{a}{\leq}$ we use $h(\mathbf{x}^{K+1}) = h\left(\sum_{i=0}^{K+1} e_{k+1,i}\mathbf{z}^i\right) \leq \sum_{i=0}^{K+1} e_{k+1,i}h(\mathbf{z}^i) = \hat{h}_{K+1}$
and in $\overset{b}{\leq}$ we use $\hat{h}_0 = h(\mathbf{x}^0)$.

Now we consider the strongly convex case. In the following, again we set $\theta_k = \theta$. Multiplying (6.51) with $(1 - \theta)^{K-k}$ and summing the results with k from 0 to K, we have

$$\sum_{k=0}^{K}(1 - \theta)^{K-k}\left\|\mathbf{w}^{j(k)} - \mathbf{y}^k\right\|^2$$

$$\leq \left(\frac{\tau^2 + \tau}{2n} + \tau\right)\sum_{k=0}^{K}\sum_{i=1}^{\min(\tau,k)}\left(\frac{i}{n} + 1\right)(1 - \theta)^{K-k}\left\|\mathbf{x}^{k-i+1} - \mathbf{y}^{k-i}\right\|^2$$

$$= \left(\frac{\tau^2 + \tau}{2n} + \tau\right)\sum_{k=0}^{K}\sum_{i=1}^{\min(\tau,k)}(1 - \theta)^{-i}\left(\frac{i}{n} + 1\right)$$

$$\times (1 - \theta)^{K-(k-i)}\left\|\mathbf{x}^{k-i+1} - \mathbf{y}^{k-i}\right\|^2$$

$$\leq \frac{1}{(1-\theta)^\tau} \left(\frac{\tau^2 + \tau}{2n} + \tau \right) \sum_{k=0}^{K} \sum_{i=1}^{\min(\tau,k)} \left(\frac{i}{n} + 1 \right)$$

$$\times (1-\theta)^{K-(k-i)} \left\| \mathbf{x}^{k-i+1} - \mathbf{y}^{k-i} \right\|^2$$

$$= \frac{1}{(1-\theta)^\tau} \left(\frac{\tau^2 + \tau}{2n} + \tau \right) \sum_{i=1}^{\tau} \left(\frac{i}{n} + 1 \right)$$

$$\times \sum_{k=i}^{K} (1-\theta)^{K-(k-i)} \left\| \mathbf{x}^{k-i+1} - \mathbf{y}^{k-i} \right\|^2$$

$$= \frac{1}{(1-\theta)^\tau} \left(\frac{\tau^2 + \tau}{2n} + \tau \right) \sum_{i=1}^{\tau} \left(\frac{i}{n} + 1 \right) \sum_{k'=0}^{K-i} (1-\theta)^{K-k'} \left\| \mathbf{x}^{k'+1} - \mathbf{y}^{k'} \right\|^2$$

$$\leq \frac{1}{(1-\theta)^\tau} \left(\frac{\tau^2 + \tau}{2n} + \tau \right) \sum_{i=1}^{\tau} \left(\frac{i}{n} + 1 \right) \sum_{k=0}^{K} (1-\theta)^{K-k} \left\| \mathbf{x}^{k+1} - \mathbf{y}^{k} \right\|^2$$

$$= \frac{[(\tau^2 + \tau)/n + 2\tau]^2}{4(1-\theta)^\tau} \sum_{k=0}^{K} (1-\theta)^{K-k} \left\| \mathbf{x}^{k+1} - \mathbf{y}^{k} \right\|^2. \tag{6.54}$$

Since we have set $\theta = \frac{-\gamma\mu + \sqrt{\gamma^2\mu^2 + 4\gamma\mu}}{2n}$, which satisfies

$$\frac{\theta^2 n^2 + (n-1)\mu\theta\gamma}{2\gamma} = (1-\theta) \left(\frac{\theta^2 n^2 + n\mu\theta\gamma}{2\gamma} \right),$$

by rearranging terms in (6.49) and using (6.50) again, we have

$$\mathbb{E}_{i_k} f(\mathbf{x}^{k+1}) + \mathbb{E}_{i_k} \hat{h}_{k+1} - F(\mathbf{x}^*) + \frac{n^2\theta^2 + n\theta\mu\gamma}{2\gamma} \mathbb{E}_{i_k} \left\| \mathbf{z}^{k+1} - \mathbf{x}^* \right\|^2$$

$$\leq (1-\theta) \left(f(\mathbf{x}^k) + \hat{h}_k - F(\mathbf{x}^*) + \frac{n^2\theta^2 + n\theta\mu\gamma}{2\gamma} \left\| \mathbf{z}^k - \mathbf{x}^* \right\|^2 \right)$$

$$- \frac{1}{\gamma} \left(\frac{1}{2} - \frac{\gamma L}{2} - \frac{C_2}{2} \right) \mathbb{E}_{i_k} \left\| \mathbf{x}^{k+1} - \mathbf{y}^k \right\|^2 + \left(\frac{\gamma L_c^2}{2nC_2} + L_c \right) \left\| \mathbf{w}^{j(k)} - \mathbf{y}^k \right\|^2. \tag{6.55}$$

From the assumption, we have $\mu/L_c \leq 1$. By the setting of γ, we have $\gamma\mu \leq 1$. So $n\theta \leq \sqrt{\gamma\mu} \leq 1$. Thus the setting of θ also satisfies the assumptions of Lemma 5.1 in Sect. 5.1.1. On the other hand, from the assumption on γ we have

$$3\gamma L_c \tau^2 \leq 2\gamma L_c + \left(\frac{3}{4} + \frac{3}{8n} \right) \gamma L_c \left[(\tau^2 + \tau)/n + 2\tau \right]^2 \leq 1. \tag{6.56}$$

Now we consider $\frac{1}{(1-\theta)^\tau}$. Without loss of generality, we assume $n \geq 2$. Then we have

$$\frac{1}{(1-\theta)^\tau} \overset{a}{\leq} \frac{1}{(1-\sqrt{\gamma\mu}/n)^\tau} \overset{b}{\leq} \frac{1}{\left[1 - \frac{1}{\tau}\sqrt{\mu/(3L_c)}/n\right]^\tau}$$

$$\overset{c}{\leq} \frac{1}{\left(1 - \frac{1}{2\sqrt{3}\tau}\right)^\tau} \leq \frac{1}{1 - \frac{1}{2\sqrt{3}}} \leq \frac{3}{2},$$

where in $\overset{a}{\leq}$ we use $\theta \leq \sqrt{\gamma\mu}/n$, in $\overset{b}{\leq}$ we use (6.56), and $\overset{c}{\leq}$ uses $\sqrt{\frac{\mu}{L_c}}/n \leq \frac{1}{n} \leq \frac{1}{2}$.

Taking expectation on (6.55), multiplying (6.55) with θ^{K-k}, and then summing the result with k from 0 to K, we have

$$\mathbb{E}f(\mathbf{x}^{K+1}) + \mathbb{E}\hat{h}_{K+1} - F(\mathbf{x}^*) + \frac{n^2\theta^2 + n\theta\mu\gamma}{2\gamma}\mathbb{E}\left\|\mathbf{z}^{K+1} - \mathbf{x}^*\right\|^2$$

$$\leq (1-\theta)^{K+1}\left(f(\mathbf{x}^0) + \hat{h}_0 - F(\mathbf{x}^*) + \frac{n^2\theta^2 + n\theta\mu\gamma}{2\gamma}\left\|\mathbf{z}^0 - \mathbf{x}^*\right\|^2\right)$$

$$-\frac{1}{\gamma}\left(\frac{1}{2} - \frac{\gamma L}{2} - \frac{C_2}{2}\right)\sum_{k=0}^{K}(1-\theta)^{K-k}\mathbb{E}\left\|\mathbf{x}^{k+1} - \mathbf{y}^k\right\|^2$$

$$+\left(\frac{\gamma L_c^2}{2nC_2} + L_c\right)\sum_{k=0}^{K}(1-\theta)^{K-k}\mathbb{E}\left\|\mathbf{w}^{j(k)} - \mathbf{y}^k\right\|^2$$

$$\overset{a}{\leq} (1-\theta)^{K+1}\left(f(\mathbf{x}^0) + \hat{h}_0 - F(\mathbf{x}^*) + \frac{n^2\theta^2 + n\theta\mu\gamma}{2\gamma}\left\|\mathbf{z}^0 - \mathbf{x}^*\right\|^2\right)$$

$$-\frac{1}{\gamma}\left\{\frac{1}{2} - \frac{\gamma L_c}{2} - \frac{C_2}{2} - \left(\frac{\gamma^2 L_c^2}{2nC_2} + \gamma L_c\right)\frac{\left[(\tau^2+\tau)/n + 2\tau\right]^2}{4(1-\theta)^\tau}\right\}$$

$$\times \sum_{k=0}^{K}(1-\theta)^{K-k}\mathbb{E}\left\|\mathbf{x}^{k+1} - \mathbf{y}^k\right\|^2,$$

where $\overset{a}{\leq}$ uses (6.54).

Setting $C_2 = \gamma L_c$, since

$$2\gamma L_c + \left(\frac{\gamma L_c}{n} + 2\gamma L_c\right)\frac{\left[(\tau^2+\tau)/n + 2\tau\right]^2}{4(1-\theta)^\tau}$$

$$\leq 2\gamma L_c + \left(\frac{\gamma L_c}{n} + 2\gamma L_c\right)\frac{3\left[(\tau^2+\tau)/n + 2\tau\right]^2}{8},$$

by the assumption of γ we have

$$\frac{1}{2} - \frac{\gamma L_c}{2} - \frac{C_2}{2} - \left(\frac{\gamma^2 L_c^2}{2nC_2} + \gamma L_c\right) \frac{\left[(\tau^2 + \tau)/n + 2\tau\right]^2}{4(1-\theta)^\tau} \geq 0.$$

Then using $h(\mathbf{x}^{K+1}) \leq \hat{h}_{K+1}$ and $\hat{h}_0 = h(\mathbf{x}^0)$, we obtain:

$$\mathbb{E}F(\mathbf{x}^{K+1}) - F(\mathbf{x}^*) + \frac{n^2\theta^2 + n\theta\mu\gamma}{2\gamma} \mathbb{E}\left\|\mathbf{z}^{K+1} - \mathbf{x}^*\right\|^2$$

$$\leq (1-\theta)^{K+1}\left(F(\mathbf{x}^0) - F(\mathbf{x}^*) + \frac{n^2\theta^2 + n\theta\mu\gamma}{2\gamma}\left\|\mathbf{z}^0 - \mathbf{x}^*\right\|^2\right). \qquad \square$$

6.2 Accelerated Distributed Algorithms

We describe accelerated distributed algorithms in this section. Distributed algorithms allow machines to handle their local data and communicate with each other to solve the coupled problems. The main challenges of distributed algorithms are:

1. Reducing communication costs. In practice, the computation time of local computation is usually significantly smaller than the communication cost. So balancing the local computation costs and communication ones becomes very important in designing an efficient distributed algorithms.
2. Improving the speed-up factor. Speed-up is the factor that the running time using a single machine divided by the running time using m machines. We say that a distributed algorithm achieves a linear speed-up if using m machines to solve can be αm times faster than using a single machine, where α is an absolute constant.

Typical organizations for communication between nodes are centralized topology where there is a central machine and all others communicate with it, and decentralized one where all the machines can only communicate with their neighbors.

6.2.1 Centralized Topology

6.2.1.1 Large Mini-Batch Algorithms

A straightforward way to implement an algorithm on a distributed system is by transporting the algorithm into a large mini-batch setting. All the machines compute the (stochastic) gradient synchronously and then send back the gradient to the central node. In this way, the algorithm is essentially identical to a serial algorithm. For infinite-sum problems, we have shown in Sect. 5.5 that the momentum technique ensures an $\Omega(\sqrt{\kappa})$ times larger mini-batch size. For finite-sum problems, the best result is obtained by Katyusha with a large mini-batch size version [2] (c.f. Algorithm 5.3 in Sect. 5.1.3). We show the convergence results as follows.

Theorem 6.3 *Suppose that $h(\mathbf{x})$ is μ-strongly convex with $\mu \leq \frac{3L}{8n}$. For Algorithm 5.3, if the step size $\gamma = \frac{1}{3L}$, $\theta_1 = \sqrt{\frac{2n\mu}{3L}}$, $\theta_3 = 1 + \frac{\mu\gamma}{\theta_1}$, the mini-batch size b satisfies $b \leq \sqrt{n}$, and $m = n/b$, then we have*

$$F(\mathbf{x}^{\text{out}}) - F(\mathbf{x}^*) \leq O\left(\left(1 + \sqrt{\mu/(Ln)}\right)^{-Sm}\right)\left(F(\mathbf{x}_0^0) - F(\mathbf{x}^*)\right),$$

where $\mathbf{x}^{\text{out}} = \frac{m\tilde{\mathbf{x}}^S + (1-2\theta_1)\mathbf{z}_m^S}{m+1-2\theta_1}$. *In other words, by setting $b = \sqrt{n}$, the total computation costs and total communication costs for Algorithm 5.3 to achieve an ϵ-accuracy are $\tilde{O}(\sqrt{n\kappa})$ and $\tilde{O}(\sqrt{\kappa})$, respectively.*

The proof of Theorem 6.3 is similar to that of Theorem 5.3 but is a little more involved by taking the mini-batch size b into account. Readers who are interested in it can refer to [2] for the details.

6.2.1.2   Dual Communication-Efficient Methods

The above mini-batch algorithms are mainly implemented on shared memory systems where the cores can access all the data. However, in some distributed systems the data are stored separately. To handle this case, we can formulate the model into a constrained problem. We consider the following objective function:

$$\mathcal{L}(\mathbf{w}) = \frac{1}{mn}\sum_{i=1}^{m}\sum_{j=1}^{n} l_{ij}(\mathbf{w}) + h(\mathbf{w}) + \frac{\lambda}{2}\|\mathbf{w}\|^2, \quad \mathbf{w} \in \mathbb{R}^d,$$

where m is the number of machines in the network and n is the number of individual functions on each machine. We assume that $h(\mathbf{w})$ is μ-strongly convex and each l_{ij} is convex and L-smooth. We also denote $\kappa = \frac{L}{\mu}$ and assume $\kappa \geq n$. For centralized algorithms, we can introduce auxiliary variables as follows:

$$\min_{\mathbf{w},\mathbf{u}} \quad \frac{1}{mn}\sum_{i=1}^{m}\sum_{j=1}^{n}\left(l_{ij}(\mathbf{w}_{ij}) + \frac{\lambda}{2}\|\mathbf{u}_i\|^2\right) + h(\mathbf{u}_0), \tag{6.57}$$

$$\text{s.t.} \quad \mathbf{w}_{ij} = \mathbf{u}_i, \quad i \in [m], \ j \in [n],$$

$$\mathbf{u}_i = \mathbf{u}_0, \quad i \in [m].$$

Let $\frac{1}{mn}\mathbf{a}_{ij} \in \mathbb{R}^d$ with $i \in [m]$ and $j \in [n]$ be the Lagrange multipliers for $\mathbf{w}_{ij} = \mathbf{u}_i$, and $\frac{1}{mn}\mathbf{b}_i \in \mathbb{R}^d$ be the Lagrange multipliers for $\mathbf{u}_i = \mathbf{u}_0$. Then we can obtain the dual problem of (6.57) as

$$
\begin{aligned}
\max_{\mathbf{a},\mathbf{b}} D(\mathbf{a}, \mathbf{b}) &= \max_{\mathbf{a},\mathbf{b}} \min_{\mathbf{w},\mathbf{u}} \left[\frac{1}{mn} \sum_{i=1}^{m} \sum_{j=1}^{n} \left(l_{ij}(\mathbf{w}_{ij}) + \langle \mathbf{a}_{ij}, \mathbf{w}_{ij} - \mathbf{u}_i \rangle + \frac{\lambda}{2} \|\mathbf{u}_i\|^2 \right) \right. \\
&\quad \left. + \frac{1}{nm} \sum_{i=1}^{m} \langle \mathbf{b}_i, \mathbf{u}_0 - \mathbf{u}_i \rangle + h(\mathbf{u}_0) \right] \\
&= \max_{\mathbf{a},\mathbf{b}} - \left[\sum_{i=1}^{m} \left(\frac{1}{mn} \sum_{j=1}^{n} l_{ij}^*(-\mathbf{a}_{ij}) + \frac{1}{2m\lambda} \left\| \frac{1}{n} \sum_{j=1}^{n} \mathbf{a}_{ij} + \frac{1}{n}\mathbf{b}_i \right\|^2 \right) \right. \\
&\quad \left. + h^* \left(-\frac{1}{nm} \sum_{i=1}^{m} \mathbf{b}_i \right) \right],
\end{aligned}
\tag{6.58}
$$

where $l_{ij}^*(\cdot)$ and $h^*(\cdot)$ are the conjugate functions of $l_{ij}(\cdot)$ and $h(\cdot)$, respectively. If $h(\mathbf{x}) \equiv 0$ we have

$$
h^*(\mathbf{x}) = \begin{cases} 0, & \mathbf{x} = \mathbf{0}, \\ +\infty, & \text{otherwise.} \end{cases}
$$

We denote $\mathbf{a}_i = [\mathbf{a}_{i1}; \cdots ; \mathbf{a}_{in}] \in \mathbb{R}^{nd}$, $\mathbf{a} = [\mathbf{a}_1; \cdots ; \mathbf{a}_m] \in \mathbb{R}^{mnd}$, and $\mathbf{b} = [\mathbf{b}_1; \cdots ; \mathbf{b}_m] \in \mathbb{R}^{md}$. To solve (6.58), when $\mathbf{b}_i$ is fixed, each $\mathbf{a}_i$ can be solved independently, so each $\mathbf{a}_i$ can be stored separately on a local machine. We refer $\mathbf{a}$ as the local variable. When each $\mathbf{a}_i$ is fixed, $\mathbf{b}_i$ reflects the "inconsistent degree" from a local solution to the global one, and solving $\mathbf{b}_i$ needs communication. So we refer $\mathbf{b}$ as the consensus variable. We can solve $\mathbf{a}$ and $\mathbf{b}$ alternately. The iteration can be written as

$$
\dot{\mathbf{a}}_i^{k+1} = \max_{\dot{\mathbf{a}}_i} D(\dot{\mathbf{a}}_i, \mathbf{b}^k), \ \dot{\mathbf{a}}_i \text{ is a subset of } \mathbf{a}_i, \ i \in [m],
$$

$$
\mathbf{b}^{k+1} = \max_{\mathbf{b}} D(\mathbf{a}^{k+1}, \mathbf{b}).
$$

The above method has been considered in [9, 18]. By designing a subproblem for $\mathbf{a}_i$, each machine can solve the local problem by efficient stochastic algorithms.

However, the above method is still suboptimal. Suppose that the sample size of $\dot{\mathbf{a}}_i^{k+1}$ is n_0, then we can only obtain the total computation costs of $O(\sqrt{n\kappa n_0})$ and communication costs of $O(\sqrt{\kappa n / n_0})$. To obtain a communication-efficient algorithm, we need choose $n_0 = n$. However, the computation costs become

$O(n\sqrt{\kappa})$, whose dependence on n is not square rooted. To obtain a faster rate, we consider the following shifted form of (6.58).

Setting $F(\mathbf{a}, \mathbf{b}) = -D(\mathbf{a}, \mathbf{b})$, we have

$$\min_{\mathbf{a}, \mathbf{b}} F(\mathbf{a}, \mathbf{b}) \equiv g_0(\mathbf{b}) + \sum_{i=1}^{m} \sum_{j=1}^{n} g_{ij}(\mathbf{a}_{ij}) + f(\mathbf{a}, \mathbf{b}), \qquad (6.59)$$

where

$$g_0(\mathbf{b}) = h^* \left(-\frac{1}{nm} \sum_{i=1}^{m} \mathbf{b}_i \right) + \sum_{i=1}^{m} \frac{\mu_1}{6mn^2} \|\mathbf{b}_i\|^2,$$

$$g_{ij}(\mathbf{a}_{ij}) = \frac{1}{mn} \left(l_{ij}^*(-\mathbf{a}_{ij}) - \frac{\mu_1}{4} \|\mathbf{a}_{ij}\|^2 \right), \quad i \in [m], \quad j \in [n],$$

$$f(\mathbf{a}, \mathbf{b}) = \sum_{i=1}^{m} f_i(\mathbf{a}_i, \mathbf{b}),$$

in which

$$f_i(\mathbf{a}_i, \mathbf{b}) = \frac{1}{mn} \sum_{j=1}^{n} \frac{\mu_1}{4} \|\mathbf{a}_{ij}\|^2 + \frac{1}{2m\lambda} \left\| \frac{1}{n} \sum_{j=1}^{n} \mathbf{a}_{ij} + \frac{1}{n} \mathbf{b}_i \right\|^2$$

$$- \frac{\mu_1}{6mn^2} \|\mathbf{b}_i\|^2, \quad i \in [m].$$

We define $\nabla_{\mathbf{a}} f(\bar{\mathbf{a}}, \bar{\mathbf{b}}) \in \mathbb{R}^{mnd}$ as the partial gradient of $f(\bar{\mathbf{a}}, \bar{\mathbf{b}})$ w.r.t. $\mathbf{a}$. Similarly, we denote $\nabla_{\mathbf{a}_i} f(\bar{\mathbf{a}}, \bar{\mathbf{b}}) \in \mathbb{R}^{nd}$, $\nabla_{\mathbf{b}_i} f(\bar{\mathbf{a}}, \bar{\mathbf{b}}) \in \mathbb{R}^{d}$, $\nabla_{\mathbf{b}} f(\bar{\mathbf{a}}, \bar{\mathbf{b}}) \in \mathbb{R}^{md}$, and $\nabla_{\mathbf{a}_{ij}} f(\bar{\mathbf{a}}, \bar{\mathbf{b}}) \in \mathbb{R}^{d}$ with $i \in [m]$ and $j \in [n]$ as the partial gradient of $f(\bar{\mathbf{a}}, \bar{\mathbf{b}})$ w.r.t. $\mathbf{a}_i$, $\mathbf{b}_i$, $\mathbf{b}$ and $\mathbf{a}_{ij}$, respectively. We have the following lemma.

Lemma 6.1 *Suppose $n \leq \frac{L}{\lambda}$ and set $\mu_1 = \frac{1}{L}$ and $L_c = \frac{1}{2mn^2\lambda} + \frac{1}{4mnL}$. Then $f(\mathbf{a}, \mathbf{b})$ is convex and has L_c-block coordinate Lipschitz continuous gradients with respect to $(\mathbf{a}_{ij}, \mathbf{b})$, i.e., for all $i \in [m]$ and $j \in [n]$, for any $\mathbf{b} \in \mathbb{R}^{md}$, and for any $\bar{\mathbf{a}} \in \mathbb{R}^{mnd}$ and $\tilde{\mathbf{a}} \in \mathbb{R}^{mnd}$ which differ only on the (i, j)-th component, i.e., $\bar{\mathbf{a}}_{k,l} = \tilde{\mathbf{a}}_{k,l}$ when $k \neq i$ or $l \neq j$, we have*

$$\|\nabla_{\mathbf{a}_{ij}} f(\tilde{\mathbf{a}}, \mathbf{b}) - \nabla_{\mathbf{a}_{ij}} f(\bar{\mathbf{a}}, \mathbf{b})\| \leq L_c \|\tilde{\mathbf{a}}_{ij} - \bar{\mathbf{a}}_{ij}\|, \qquad (6.60)$$

and for any $\mathbf{a} \in \mathbb{R}^{mnd}$, $\tilde{\mathbf{b}} \in \mathbb{R}^{md}$, and $\bar{\mathbf{b}} \in \mathbb{R}^{md}$, we have

$$\|\nabla_{\mathbf{b}} f(\mathbf{a}, \tilde{\mathbf{b}}) - \nabla_{\mathbf{b}} f(\mathbf{a}, \bar{\mathbf{b}})\| \leq L_c \|\tilde{\mathbf{b}} - \bar{\mathbf{b}}\|. \qquad (6.61)$$

Moreover, $g_0(\mathbf{b})$ is $\frac{1}{6mn^2 L}$-strongly convex and $g_{ij}(\mathbf{a}_{ij})$ ($j \in [n]$, $i \in [m]$) are $\frac{3}{4nmL}$-strongly convex.

Proof By checking, we can find that (6.60) is right. Because $\frac{\mu_1}{6mn^2} = \frac{1}{6mn^2L} \le \frac{1}{4mnL}$, (6.61) is right. Also, it is easy to prove that $g_0(\mathbf{b})$ is $\frac{\mu_1}{6mn^2}$-strongly convex and $g_{ij}(\mathbf{a}_{ij})$ with $i \in [m]$ and $j \in [n]$ are $\frac{3\mu_1}{4nm}$-strongly convex.

We now prove that $f(\mathbf{a}, \mathbf{b})$ is convex. Consider the following inequality:

$$(1 + \eta)a^2 + \left(1 + \frac{1}{\eta}\right) b^2 \ge (a + b)^2.$$

Dividing $1 + \eta$ on both sides, we have

$$a^2 \ge \frac{(a + b)^2}{1 + \eta} - \frac{1}{\eta}b^2. \tag{6.62}$$

Then setting

$$\eta = \frac{2}{\mu_1 \lambda} = \frac{2L}{\lambda} \ge 2n \ge 2, \tag{6.63}$$

we have

$$\frac{1}{mn} \sum_{j=1}^{n} \frac{\mu_1}{2} \|\mathbf{a}_{ij}\|^2 + \frac{1}{2m\lambda} \left\| \frac{1}{n} \sum_{j=1}^{n} \mathbf{a}_{ij} + \frac{1}{n} \mathbf{b}_i \right\|^2$$

$$\overset{a}{\ge} \frac{1}{mn} \sum_{j=1}^{n} \frac{\mu_1}{2} \|\mathbf{a}_{ij}\|^2 + \frac{1}{2m\lambda(1 + \eta)} \left\| \frac{1}{n} \mathbf{b}_i \right\|^2 - \frac{1}{2m\lambda\eta} \left\| \frac{1}{n} \sum_{j=1}^{n} \mathbf{a}_{ij} \right\|^2$$

$$\ge \frac{1}{mn} \sum_{j=1}^{n} \frac{\mu_1}{2} \|\mathbf{a}_{ij}\|^2 + \frac{1}{2m\lambda(1 + \eta)} \left\| \frac{1}{n} \mathbf{b}_i \right\|^2 - \frac{1}{2mn\lambda\eta} \sum_{j=1}^{n} \|\mathbf{a}_{ij}\|^2$$

$$\overset{b}{=} \frac{1}{mn} \sum_{j=1}^{n} \frac{\mu_1}{2} \|\mathbf{a}_{ij}\|^2 + \frac{1}{2m\lambda(1 + \eta)} \left\| \frac{1}{n} \mathbf{b}_i \right\|^2 - \frac{1}{2mn} \sum_{j=1}^{n} \frac{\mu_1}{2} \|\mathbf{a}_{ij}\|^2$$

$$\overset{c}{\ge} \frac{1}{2mn} \sum_{j=1}^{n} \frac{\mu_1}{2} \|\mathbf{a}_{ij}\|^2 + \frac{1}{2m(\frac{\eta\lambda}{2} + \eta\lambda)} \left\| \frac{1}{n} \mathbf{b}_i \right\|^2$$

$$\overset{d}{=} \frac{1}{mn} \sum_{j=1}^{n} \frac{\mu_1}{4} \|\mathbf{a}_{ij}\|^2 + \frac{\mu_1}{6m} \left\| \frac{1}{n} \mathbf{b}_i \right\|^2,$$

where $\overset{a}{\ge}$ uses (6.62), and $\overset{b}{=}$, $\overset{c}{\ge}$, and $\overset{d}{\ge}$ all use (6.63).

Then by summing $i = 1$ to m, we obtain that $f(\mathbf{a}, \mathbf{b})$ is convex as it is a nonnegative quadratic function. $\square$

Algorithm 6.3 Distributed stochastic communication accelerated dual (DSCAD)

Input θ_k, p_b, and L_2. Set $\mathbf{a}^0 = \hat{\mathbf{a}}^0 = \mathbf{0}$, $\mathbf{b}^0 = \hat{\mathbf{b}}^0 = \mathbf{0}$, and $p_a = \frac{1-p_b}{n}$.

1 **for** $k = 0$ **to** K **do**

2 $\quad \tilde{\mathbf{a}}^k = (1 - \theta_k)\mathbf{a}^k + \theta_k \hat{\mathbf{a}}^k$,

3 $\quad \tilde{\mathbf{b}}^k = (1 - \theta_k)\mathbf{b}^k + \theta_k \hat{\mathbf{b}}^k$,

4 $\quad$ Uniformly sample a random number q from $[0, 1]$,

5 $\quad$ If $q \leq p_b$, $\qquad \diamond$ Communication

7 $\quad\quad \hat{\mathbf{b}}^{k+1} = \mathrm{argmin}_{\mathbf{b}} \left(g_0(\mathbf{b}) + \left\langle \nabla_{\mathbf{b}} f(\tilde{\mathbf{a}}^k, \tilde{\mathbf{b}}^k), \mathbf{b} \right\rangle + \frac{\theta_k L_2}{2p_b} \left\| \mathbf{b} - \hat{\mathbf{b}}^k \right\|^2 \right)$.

8 $\quad$ Else $\qquad \diamond$ Local Update

9 $\quad\quad$ For each machine i with $i \in [m]$, randomly sample an index $j(i)$,

10 $\quad\quad \hat{\mathbf{a}}_{i,j(i)}^{k+1} = \mathrm{argmin}_{\mathbf{a}_{i,j(i)}} \left(g_{i,j(i)}(\mathbf{a}_{i,j(i)}) + \left\langle \nabla_{\mathbf{a}_{i,j(i)}} f_i(\tilde{\mathbf{a}}_i^k, \tilde{\mathbf{b}}^k), \mathbf{a}_{i,j(i)} \right\rangle + \frac{\theta_k L_2}{2p_a} \left\| \mathbf{a}_{i,j(i)} - \hat{\mathbf{a}}_{i,j(i)}^k \right\|^2 \right)$.

11 $\quad$ End If

12 $\quad \mathbf{a}^{k+1} = \tilde{\mathbf{a}}^k + \frac{\theta_k}{p_a}(\hat{\mathbf{a}}^{k+1} - \hat{\mathbf{a}}^k)$,

13 $\quad \mathbf{b}^{k+1} = \tilde{\mathbf{b}}^k + \frac{\theta_k}{p_b}(\hat{\mathbf{b}}^{k+1} - \hat{\mathbf{b}}^k)$.

14 **end for** k

Output $\mathbf{a}^{K+1}$ and $\mathbf{b}^{K+1}$.

Lemma 6.1 suggests that $f(\mathbf{a}, \mathbf{b})$ has the same block Lipschitz constants and the strongly convex moduli of $g_{ij}(\mathbf{a}_{ij})$ are $O(n)$ times larger than that of $g_0(\mathbf{b})$. We can update $\mathbf{a}$ and $\mathbf{b}$ with a nonuniform probability, which is known as importance sampling in [3, 17]. During the updates, if $\mathbf{a}$ is chosen, we solve $\mathbf{b}$ with communication. Otherwise, for each machine i, we randomly choose a sample $j(i)$ and update $\mathbf{a}_{i,j(i)}$. By integrating the momentum technique [7, 11], we obtain the framework called Distributed Stochastic Communication Accelerated Dual (DSCAD) and shown in Algorithm 6.3. The main improvement in the algorithm is the variants that the machines are scheduled to do local communication or communicate with each other under a certain probability. In this way, in each round of communication, each machine is not solving a local subproblem but instead concurrently solving the original problem (6.59). The algorithm ensures convergence with computation costs of $T_1 = O(\sqrt{n\kappa})$ and communication costs of $T_2 = \tilde{O}(\sqrt{\kappa})$.

DSCAD can also be implemented in decentralized systems. So we leave the proof of the convergence result to the next section.

6.2.2 Decentralized Topology

For decentralized topology, the machines are connected by an undirected network and cooperatively solve a joint problem. We represent the topology of the network as a graph $\mathfrak{g} = \{V, E\}$, where V and E are the node and the edge sets, respectively. Each node $v \in V$ represents a machine in the network and $e_{ij} = (i, j) \in E$ indicates that nodes i and j are connected. Let the Laplacian matrix (Definition A.2) of the graph $\mathfrak{g}$ be $\mathbf{L}$.

The early works for decentralized algorithms include typically decentralized gradient descent [16], which exhibits sublinear convergence rate even if each $f_i(\mathbf{x})$

is strongly convex and has Lipschitz continuous gradients. More recently, a number of methods of linear convergence rate have been designed, such as EXTRA [15] and augmented Lagrangians [6], and a more recent work [14] shows that a little variants of decentralized gradient descent can achieve a linear convergence rate and can further obtain the optimal convergence rate with iteration costs of $\tilde{O}(\sqrt{\kappa})$, and $\tilde{O}(\sqrt{\kappa \kappa_g})$ by fusing with the momentum technique, where κ is the condition number of the local function and κ_g is the eigengap of the gossip matrix (matrix $\mathbf{A}$ in (6.64)) used for communication. We consider the following optimization problem:

$$\min_{\mathbf{w}} \ \mathcal{L}_2(\mathbf{w}) \equiv \frac{1}{mn} \sum_{i=1}^{m} \sum_{j=1}^{n} \left(l_{ij}(\mathbf{w}_i) + \frac{\lambda}{2} h(\mathbf{w}) \right), \tag{6.64}$$

s.t. $\mathbf{Aw} = \mathbf{0}$,

where $\mathbf{w} = [\mathbf{w}_1; \cdots ; \mathbf{w}_m] \in \mathbb{R}^{md}$, $\mathbf{A} = (\mathbf{L}/\|\mathbf{L}\|)^{1/2} \otimes \mathbf{I}_d \in \mathbb{R}^{dm \times dm}$ is the gossip matrix, $\otimes$ indicates the Kronecker product, $\mathbf{I}_d$ is the identity matrix with size d, $\mathbf{w}_i \in \mathbb{R}^d$, and $\lambda \geq 0$. $\mathbf{A}$ is a symmetric matrix and the choice of $\mathbf{A}$ ensures that computing $\mathbf{A}^2 \mathbf{w}$ needs one time of communication [14, 15]. For simplicity, we assume that $h(\mathbf{w}) = \|\mathbf{w}\|^2$.

Then by introducing auxiliary variables to split the loss and the regularization terms, we have

$$\min_{\mathbf{w}, \mathbf{u}} \ \frac{1}{mn} \sum_{i=1}^{m} \sum_{j=1}^{n} \left(l_{ij}(\mathbf{w}_{ij}) + \frac{\lambda}{2} \|\mathbf{u}_i\|^2 \right)$$

$$s.t. \ \mathbf{w}_{ij} = \mathbf{u}_i, \quad i \in [m], \quad j \in [n],$$

$$\mathbf{Au} = \mathbf{0},$$

where $\mathbf{u} = [\mathbf{u}_1; \cdots ; \mathbf{u}_m]$. By introducing the dual variables $\frac{1}{mn} \mathbf{a}_{ij}$ with $i \in [m]$ and $j \in [n]$ for $\mathbf{w}_{ij} = \mathbf{u}_i$, and $\frac{1}{mn} \mathbf{b}$ for $\mathbf{Au} = \mathbf{0}$, we can derive the dual problem as

$$\max_{\mathbf{a}, \mathbf{b}} \min_{\mathbf{w}, \mathbf{u}} \left[\frac{1}{mn} \sum_{i=1}^{m} \sum_{j=1}^{n} \left(l_{ij}(\mathbf{w}_{ij}) + \langle \mathbf{a}_{ij}, \mathbf{w}_{ij} - \mathbf{u}_i \rangle + \frac{\lambda}{2} \|\mathbf{u}_i\|^2 \right) + \frac{1}{nm} \langle \mathbf{A}^T \mathbf{b}, \mathbf{u} \rangle \right]$$

$$= \max_{\mathbf{a}, \mathbf{b}} \min_{\mathbf{w}, \mathbf{u}} \left[\frac{1}{mn} \sum_{i=1}^{m} \sum_{j=1}^{n} \left(l_{ij}(\mathbf{w}_{ij}) + \langle \mathbf{a}_{ij}, \mathbf{w}_{ij} - \mathbf{u}_i \rangle + \frac{\lambda}{2} \|\mathbf{u}_i\|^2 \right) \right.$$

$$\left. + \frac{1}{nm} \sum_{i=1}^{m} \langle \mathbf{A}_i^T \mathbf{b}, \mathbf{u}_i \rangle \right]$$

$$= \max_{\mathbf{a}, \mathbf{b}} - \left(\frac{1}{mn} \sum_{i=1}^{m} \sum_{j=1}^{n} l_{ij}^*(-\mathbf{a}_{ij}) + \frac{1}{2m\lambda} \left\| \frac{1}{n} \sum_{j=1}^{n} \mathbf{a}_{ij} + \frac{1}{n} \mathbf{A}_i^T \mathbf{b} \right\|^2 \right), \tag{6.65}$$

where we use $\mathbf{A}$ being symmetric and $\mathbf{A}_i^T$ denotes the i-th row-block of $\mathbf{A}$. We can transform (6.65) into a shifted form that is similar to (6.59) by setting

$$g_0(\mathbf{b}) = \sum_{i=1}^{m} \frac{\mu_1 \mu_2}{6mn^2} \|\mathbf{b}_i\|^2,$$

$$g_{ij}(\mathbf{a}_{ij}) = \frac{1}{mn} \left(l_{ij}^*(-\mathbf{a}_{ij}) - \frac{\mu_1}{4} \|\mathbf{a}_{ij}\|^2 \right), \quad i \in [m], \quad j \in [n],$$

$$f_i(\mathbf{a}_i, \mathbf{b}) = \frac{1}{mn} \sum_{j=1}^{n} \frac{\mu_1}{4} \|\mathbf{a}_{ij}\|^2 + \frac{1}{2m\lambda} \left\| \frac{1}{n} \sum_{j=1}^{n} \mathbf{a}_{ij} + \frac{1}{n} \mathbf{A}_i^T \mathbf{b} \right\|^2$$

$$- \frac{\mu_1 \mu_2}{6mn^2} \|\mathbf{b}_i\|^2, \ i \in [m].$$

The above has the similar form as (6.59) and we can solve the transformed problem by Algorithm 6.3.

Lemma 6.2 *Suppose* $n \leq \frac{L}{\lambda}$ *and set* $u_1 = \frac{1}{L}$ *and* $\mu_2 = 1/\kappa_g$. *Then* $f(\mathbf{a}, \mathbf{b})$ *has* L_c-*block coordinate Lipschitz continuous gradients w.r.t.* $(\mathbf{a}_{ij}, \mathbf{b})$ *(see (6.60) and (6.61)), where* $L_c = \frac{1}{2mn^2\lambda} + \frac{1}{4mnL}$. $f(\mathbf{a}, \mathbf{b})$ *is convex w.r.t.* $(\mathbf{a}, \mathbf{b}) \in \mathbb{R}^{mnd} \times$ *Span*$(\mathbf{A})$. $g_0(\mathbf{b})$ *is* $\frac{1}{6mn^2\kappa_g L}$-*strongly convex.* $g_{ij}(\mathbf{a}_{ij})$ *with* $i \in [m]$ *and* $j \in [n]$ *are* $\frac{1}{4nmL}$-*strongly convex.*

Proof By checking, we can find that (6.60) is right. Using the fact that $\sum_{i=1}^{m} \mathbf{A}_i \mathbf{A}_i^T = \mathbf{A}^2$ and $\|\mathbf{A}\| = 1$, the block Lipschitz constant of $\mathbf{b}$ for $f(\mathbf{a}, \mathbf{b})$ is less than $\frac{1}{2mn^2\lambda} + \frac{1}{4mnL}$, so (6.61) is right. Also, it is easy to prove that $g_0(\mathbf{b})$ is $\frac{\mu_1}{3mn^2}$-strongly convex and $g_{ij}(\mathbf{a}_{ij})$ are $\frac{1}{4nmL}$-strongly convex, $i \in [m], j \in [n]$.

Now we prove that $f(\mathbf{a}, \mathbf{b})$ is convex w.r.t. $(\mathbf{a}, \mathbf{b}) \in \mathbb{R}^{mnd} \times$ Span$(\mathbf{A})$. Applying (6.62) and using

$$\eta = \frac{2}{\mu_1 \lambda} = \frac{2L}{\lambda} \geq 2n \geq 2, \tag{6.66}$$

we have

$$\frac{1}{mn} \sum_{j=1}^{n} \frac{\mu_1}{2} \|\mathbf{a}_{ij}\|^2 + \frac{1}{2m\lambda} \left\| \frac{1}{n} \sum_{j=1}^{n} \mathbf{a}_{ij} + \frac{1}{n} \mathbf{A}_i^T \mathbf{b} \right\|^2$$

$$\geq \frac{1}{mn} \sum_{j=1}^{n} \frac{\mu_1}{2} \|\mathbf{a}_{ij}\|^2 + \frac{1}{2m\lambda(1+\eta)} \left\| \frac{1}{n} \mathbf{A}_i^T \mathbf{b} \right\|^2 - \frac{1}{2mn} \sum_{j=1}^{n} \frac{\mu_1}{2} \|\mathbf{a}_{ij}\|^2$$

$$\overset{a}{\geq} \frac{1}{mn}\sum_{j=1}^{n}\frac{\mu_1}{4}\|\mathbf{a}_{ij}\|^2 + \frac{1}{2m(\frac{\eta\lambda}{2}+\eta\lambda)}\left\|\frac{1}{n}\mathbf{A}_i^T\mathbf{b}\right\|^2$$

$$= \frac{1}{mn}\sum_{j=1}^{n}\frac{\mu_1}{4}\|\mathbf{a}_{ij}\|^2 + \frac{\mu_1}{3m}\left\|\frac{1}{n}\mathbf{A}_i^T\mathbf{b}\right\|^2, \tag{6.67}$$

where $\overset{a}{\geq}$ uses (6.66).

Summing (6.67) with $i = 1, \cdots, m$, we have

$$\frac{1}{mn}\sum_{i=1}^{m}\sum_{j=1}^{n}\frac{\mu_1}{2}\|\mathbf{a}_{ij}\|^2 + \frac{1}{2m\lambda}\sum_{i=1}^{m}\left\|\frac{1}{n}\sum_{j=1}^{n}\mathbf{a}_{ij} + \frac{1}{n}\mathbf{A}_i^T\mathbf{b}\right\|^2$$

$$\geq \frac{1}{mn}\sum_{i=1}^{m}\sum_{j=1}^{n}\frac{\mu_1}{4}\|\mathbf{a}_{ij}\|^2 + \frac{\mu_1}{3m}\left\|\frac{1}{n}\mathbf{A}\mathbf{b}\right\|^2$$

$$\overset{a}{\geq} \frac{1}{mn}\sum_{i=1}^{m}\sum_{j=1}^{n}\frac{\mu_1}{4}\|\mathbf{a}_{ij}\|^2 + \frac{\mu_1}{3mn^2\kappa_g}\|\mathbf{b}\|^2,$$

where in $\overset{a}{=}$ we use $\|\mathbf{A}\mathbf{b}\|^2 \geq \frac{\|\mathbf{b}\|^2}{\kappa_g}$ if $\mathbf{b} \in \mathrm{Span}\mathbf{A}$. So $f(\mathbf{a}, \mathbf{b})$ is convex as it is a nonnegative quadratic function. $\square$

We give a unified convergence results for Algorithm 6.3.

Theorem 6.4 *For Algorithm 6.3, suppose that $f(\mathbf{a}, \mathbf{b})$ is convex and has L_2-block coordinate Lipschitz continuous gradient, $g_0(\mathbf{b})$ is u_3-strongly convex, $g_{ij}(\mathbf{a}_{ij})$ with $i \in [m]$ and $j \in [n]$ are u_4-strongly convex, and $\max(\mu_3, \mu_4) \leq L_2$. Then by setting $p_b = \frac{1}{1+n\sqrt{\frac{\mu_3}{\mu_4}}}$ and $\theta_k \equiv \theta = \frac{p_b\sqrt{\mu_3/L_2}}{2}$, we have*

$$\mathbb{E}\left(F(\mathbf{a}^{k+1}, \mathbf{b}^{k+1}) - F(\mathbf{a}^*, \mathbf{b}^*) + \frac{\theta^2 L_2 + \theta p_b\mu_3}{2p_b^2}\mathbb{E}\left\|\hat{\mathbf{b}}^{k+1} - \mathbf{b}^*\right\|^2\right.$$

$$\left. + \frac{\theta^2 L_2 + \theta p_a\mu_4}{2p_a^2}\mathbb{E}\left\|\hat{\mathbf{a}}^{k+1} - \mathbf{a}^*\right\|^2\right)$$

$$\leq (1-\theta)^k\left(F(\mathbf{a}^0, \mathbf{b}^0) - F(\mathbf{a}^*, \mathbf{b}^*) + \frac{\theta^2 L_2 + \theta p_b\mu_3}{2p_b^2}\mathbb{E}\left\|\hat{\mathbf{b}}^0 - \mathbf{b}^*\right\|^2\right.$$

$$\left. + \frac{\theta^2 L_2 + \theta p_a\mu_4}{2p_a^2}\mathbb{E}\left\|\hat{\mathbf{a}}^0 - \mathbf{a}^*\right\|^2\right).$$

Proof We use $\mathbf{b}_c^{k+1}$ and $\hat{\mathbf{b}}_c^{k+1}$ to denote the result of $\mathbf{b}^{k+1}$ and $\hat{\mathbf{b}}^{k+1}$, respectively, if $\mathbf{b}$ is chosen to update at iteration k (i.e., $q \leq pb$). Similarly, $\mathbf{a}_{i,j(i),c}^{k+1}$ and $\hat{\mathbf{a}}_{i,j(i),c}^{k+1}$ denote the result of $\mathbf{a}_{i,j(i)}^{k+1}$ and $\hat{\mathbf{a}}_{i,j(i)}^{k+1}$, respectively, if $\mathbf{a}_{i,j(i)}$ is chosen update at iteration k. Then by the optimality condition of $\hat{\mathbf{b}}^{k+1}$ in Step 7, we have

$$\partial g_0(\hat{\mathbf{b}}_c^{k+1}) + \nabla_{\mathbf{b}} f(\tilde{\mathbf{a}}^k, \tilde{\mathbf{b}}^k) + \frac{\theta L_2}{p_b}(\hat{\mathbf{b}}_c^{k+1} - \hat{\mathbf{b}}^k) = \mathbf{0}. \tag{6.68}$$

From Step 13, we have

$$\partial g_0(\hat{\mathbf{b}}_c^{k+1}) + \nabla_{\mathbf{b}} f(\tilde{\mathbf{a}}^k, \tilde{\mathbf{b}}^k) + L_2(\mathbf{b}_c^{k+1} - \tilde{\mathbf{b}}^k) = \mathbf{0}. \tag{6.69}$$

If $\mathbf{b}$ is chosen to update at iteration k (i.e., $q \leq p_b$), we have $\mathbf{a}^{k+1} = \tilde{\mathbf{a}}^k$. Since $f(\mathbf{a}, \mathbf{b})$ has L_2-block coordinate Lipschitz continuous gradient w.r.t. $\mathbf{b}$, we have

$$f(\mathbf{a}^{k+1}, \mathbf{b}^{k+1}) \leq f(\tilde{\mathbf{a}}^k, \tilde{\mathbf{b}}^k) + \left\langle \nabla_{\mathbf{b}} f(\tilde{\mathbf{a}}^k, \tilde{\mathbf{b}}^k), \mathbf{b}_c^{k+1} - \tilde{\mathbf{b}}^k \right\rangle + \frac{L_2}{2} \left\| \mathbf{b}_c^{k+1} - \tilde{\mathbf{b}}^k \right\|^2. \tag{6.70}$$

Substituting (6.69) into (6.70), we have

$$f(\mathbf{a}^{k+1}, \mathbf{b}^{k+1}) \leq f(\tilde{\mathbf{a}}^k, \tilde{\mathbf{b}}^k) - \left\langle \partial g_0(\hat{\mathbf{b}}_c^{k+1}), \mathbf{b}_c^{k+1} - \tilde{\mathbf{b}}^k \right\rangle - \frac{L_2}{2} \left\| \mathbf{b}_c^{k+1} - \tilde{\mathbf{b}}^k \right\|^2. \tag{6.71}$$

Similar to (6.69), we have

$$\partial g_{i,j(i)} \left(\hat{\mathbf{a}}_{i,j(i),c}^{k+1} \right) + \nabla_{\mathbf{a}_{i,j(i)}} f_i(\tilde{\mathbf{a}}_i^k, \tilde{\mathbf{b}}^k) + L_2 \left(\mathbf{a}_{i,j(i),c}^{k+1} - \tilde{\mathbf{a}}_{i,j(i)}^k \right) = \mathbf{0}. \tag{6.72}$$

If $\mathbf{a}$ is chosen to update (i.e., $q > p_b$), we have $\mathbf{b}^{k+1} = \tilde{\mathbf{b}}^k$. For each f_i, suppose that $j(i)$ is chosen. Since $f_i(\mathbf{a}_i, \mathbf{b})$ has L_2-coordinate Lipschitz continuous gradient w.r.t. $\mathbf{a}_{i,j(i)}$, we have

$$f_i(\mathbf{a}^{k+1}, \mathbf{b}^{k+1})$$
$$\leq f_i(\tilde{\mathbf{a}}^k, \tilde{\mathbf{b}}^k) + \left\langle \nabla_{\mathbf{a}_{j(i)}} f_i(\tilde{\mathbf{a}}_i^k, \tilde{\mathbf{b}}^k), \mathbf{a}_{i,j(i),c}^{k+1} - \tilde{\mathbf{a}}_{i,j(i)}^k \right\rangle + \frac{L_2}{2} \left\| \mathbf{a}_{i,j(i),c}^{k+1} - \tilde{\mathbf{a}}_{i,j(i)}^k \right\|^2$$
$$\overset{a}{=} f_i(\tilde{\mathbf{a}}^k, \tilde{\mathbf{b}}^k) - \left\langle \partial g_{i,j(i)}(\hat{\mathbf{a}}_{i,j(i),c}^{k+1}), \mathbf{a}_{i,j(i),c}^{k+1} - \tilde{\mathbf{a}}_{i,j(i)}^k \right\rangle - \frac{L_2}{2} \left\| \mathbf{a}_{i,j(i),c}^{k+1} - \tilde{\mathbf{a}}_{i,j(i)}^k \right\|^2,$$

where $\stackrel{a}{=}$ uses (6.72). Taking expectation only on $j(i)$ under the condition that $\mathbf{a}$ is chosen at iteration k, we have

$$\mathbb{E}_a f_i(\mathbf{a}^{k+1}, \mathbf{b}^{k+1})$$
$$\leq f_i(\tilde{\mathbf{a}}^k, \tilde{\mathbf{b}}^k) - \frac{1}{n} \sum_{j=1}^{n} \left\langle \partial g_{ij}(\hat{\mathbf{a}}_{i,j,c}^{k+1}), \mathbf{a}_{i,j,c}^{k+1} - \tilde{\mathbf{a}}_{ij}^k \right\rangle - \frac{1}{n} \sum_{j=1}^{n} \frac{L_2}{2} \left\| \mathbf{a}_{i,j,c}^{k+1} - \tilde{\mathbf{a}}_{ij}^k \right\|^2.$$

Summing $i = 1$ to m, we have

$$\mathbb{E}_a f(\mathbf{a}^{k+1}, \mathbf{b}^{k+1}) \leq f(\tilde{\mathbf{a}}^k, \tilde{\mathbf{b}}^k) - \frac{1}{n} \sum_{i=1}^{m} \sum_{j=1}^{n} \left\langle \partial g_{ij}(\hat{\mathbf{a}}_{i,j,c}^{k+1}), \mathbf{a}_{i,j,c}^{k+1} - \tilde{\mathbf{a}}_{ij}^k \right\rangle$$
$$- \frac{1}{n} \sum_{i=1}^{m} \sum_{j=1}^{n} \frac{L_2}{2} \left\| \mathbf{a}_{i,j,c}^{k+1} - \tilde{\mathbf{a}}_{ij}^k \right\|^2.$$

Then by taking expectation on the random choice of $\mathbf{a}$ and $\mathbf{b}$ at iteration k, we have

$$\mathbb{E}_k f(\mathbf{a}^{k+1}, \mathbf{b}^{k+1})$$
$$\leq f(\tilde{\mathbf{a}}^k, \tilde{\mathbf{b}}^k) - p_b \left\langle \partial g_0(\hat{\mathbf{b}}_c^{k+1}), \mathbf{b}_c^{k+1} - \tilde{\mathbf{b}}^k \right\rangle$$
$$- \sum_{i=1}^{m} \sum_{j=1}^{n} p_a \left\langle \partial g_{ij}(\hat{\mathbf{a}}_{i,j,c}^{k+1}), \mathbf{a}_{i,j,c}^{k+1} - \tilde{\mathbf{a}}_{ij}^k \right\rangle$$
$$- \frac{L_2 p_b}{2} \left\| \mathbf{b}_c^{k+1} - \tilde{\mathbf{b}}^k \right\|^2 - \sum_{i=1}^{m} \sum_{j=1}^{n} \frac{L_2 p_a}{2} \left\| \mathbf{a}_{i,j,c}^{k+1} - \tilde{\mathbf{a}}_{ij}^k \right\|^2, \qquad (6.73)$$

where we use $p_a = \frac{1-p_b}{n}$.

For $\left\| \hat{\mathbf{b}}_c^{k+1} - \mathbf{b}^* \right\|^2$, we have

$$\frac{1}{2} \left\| \hat{\mathbf{b}}_c^{k+1} - \mathbf{b}^* \right\|^2$$
$$= \frac{1}{2} \left\| \hat{\mathbf{b}}_c^{k+1} - \hat{\mathbf{b}}^k + \hat{\mathbf{b}}^k - \mathbf{b}^* \right\|^2$$
$$= \frac{1}{2} \left\| \hat{\mathbf{b}}_c^{k+1} - \hat{\mathbf{b}}^k \right\|^2 + \frac{1}{2} \left\| \hat{\mathbf{b}}^k - \mathbf{b}^* \right\|^2 + \left\langle \hat{\mathbf{b}}_c^{k+1} - \hat{\mathbf{b}}^k, \hat{\mathbf{b}}^k - \mathbf{b}^* \right\rangle$$
$$\stackrel{a}{=} \frac{1}{2} \left\| \hat{\mathbf{b}}_c^{k+1} - \hat{\mathbf{b}}^k \right\|^2 + \frac{1}{2} \left\| \hat{\mathbf{b}}^k - \mathbf{b}^* \right\|^2$$
$$- \frac{p_b}{\theta L_2} \left\langle \partial g_0(\hat{\mathbf{b}}_c^{k+1}) + \nabla_{\mathbf{b}} f(\tilde{\mathbf{a}}^k, \tilde{\mathbf{b}}^k), \hat{\mathbf{b}}^k - \mathbf{b}^* \right\rangle, \qquad (6.74)$$

where $\overset{a}{=}$ uses (6.68).

Considering the expectation of $\left\|\hat{\mathbf{b}}^{k+1} - \mathbf{b}^*\right\|^2$ under the random numbers in iteration k, we have

$$\mathbb{E}_k \left\|\hat{\mathbf{b}}^{k+1} - \mathbf{b}^*\right\|^2 = p_b \left\|\hat{\mathbf{b}}_c^{k+1} - \mathbf{b}^*\right\|^2 + (1 - p_b) \left\|\hat{\mathbf{b}}^k - \mathbf{b}^*\right\|^2. \qquad (6.75)$$

Multiplying (6.75) by $L_2 \theta^2 / (2 p_b^2)$ and substituting (6.74) into it, we have

$$\frac{L_2 \theta^2}{2 p_b^2} \mathbb{E}_k \left\|\hat{\mathbf{b}}^{k+1} - \mathbf{b}^*\right\|^2 - \frac{L_2 \theta^2}{2 p_b^2} \left\|\hat{\mathbf{b}}^k - \mathbf{b}^*\right\|^2$$

$$= \frac{L_2 \theta^2}{2 p_b} \left\|\hat{\mathbf{b}}_c^{k+1} - \hat{\mathbf{b}}^k\right\|^2 - \left\langle \partial g_0(\hat{\mathbf{b}}_c^{k+1}) + \nabla_\mathbf{b} f(\tilde{\mathbf{a}}^k, \tilde{\mathbf{b}}^k), \theta \hat{\mathbf{b}}^k - \theta \mathbf{b}^* \right\rangle. \qquad (6.76)$$

From Step 3 of Algorithm 6.3, we have

$$\theta \hat{\mathbf{b}}^k = \tilde{\mathbf{b}}^k - (1 - \theta) \mathbf{b}^k, \qquad (6.77)$$

and from Step 13 of Algorithm 6.3, we have

$$\mathbf{b}_c^{k+1} - \tilde{\mathbf{b}}^k = \frac{\theta}{p_b} (\hat{\mathbf{b}}_c^{k+1} - \hat{\mathbf{b}}^k). \qquad (6.78)$$

So we have

$$\frac{L_2 \theta^2}{2 p_b^2} \mathbb{E}_k \left\|\hat{\mathbf{b}}^{k+1} - \mathbf{b}^*\right\|^2 - \frac{L_2 \theta^2}{2 p_b^2} \left\|\hat{\mathbf{b}}^k - \mathbf{b}^*\right\|^2$$

$$\overset{a}{=} \frac{L_2 p_b}{2} \left\|\mathbf{b}_c^{k+1} - \tilde{\mathbf{b}}^k\right\|^2 - \left\langle \partial g_0(\hat{\mathbf{b}}_c^{k+1}), \theta \hat{\mathbf{b}}_c^{k+1} - p_b(\mathbf{b}_c^{k+1} - \tilde{\mathbf{b}}^k) - \theta \mathbf{b}^* \right\rangle$$

$$\quad - \left\langle \nabla_\mathbf{b} f(\tilde{\mathbf{a}}^k, \tilde{\mathbf{b}}^k), \theta \hat{\mathbf{b}}^k - \theta \mathbf{b}^* \right\rangle.$$

$$\overset{b}{=} \frac{L_2 p_b}{2} \left\|\mathbf{b}_c^{k+1} - \tilde{\mathbf{b}}^k\right\|^2 - \left\langle \partial g_0(\hat{\mathbf{b}}_c^{k+1}), \theta \hat{\mathbf{b}}_c^{k+1} - p_b(\mathbf{b}_c^{k+1} - \tilde{\mathbf{b}}^k) - \theta \mathbf{b}^* \right\rangle$$

$$\quad - \left\langle \nabla_\mathbf{b} f(\tilde{\mathbf{a}}^k, \tilde{\mathbf{b}}^k), \tilde{\mathbf{b}}^k - (1 - \theta) \mathbf{b}^k - \theta \mathbf{b}^* \right\rangle, \qquad (6.79)$$

where $\overset{a}{=}$ uses (6.76) and (6.78) and $\overset{b}{=}$ uses (6.77).

Using the same technique on $\mathbf{a}_{ij}^k$, we have

$$\frac{L_2\theta^2}{2p_a^2}\mathbb{E}_k\left\|\hat{\mathbf{a}}_{ij}^{k+1} - \mathbf{a}_{ij}^*\right\|^2 - \frac{L_2\theta^2}{2p_a^2}\left\|\hat{\mathbf{a}}_{ij}^k - \mathbf{a}_{ij}^*\right\|^2$$

$$= \frac{L_2p_a}{2}\left\|\mathbf{a}_{i,j,c}^{k+1} - \tilde{\mathbf{a}}_{ij}^k\right\|^2 - \left\langle\partial g_{ij}(\hat{\mathbf{a}}_{i,j,c}^{k+1}), \theta\hat{\mathbf{a}}_{i,j,c}^{k+1} - p_a(\mathbf{a}_{i,j,c}^{k+1} - \tilde{\mathbf{a}}_{ij}^k) - \theta\mathbf{a}_{ij}^*\right\rangle$$

$$- \left\langle\nabla_{\mathbf{a}_{i,j(i)}}f_i(\tilde{\mathbf{a}}^k, \tilde{\mathbf{b}}^k), \tilde{\mathbf{a}}_{ij}^k - (1-\theta)\mathbf{a}_{ij}^k - \theta\mathbf{a}_{ij}^*\right\rangle. \tag{6.80}$$

Then by summing $i = 1, \cdots, m$ and $j = 1, \cdots, n$ for (6.80), and adding (6.79) and (6.73) we have

$$\mathbb{E}_k f(\mathbf{a}^{k+1}, \mathbf{b}^{k+1}) + \frac{\theta^2 L_2}{2p_b^2}\mathbb{E}_k\left\|\hat{\mathbf{b}}^{k+1} - \mathbf{b}^*\right\|^2 - \frac{\theta^2 L_2}{2p_b^2}\left\|\hat{\mathbf{b}}^k - \mathbf{b}^*\right\|^2$$

$$+ \frac{\theta^2 L_2}{2p_a^2}\mathbb{E}\left\|\hat{\mathbf{a}}^{k+1} - \mathbf{a}^*\right\|^2 - \frac{\theta^2 L_2}{2p_a^2}\left\|\hat{\mathbf{a}}^k - \mathbf{a}^*\right\|^2$$

$$\leq f(\tilde{\mathbf{a}}^k, \tilde{\mathbf{b}}^k) - \theta\left\langle\partial g_0(\hat{\mathbf{b}}_c^{k+1}), \hat{\mathbf{b}}_c^{k+1} - \mathbf{b}^*\right\rangle$$

$$- \sum_{i=1}^m\sum_{j=1}^n \theta\left\langle\partial g_{ij}(\hat{\mathbf{a}}_{i,j,c}^{k+1}), \hat{\mathbf{a}}_{i,j,c}^{k+1} - \mathbf{a}_{ij}^*\right\rangle$$

$$- \sum_{i=1}^m\sum_{j=1}^n\left\langle\nabla_{\mathbf{a}_{i,j(i)}}f_i(\tilde{\mathbf{a}}_i^k, \tilde{\mathbf{b}}^k), \tilde{\mathbf{a}}_{ij}^k - (1-\theta)\mathbf{a}_{ij}^k - \theta\mathbf{a}_{ij}^*\right\rangle$$

$$- \left\langle\nabla_{\mathbf{b}}f(\tilde{\mathbf{a}}^k, \tilde{\mathbf{b}}^k), \tilde{\mathbf{b}}^k - (1-\theta)\mathbf{b}^k - \theta\mathbf{b}^*\right\rangle. \tag{6.81}$$

By the μ_3-strongly convexity of g_0, we have

$$- \left\langle\partial g_0(\hat{\mathbf{b}}_c^{k+1}), \hat{\mathbf{b}}_c^{k+1} - \mathbf{b}^*\right\rangle$$

$$\leq -g_0(\hat{\mathbf{b}}_c^{k+1}) + g_0(\mathbf{b}^*) - \frac{\mu_3}{2}\left\|\hat{\mathbf{b}}_c^{k+1} - \mathbf{b}^*\right\|^2$$

$$\overset{a}{=} -g_0(\hat{\mathbf{b}}_c^{k+1}) + g_0(\mathbf{b}^*) - \frac{\mu_3}{2p_b}\mathbb{E}_k\left\|\hat{\mathbf{b}}^{k+1} - \mathbf{b}^*\right\|^2 + \frac{\mu_3(1-p_b)}{2p_b}\left\|\hat{\mathbf{b}}^k - \mathbf{b}^*\right\|^2, \tag{6.82}$$

where $\overset{a}{=}$ uses (6.75).

Similarly, for $\mathbf{a}_{ij}$ we have

$$
-\left\langle \partial g_{ij}(\hat{\mathbf{a}}_{i,j,c}^{k+1}), \hat{\mathbf{a}}_{i,j,c}^{k+1} - \mathbf{a}_{ij}^* \right\rangle
$$

$$
\leq -g_{ij}(\hat{\mathbf{a}}_{i,j,c}^{k+1}) + g_{ij}(\mathbf{a}_{ij}^*) - \frac{\mu_4}{2} \left\| \hat{\mathbf{a}}_{i,j,c}^{k+1} - \mathbf{a}_{ij}^* \right\|^2
$$

$$
= -g_{ij}(\hat{\mathbf{a}}_{i,j,c}^{k+1}) + g_{ij}(\mathbf{a}_{ij}^*) - \frac{\mu_4}{2 p_a} \mathbb{E}_k \left\| \hat{\mathbf{a}}_{ij}^{k+1} - \mathbf{a}_{ij}^* \right\|^2
$$

$$
+ \frac{\mu_4(1 - p_a)}{2 p_a} \left\| \hat{\mathbf{a}}_{ij}^{k+1} - \mathbf{a}_{ij}^* \right\|^2 . \tag{6.83}
$$

Next, we have

$$
-\left\langle \nabla_{\mathbf{b}} f(\tilde{\mathbf{a}}^k, \tilde{\mathbf{b}}^k), \tilde{\mathbf{b}}^k - (1-\theta)\mathbf{b}^k - \theta\mathbf{b}^* \right\rangle
$$

$$
- \sum_{i=1}^{m} \sum_{j=1}^{n} \left\langle \nabla_{\mathbf{a}_{i,j(i)}} f_i(\tilde{\mathbf{a}}^k, \tilde{\mathbf{b}}^k), \tilde{\mathbf{a}}_{ij}^k - (1-\theta)\mathbf{a}_{ij}^k - \theta\mathbf{a}_{ij}^* \right\rangle
$$

$$
= -\left\langle \nabla_{\mathbf{b}} f(\tilde{\mathbf{a}}^k, \tilde{\mathbf{b}}^k), \tilde{\mathbf{b}}^k - (1-\theta)\mathbf{b}^k - \theta\mathbf{b}^* \right\rangle
$$

$$
- \left\langle \nabla_{\mathbf{a}} f(\tilde{\mathbf{a}}^k, \tilde{\mathbf{b}}^k), \tilde{\mathbf{a}}^k - (1-\theta)\mathbf{a}^k - \theta\mathbf{a}^* \right\rangle
$$

$$
\overset{a}{\leq} -f(\tilde{\mathbf{a}}^k, \tilde{\mathbf{b}}^k) + (1-\theta)f(\mathbf{a}^k, \mathbf{b}^k) + \theta f(\mathbf{a}^*, \mathbf{b}^*), \tag{6.84}
$$

where in $\overset{a}{\leq}$ we use the convexity of $f(\mathbf{a}, \mathbf{b})$.

Substituting (6.82), (6.83), and (6.84) into (6.81), we have

$$
\mathbb{E}_k f(\mathbf{a}^{k+1}, \mathbf{b}^{k+1}) + \theta g_0(\hat{\mathbf{b}}_c^{k+1}) + \theta \sum_{i=1}^{m} \sum_{j=1}^{n} g_{ij}(\hat{\mathbf{a}}_{i,j,c}^{k+1})
$$

$$
+ \frac{\theta^2 L_2 + \theta p_b \mu_3}{2 p_b^2} \mathbb{E}_k \left\| \hat{\mathbf{b}}^{k+1} - \mathbf{b}^* \right\|^2 - \frac{\theta^2 L_2 + \theta \mu_3 p_b(1 - p_b)}{2 p_b^2} \left\| \hat{\mathbf{b}}^k - \mathbf{b}^* \right\|^2
$$

$$
+ \frac{\theta^2 L_2 + \theta p_a \mu_4}{2 p_a^2} \mathbb{E}_k \left\| \hat{\mathbf{a}}^{k+1} - \mathbf{a}^* \right\|^2 - \frac{\theta^2 L_2 + \theta \mu_4 p_a(1 - p_a)}{2 p_a^2} \left\| \hat{\mathbf{a}}^k - \mathbf{a}^* \right\|^2
$$

$$
\leq (1-\theta)f(\mathbf{a}^k, \mathbf{b}^k) + \theta F(\mathbf{a}^*, \mathbf{b}^*).
$$

Using the definition of $\hat{g}^k$ in Lemma 6.3 (see the end of the proof), we have

$$
\mathbb{E}_k\left(f(\mathbf{a}^{k+1},\mathbf{b}^{k+1})+\hat{g}^{k+1}\right)
$$

$$
\leq (1-\theta)\left(f(\mathbf{a}^k,\mathbf{b}^k)+\hat{g}^k\right)+\theta F(\mathbf{a}^*,\mathbf{b}^*)
$$

$$
-\frac{\theta^2 L_2+\theta p_b\mu_3}{2p_b^2}\mathbb{E}_k\left\|\hat{\mathbf{b}}^{k+1}-\mathbf{b}^*\right\|^2+\frac{\theta^2 L_2+\theta\mu_3 p_b(1-p_b)}{2p_b^2}\left\|\hat{\mathbf{b}}^k-\mathbf{b}^*\right\|^2
$$

$$
-\frac{\theta^2 L_2+\theta p_a\mu_4}{2p_a^2}\mathbb{E}_k\left\|\hat{\mathbf{a}}^{k+1}-\mathbf{a}^*\right\|^2+\frac{\theta^2 L_2+\theta\mu_4 p_a(1-p_a)}{2p_a^2}\left\|\hat{\mathbf{a}}^k-\mathbf{a}^*\right\|^2.
$$

From the setting of p_b, we have $\frac{p_b}{p_a}=\frac{\sqrt{\mu_4}}{\sqrt{\mu_3}}$, then $\theta=\frac{p_b\sqrt{u_3/L_2}}{2}=\frac{p_a\sqrt{u_4/L_2}}{2}$. So we can have

$$
\frac{\theta^2 L_2+\theta p_b\mu_3}{2p_b^2}(1-\theta)\geq\frac{\theta^2 L_2+\theta\mu_3 p_b(1-p_b)}{2p_b^2}, \tag{6.85}
$$

and

$$
\frac{\theta^2 L_2+\theta p_a\mu_4}{2p_a^2}(1-\theta)\geq\frac{\theta^2 L_2+\theta\mu_4 p_a(1-p_a)}{2p_a^2}. \tag{6.86}
$$

Indeed, by solving (6.85) we have

$$
\theta\leq\frac{p_b\left(-\mu_3/L_2+\sqrt{\frac{\mu_3^2}{L_2^2}+4\frac{\mu_3}{L_2}}\right)}{2}. \tag{6.87}
$$

Using $u_3/L_2\leq 1$, we can check that $\theta=\frac{p_b\sqrt{u_3/L_2}}{2}$ satisfies (6.87). Also using $u_4/L_2\leq 1$, we have that $\theta=\frac{p_a\sqrt{u_4/L_2}}{2}$ satisfies (6.86). Then we have

$$
\mathbb{E}_k\left(f(\mathbf{a}^{k+1},\mathbf{b}^{k+1})+\hat{g}^{k+1}-F(\mathbf{a}^*,\mathbf{b}^*)\right)
$$

$$
+\mathbb{E}_k\left(\frac{\theta^2 L_2+\theta p_b\mu_3}{2p_b^2}\left\|\hat{\mathbf{b}}^{k+1}-\mathbf{b}^*\right\|^2+\frac{\theta^2 L_2+\theta p_a\mu_4}{2p_a^2}\left\|\hat{\mathbf{a}}^{k+1}-\mathbf{a}^*\right\|^2\right)
$$

$$
\leq(1-\theta)\left(f(\mathbf{a}^k,\mathbf{b}^k)+\hat{g}^k-F(\mathbf{a}^*,\mathbf{b}^*)\right.
$$

$$
\left.+\frac{\theta^2 L_2+\theta p_b\mu_3}{2p_b^2}\left\|\hat{\mathbf{b}}^k-\mathbf{b}^*\right\|^2+\frac{\theta^2 L_2+\theta p_a\mu_4}{2p_a^2}\left\|\hat{\mathbf{a}}^k-\mathbf{a}^*\right\|^2\right).
$$

By taking full expectation, we have

$$\mathbb{E}\left(f(\mathbf{a}^{k+1}, \mathbf{b}^{k+1}) + \hat{g}^{k+1} - F(\mathbf{a}^*, \mathbf{b}^*)\right)$$

$$+ \left(\frac{\theta^2 L_2 + \theta p_b \mu_3}{2p_b^2} \mathbb{E}\left\|\hat{\mathbf{b}}^{k+1} - \mathbf{b}^*\right\|^2 + \frac{\theta^2 L_2 + \theta p_a \mu_4}{2p_a^2} \mathbb{E}\left\|\hat{\mathbf{a}}^{k+1} - \mathbf{a}^*\right\|^2\right)$$

$$\leq (1 - \theta)^k \left(f(\mathbf{a}^0, \mathbf{b}^0) + \hat{g}^0 - F(\mathbf{a}^*, \mathbf{b}^*)\right.$$

$$\left. + \frac{\theta^2 L_2 + \theta p_b \mu_3}{2p_b^2} \left\|\hat{\mathbf{b}}^0 - \mathbf{b}^*\right\|^2 + \frac{\theta^2 L_2 + \theta p_a \mu_4}{2p_a^2} \left\|\hat{\mathbf{a}}^0 - \mathbf{a}^*\right\|^2\right).$$

Then using the convexity of g_0 we have

$$g_0(\mathbf{b}^{k+1}) = g_0\left(\sum_{i=0}^{k+1} e_{k+1,i,1} \hat{\mathbf{b}}^i\right) \leq \sum_{i=0}^{k+1} e_{k+1,i,1} g_0(\hat{\mathbf{b}}^i) = \hat{g}_0^{k+1}.$$

Also for any $\mathbf{a}_{ij}$, we have

$$g_{ij}(\mathbf{a}_{ij}^{k+1}) = g_{ij}\left(\sum_{q=0}^{k+1} e_{k+1,q,2} \hat{\mathbf{a}}_{ij}^q\right) \leq \sum_{q=0}^{k+1} e_{k+1,q,2} g_{ij}(\hat{\mathbf{a}}_{ij}^q) = \hat{g}_{ij}^{k+1},$$

and $\hat{g}^0 = g_0(\hat{\mathbf{b}}^0) + \sum_{i=1}^m \sum_{j=1}^n g_{ij}(\hat{\mathbf{a}}_{ij}^0)$. We obtain Theorem 6.4. $\qquad\square$

The following lemma is a straightforward extension of Lemma 5.1 in Sect. 5.1.1.

Lemma 6.3 *From Algorithm 6.3 with the parameters set in Theorem 6.4, we have that $\mathbf{b}^k$ is a convex combination of $\{\hat{\mathbf{b}}^i\}_{i=0}^k$, i.e., $\mathbf{b}^k = \sum_{i=0}^k e_{k,i,1} \hat{\mathbf{b}}^i$, where $e_{0,0,1} = 1$, $e_{1,0,1} = 1 - \theta/p_b$, $e_{1,1,1} = \theta/p_b$. And for $k > 1$, we have*

$$e_{k,i,1} = \begin{cases} (1 - \theta)e_{k-1,i,1}, & i \leq k - 2, \\ (1 - \theta)\theta/p_b + \theta - \theta/p_b, & i = k - 1, \\ \theta/p_b, & i = k. \end{cases} \tag{6.88}$$

Also, $\mathbf{a}^k$ is a convex combination of $\{\hat{\mathbf{a}}^i\}_{i=0}^k$, i.e., $\mathbf{a}_j^k = \sum_{i=0}^k e_{k,i,2} \hat{\mathbf{a}}_j^i$, with $e_{0,0,2} = 1$, $e_{1,0,2} = 1 - \theta/p_a$, $e_{1,1,2} = \theta/p_a$. And for $k > 1$, we have

$$e_{k,i,2} = \begin{cases} (1 - \theta)e_{k-1,i,2}, & i \leq k - 2, \\ (1 - \theta)\theta/p_a + \theta - \theta/p_a, & i = k - 1, \\ \theta/p_a, & i = k. \end{cases}$$

Set $\hat{g}_0^k = \sum_{q=0}^k e_{k,q,1} g_0(\hat{\mathbf{b}}^q)$, $\hat{g}_{ij}^k = \sum_{q=0}^k e_{k,q,2} g_{ij}(\hat{\mathbf{a}}_{ij}^q)$, *and* $\hat{g}^k = \hat{g}_0^k + \sum_{i=1}^m \sum_{j=1}^n \hat{g}_{ij}^k$, *we have*

$$\mathbb{E}_k(\hat{g}_0^{k+1}) = (1 - \theta)\hat{g}_0^k + \theta g_0(\hat{\mathbf{b}}_c^{k+1}), \tag{6.89}$$

and

$$\mathbb{E}_k(\hat{g}_{ij}^{k+1}) = (1 - \theta)\hat{g}_{ij}^k + \theta g_{ij}(\hat{\mathbf{a}}_{i,j,c}^{k+1}),$$

with $i \in [m]$ and $j \in [n]$, where $\mathbb{E}_k$ denotes that the expectation is only taken on the random number in the k-th iteration under the condition that $\mathbf{a}^k$ and $\mathbf{b}^k$ are known, $\hat{\mathbf{b}}_c^{k+1}$ denotes the result of $\hat{\mathbf{b}}^{k+1}$ if $\mathbf{b}$ is chosen to update at iteration k, and $\hat{\mathbf{a}}_{i,j,c}^{k+1}$ denotes the result of $\hat{\mathbf{a}}_{ij}^{k+1}$ if $\mathbf{a}_{ij}$ is chosen to update at iteration k.

Proof We consider $e_{k,i,j}$ first. When $k = 0$ and 1, it is true that $e_{0,0,1} = 1$ and $e_{0,0,2} = 1$. We first prove (6.88). Assume for k, (6.88) holds. From Steps 3 and 13, we have

$$\mathbf{b}^{k+1} = (1 - \theta)\mathbf{b}^k + \theta\hat{\mathbf{b}}^k + \theta/p_b(\hat{\mathbf{b}}^{k+1} - \hat{\mathbf{b}}^k)$$

$$= (1 - \theta)\sum_{i=0}^k e_{k,i,1}\hat{\mathbf{b}}^i + \theta\hat{\mathbf{b}}^k + \theta/p_b(\hat{\mathbf{b}}^{k+1} - \hat{\mathbf{b}}^k)$$

$$= (1 - \theta)\sum_{i=0}^{k-1} e_{k,i,1}\hat{\mathbf{b}}^i + \left[(1 - \theta)e_{k,k,1} + \theta - \theta/p_b\right]\hat{\mathbf{b}}^k + \theta/p_b\hat{\mathbf{b}}^{k+1}.$$

Comparing the results, we obtain (6.88). To prove convex combination, it is easy to prove that the weights sum to 1. We then prove that $e_{k,i,1} \geq 0$ for all $k \geq 0$ and $0 \leq i \leq k$. When $k = 0$ and $k = 1$, we have $e_{0,0,1} = 1 \geq 0$, $e_{1,0,1} = 1 - \frac{\theta}{p_b} = 1 - \frac{\sqrt{\mu_4/L_2}}{2} \geq 0$, $e_{1,1,1} = \frac{\theta}{p_b} \geq 0$. When $k \geq 1$, suppose that at k, $e_{k,i,1} \geq 0$ with $0 \leq i \leq k$. Then we have $e_{k+1,i,1} = (1 - \theta)e_{k,i,1} \geq 0$ $(i \leq k - 1)$, $e_{k+1,k+1,1} = \theta/p_b \geq 0$, and

$$e_{k+1,k,1} = (1 - \theta)\theta/p_b + \theta - \theta/p_b = \theta(1 - \theta/p_b) = \theta\left(1 - \frac{\sqrt{\mu_4/L_2}}{2}\right) \geq 0.$$

We then prove (6.89), we have

$$\mathbb{E}_k \hat{g}_0^{k+1} \overset{a}{=} \sum_{i=0}^{k} e_{k+1,i,1} g_0(\hat{\mathbf{b}}^i) + (\theta/p_b)\mathbb{E}_k g_0(\hat{\mathbf{b}}^{k+1})$$

$$\overset{b}{=} \sum_{i=0}^{k} e_{k+1,i,1} g_0(\hat{\mathbf{b}}^i) + \theta\left(g_0(\hat{\mathbf{b}}_c^{k+1}) + \frac{1-p_b}{p_b} g_0(\hat{\mathbf{b}}^k)\right)$$

$$= \sum_{i=0}^{k} e_{k+1,i,1} g_0(\hat{\mathbf{b}}^i) + \theta g_0(\hat{\mathbf{b}}_c^{k+1}) + (1/p_b - 1)\theta g_0(\hat{\mathbf{b}}^k)$$

$$\overset{c}{=} \sum_{i=0}^{k-1} e_{k+1,i,1} g_0(\hat{\mathbf{b}}^i) + [(1-\theta)\theta/p_b + \theta - \theta/p_b] g_0(\hat{\mathbf{b}}^k)$$

$$+(1/p_b - 1)\theta g_0(\hat{\mathbf{b}}^k) + \theta g_0(\hat{\mathbf{b}}_c^{k+1})$$

$$= \sum_{i=0}^{k-1} e_{k+1,i,1} g_0(\hat{\mathbf{b}}^i) + (1-\theta)\theta/p_b g_0(\hat{\mathbf{b}}^k) + \theta g_0(\hat{\mathbf{b}}_c^{k+1})$$

$$\overset{d}{=} \sum_{i=0}^{k-1}(1-\theta)e_{k,i,1} g_0(\hat{\mathbf{b}}^i) + (1-\theta)e_{k,k,1} g_0(\hat{\mathbf{b}}^k) + \theta g_0(\hat{\mathbf{b}}_c^{k+1})$$

$$= (1-\theta)\sum_{i=0}^{k} e_{k,i,1} g_0(\hat{\mathbf{b}}^i) + \theta g_0(\hat{\mathbf{b}}_c^{k+1}) = (1-\theta)\hat{g}_0^k + \theta g_0(\hat{\mathbf{b}}_c^{k+1}),$$

where in $\overset{a}{=}$ we use $e_{k+1,k+1,1} = \theta/p_b$, in $\overset{b}{=}$ we use

$$\mathbb{E}_k g_0(\hat{\mathbf{b}}^{k+1}) = p_b g_0(\hat{\mathbf{b}}_c^{k+1}) + (1 - p_b)g_0(\hat{\mathbf{b}}^k),$$

in $\overset{c}{=}$ we use $e_{k+1,k,1} = (1-\theta)\theta/p_b + \theta - \theta/p_b$, and in $\overset{d}{=}$ we use $e_{k+1,i,1} = (1-\theta)e_{k,i,1}$ for $i \le k-1$ and $e_{k,k,1} = \theta/p_b$. By the same way, we can prove the result for $\mathbf{a}$. □

With Theorem 6.4 in hand, we can compute the communication and the iteration costs of Algorithm 6.3, which are stated in the following theorems.

Theorem 6.5 *Assume $\frac{L}{\lambda} \ge n$. Set $\mu_1 = \frac{1}{L}$, $L_2 = \frac{1}{2mn^2\lambda} + \frac{1}{4mnL}$, $\mu_3 = \frac{1}{6mn^2L}$, and $\mu_4 = \frac{3}{4mnL}$. It takes $\tilde{O}\left(\sqrt{nL/\lambda}\right)$ iterations to obtain an ϵ-accuracy solution for problem (6.58) satisfying $D(\mathbf{a}^*, \mathbf{b}^*) - D(\mathbf{a}^k, \mathbf{b}^k) \le \epsilon$. The communication and the iteration costs are $\tilde{O}\left(\sqrt{nL/\lambda}\right)$ and $\tilde{O}\left(\sqrt{L/\lambda}\right)$, respectively.*

Also for the decentralized case, for problem (6.65), because $g_0(\mathbf{b}) = \frac{\mu_1\mu_2}{6mn^2}\|\mathbf{b}\|^2$, from Steps 3, 7, and 13 of Algorithm 6.3 and $\mathbf{b}^0 = \hat{\mathbf{b}}^0 = \mathbf{0} \in \text{Span}(\mathbf{A})$, we can obtain that $\mathbf{b}^k \in \text{Span}(\mathbf{A})$ and $\hat{\mathbf{b}}^k \in \text{Span}(\mathbf{A})$ for all $k \geq 0$. Thus we have

Theorem 6.6 *Assume $\frac{L}{\lambda} \geq n$. Set $\mu_1 = \frac{1}{L}$, $L_2 = \frac{1}{2mn^2\lambda} + \frac{1}{4mnL}$, $\mu_3 = \frac{1}{6mn^2\kappa_g L}$, and $\mu_4 = \frac{1}{4mnL}$. It takes $\tilde{O}\left(\left(\sqrt{\kappa_g} + \sqrt{n}\right)\sqrt{L/\lambda}\right)$ iterations to obtain an ϵ-accuracy solution for problem (6.65) satisfying $D(\mathbf{a}^*, \mathbf{b}^*) - D(\mathbf{a}^k, \mathbf{b}^k) \leq \epsilon$. The communication and the iteration costs are $\tilde{O}\left(\sqrt{nL/\lambda}\right)$ and $\tilde{O}\left(\sqrt{\kappa_g L/\lambda}\right)$, respectively.*

References

1. A. Agarwal, J.C. Duchi, Distributed delayed stochastic optimization, in *Advances in Neural Information Processing Systems*, Granada, vol. 24 (2011), pp. 873–881
2. Z. Allen-Zhu, Katyusha: the first truly accelerated stochastic gradient method, in *Proceedings of the 49th Annual ACM SIGACT Symposium on Theory of Computing*, Montreal, (2017), pp. 1200–1206
3. Z. Allen-Zhu, Z. Qu, P. Richtárik, Y. Yuan, Even faster accelerated coordinate descent using non-uniform sampling, in *Proceedings of the 33th International Conference on Machine Learning*, New York, (2016), pp. 1110–1119
4. C. Fang, Z. Lin, Parallel asynchronous stochastic variance reduction for nonconvex optimization, in *Proceedings of the 31th AAAI Conference on Artificial Intelligence*, San Francisco, (2017), pp. 794–800
5. C. Fang, Y. Huang, Z. Lin, Accelerating asynchronous algorithms for convex optimization by momentum compensation (2018). Preprint. arXiv:1802.09747
6. D. Jakovetić, J.M. Moura, J. Xavier, Linear convergence rate of a class of distributed augmented Lagrangian algorithms. IEEE Trans. Automat. Contr. **60**(4), 922–936 (2014)
7. Q. Lin, Z. Lu, L. Xiao, An accelerated proximal coordinate gradient method, in *Advances in Neural Information Processing Systems*, Montreal, vol. 27 (2014), pp. 3059–3067
8. J. Liu, S.J. Wright, C. Ré, V. Bittorf, S. Sridhar, An asynchronous parallel stochastic coordinate descent algorithm. J. Mach. Learn. Res. **16**(1), 285–322 (2015)
9. C. Ma, V. Smith, M. Jaggi, M.I. Jordan, P. Richtarik, M. Takac, Adding vs. averaging in distributed primal-dual optimization, arXiv preprint, arXiv:1502.03508 (2015)
10. H. Mania, X. Pan, D. Papailiopoulos, B. Recht, K. Ramchandran, M.I. Jordan, Perturbed iterate analysis for asynchronous stochastic optimization. SIAM J. Optim. **27**(4), 2202–2229 (2017)
11. Y. Nesterov, A method for unconstrained convex minimization problem with the rate of convergence $O(1/k^2)$. Sov. Math. Dokl. **27**(2), 372–376 (1983)
12. B. Recht, C. Re, S. Wright, F. Niu, HOGWILD!: a lock-free approach to parallelizing stochastic gradient descent, in *Advances in Neural Information Processing Systems*, Granada, vol. 24 (2011), pp. 693–701
13. S.J. Reddi, A. Hefny, S. Sra, B. Poczos, A.J. Smola, On variance reduction in stochastic gradient descent and its asynchronous variants, in *Advances in Neural Information Processing Systems*, Montreal, vol. 28 (2015), pp. 2647–2655
14. K. Seaman, F. Bach, S. Bubeck, Y.T. Lee, L. Massoulié, Optimal algorithms for smooth and strongly convex distributed optimization in networks, in *Proceedings of the 34th International Conference on Machine Learning*, Sydney, (2017), pp. 3027–3036
15. W. Shi, Q. Ling, G. Wu, W. Yin, EXTRA: an exact first-order algorithm for decentralized consensus optimization. SIAM J. Optim. **25**(2), 944–966 (2015)

16. K. Yuan, Q. Ling, W. Yin, On the convergence of decentralized gradient descent. SIAM J. Optim. **26**(3), 1835–1854 (2016)
17. P. Zhao, T. Zhang, Stochastic optimization with importance sampling for regularized loss minimization, in *Proceedings of the 32th International Conference on Machine Learning*, Lille, (2015), pp. 1–9
18. S. Zheng, J. Wang, F. Xia, W. Xu, T. Zhang, A general distributed dual coordinate optimization framework for regularized loss minimization. J. Mach. Learn. Res. **18**(115), 1–52 (2017)

Chapter 7
Conclusions

In the previous chapters, we have introduced many representative accelerated first-order algorithms used or investigated by the machine learning community. It is inevitable that our review is incomplete and biased. Moreover, new accelerated algorithms still emerge when this book was under preparation, but we had to leave them out.

Although accelerated algorithms are very attractive in theory, in practice whether we should use them depends on many practical factors. For example, when using the warm start technique to solve a series of LASSO problems with different penalties, the advantages of accelerated algorithms may diminish when the grid of penalties is fine enough. For low-rank problems [7], the inter-/extra-polation may destroy the low-rank structure of matrices and make the related SVD (used in the singular value thresholding) computation much more expensive, counteracting the benefit of requiring less iterations. Since most accelerated algorithms only consider the L-smoothness and μ-strong convexity of the objective functions, and do not consider other characteristics of the problems, for some problems the unaccelerated algorithms and accelerated ones can actually converge equally fast, or even faster than the predicted rate (such as the problems with the objective functions being only restricted strongly convex, meaning that strongly convex only on a subset of the domain, and the "well-behaved" learning problems, meaning that the signal to noise ratio is decently high, the correlations between predictor variables are under control, and the number of predictors is larger than that of observations). One remarkable example is that although there have been several algorithms for nonconvex problems that are proven to converge faster to stationary points than gradient descent does, in reality when training deep neural networks they still cannot beat gradient descent.

On the other hand, optimization actually involves multiple aspects of computation. If some details of computation are not treated appropriately, accelerated algorithms can be slow. For example, the singular value thresholding (SVT) operator is often encountered when solving low-rank models. However, if naively implementing it, full SVD will be needed and the computation can be extremely expensive.

© Springer Nature Singapore Pte Ltd. 2020
Z. Lin et al., *Accelerated Optimization for Machine Learning*,
https://doi.org/10.1007/978-981-15-2910-8_7

Actually, very often SVT could be done by partial SVD [7], whose computation can be much cheaper than that of full SVD. For distributed optimization, some pre-processing such as balancing the data according to the computing power and communication bandwidth (if possible) can help a lot.

Although many intricate accelerated algorithms have been proposed, there are fundamental complexity lower bounds that cannot be broken if the algorithms are designed in the traditional ways. Some algorithms have achieved the complexity lower bounds, if looking at the orders and ignoring the constant factors (e.g., [2, 5, 6, 9]). So it is not quite exciting to improve the constants. Recently, there has been some work on using machine learning techniques to boost convergence in optimization algorithms, which seem to break the complexity lower bounds when running on test data [1, 3, 8, 11, 12]. Although promising experimental results have been shown, rare algorithms have theoretical guarantee. Thus most learning-based optimization algorithms remain heuristic. Chen et al. might be the first to provide convergence guarantees [1]. However, their proof is only valid for the LASSO problem. Learning-based optimization algorithms for general problems, rather than a particular problem, that have convergence guarantees are scarce. Liu et al. [8] and Xie et al. [10], which aim at solving nonconvex inverse problems and linearly constrained separable convex problems, respectively, are among the very limited literatures. It is not surprising that when considering characteristics of data, optimization can be accelerated. An example from the traditional algorithms is assuming that the dictionary matrix in the LASSO problem has the Restricted Isometric Property or the sampling operator in the matrix completion problem has the Matrix Restricted Isometric Property [4]. Learning-based optimization just aims at describing the characteristics of data more accurately, but currently only using samples themselves rather than mathematical properties. Although learning-based optimization is still in its infant age, we believe that it will bring another wave of acceleration in the future.

References

1. X. Chen, J. Liu, Z. Wang, W. Yin, Theoretical linear convergence of unfolded ISTA and its practical weights and thresholds, in *Advances in Neural Information Processing Systems*, Montreal, vol. 31 (2018), pp. 9079–9089
2. C. Fang, C.J. Li, Z. Lin, T. Zhang, SPIDER: near-optimal non-convex optimization via stochastic path-integrated differential estimator, in *Advances in Neural Information Processing Systems*, Montreal, vol. 31 (2018), pp. 689–699
3. K. Gregor, Y. LeCun, Learning fast approximations of sparse coding, in *Proceedings of the 27th International Conference on Machine Learning*, Haifa, (2010), pp. 399–406
4. M.-J. Lai, W. Yin, Augmented ℓ_1 and nuclear-norm models with a globally linearly convergent algorithm. SIAM J. Imag. Sci. **6**(2), 1059–1091 (2013)
5. G. Lan, Y. Zhou, An optimal randomized incremental gradient method. Math. Program. **171**(1–2), 167–215 (2018)
6. H. Li, Z. Lin, Accelerated alternating direction method of multipliers: an optimal $O(1/K)$ nonergodic analysis. J. Sci. Comput. **79**(2), 671–699 (2019)

7. Z. Lin, H. Zhang, *Low-Rank Models in Visual Analysis: Theories, Algorithms, and Applications* (Academic, New York, 2017)
8. R. Liu, S. Cheng, Y. He, X. Fan, Z. Lin, Z. Luo, On the convergence of learning-based iterative methods for nonconvex inverse problems. IEEE Trans. Pattern Anal. Mach. Intell. (2020). https://doi.org/10.1109/TPAMI.2019.2920591
9. K. Seaman, F. Bach, S. Bubeck, Y.T. Lee, L. Massoulié, Optimal algorithms for smooth and strongly convex distributed optimization in networks, in *Proceedings of the 34th International Conference on Machine Learning*, Sydney, (2017), pp. 3027–3036
10. X. Xie, J. Wu, G. Liu, Z. Zhong, Z. Lin, Differentiable linearized ADMM, in *Proceedings of the 36th International Conference on Machine Learning*, Long Beach, (2019), pp. 6902–6911
11. Y. Yang, J. Sun, H. Li, Z. Xu, Deep ADMM-Net for compressive sensing MRI, in *Advances in Neural Information Processing Systems*, Barcelona, vol. 29 (2016), pp. 10–18
12. J. Zhang, B. Ghanem, ISTA-Net: interpretable optimization-inspired deep network for image compressive sensing, in *Proceedings of the IEEE Conference on Computer Vision and Pattern Recognition*, Salt Lake, (2018), pp. 1828–1837

Appendix A
Mathematical Preliminaries

In this appendix, we list the conventions of notations and some basic definitions and facts that are used in the book.

A.1 Notations

Notations	Meanings
Normal font, e.g., s	A scalar
Bold lowercase, e.g., $\mathbf{v}$	A vector
Bold capital, e.g., $\mathbf{M}$	A matrix
Calligraphic capital, e.g., $\mathcal{T}$	A subspace, an operator, or a set
$\mathbb{R}, \mathbb{R}^+$	Set of real numbers, set of nonnegative real numbers
$\mathbb{Z}^+$	Set of nonnegative intergers
$[n]$	$\{1, 2, \cdots, n\}$
$\mathbb{E}X$	Expectation of random variable (or random vector) X
$\mathbf{I}, \mathbf{0}, \mathbf{1}$	The identity matrix, all-zero matrix or vector, and all-one vector
$\mathbf{x} \geq \mathbf{y}$	$\mathbf{x} - \mathbf{y}$ is a nonnegative vector
$\mathbf{X} \succeq \mathbf{Y}$	$\mathbf{X} - \mathbf{Y}$ is a positive semi-definite matrix
$f(N) = O(g(N))$	$\exists a > 0$, such that $\frac{f(N)}{g(N)} \leq a$ for all $N \in \mathbb{Z}^+$
$f(N) = \tilde{O}(g(N))$	$\exists a > 0$, such that $\frac{\tilde{f}(N)}{g(N)} \leq a$ for all $N \in \mathbb{Z}^+$, where $\tilde{f}(N)$ is the function ignoring poly-logarithmic factors in $f(N)$

© Springer Nature Singapore Pte Ltd. 2020
Z. Lin et al., *Accelerated Optimization for Machine Learning*,
https://doi.org/10.1007/978-981-15-2910-8

$f(N) = \Omega(g(N))$	$\exists a > 0$, such that $\frac{f(N)}{g(N)} \geq a$ for all $N \in \mathbb{Z}^+$
$\mathbf{x}_i$	The i-th vector in a sequence or the i-th coordinate of $\mathbf{x}$
$\nabla f(\mathbf{x})$	Gradient of f at $\mathbf{x}$
$\nabla_i f(\mathbf{x})$	$\frac{\partial f}{\partial \mathbf{x}_i}$
$\mathbf{X}_{:j}$	The j-th column of matrix $\mathbf{X}$
$\mathbf{X}_{ij}$	The entry at the i-th row and the j-th column of $\mathbf{X}$
$\mathbf{X}^T$	Transpose of matrix $\mathbf{X}$
$\mathrm{Diag}(\mathbf{x})$	Diagonal matrix whose diagonal entries are entries of vector $\mathbf{x}$
$\sigma_i(\mathbf{X})$	The i-th largest singular value of matrix $\mathbf{X}$
$\lambda_i(\mathbf{X})$	The i-th largest eigenvalue of matrix $\mathbf{X}$
$|\mathbf{X}|$	Matrix whose (i, j)-th entry is $|\mathbf{X}_{ij}|$
$\mathrm{Span}(\mathbf{X})$	The subspace spanned by the columns of $\mathbf{X}$
$\|\cdot\|$	Operator norm of an operator or a matrix
$\|\cdot\|_2$ or $\|\cdot\|$	ℓ_2 norm of vectors, $\|\mathbf{v}\|_2 = \sqrt{\sum_i \mathbf{v}_i^2}$; $\|\cdot\|$ is also used for general norm of vectors
$\|\cdot\|_*$	Nuclear norm of matrices, the sum of singular values
$\|\cdot\|_0$	ℓ_0 pseudo-norm, number of nonzero entries
$\|\cdot\|_1$	ℓ_1 norm, $\|\mathbf{X}\|_1 = \sum_{i,j} |\mathbf{X}_{ij}|$
$\|\cdot\|_p$	ℓ_p norm, $\|\mathbf{X}\|_p = \left(\sum_{i,j} |\mathbf{X}_{ij}|^p \right)^{1/p}$
$\|\cdot\|_F$	Frobenius norm of a matrix, $\|\mathbf{X}\|_F = \sqrt{\sum_{i,j} \mathbf{X}_{ij}^2}$
$\|\cdot\|_\infty$	ℓ_∞ norm, $\|\mathbf{X}\|_\infty = \max_{ij} |\mathbf{X}_{ij}|$
$\mathrm{conv}(\mathcal{X})$	Convex hull of set $\mathcal{X}$
∂f	Subgradient (resp. supergradient) of a convex (resp concave) function f
f^*	Optimum value of $f(\mathbf{x})$, where $\mathbf{x}$ varies in $\mathrm{dom} f$ and the constraints
$f^*(\mathbf{x})$	The conjugate function of $f(\mathbf{x})$
$\mathrm{Prox}_{\alpha f}(\cdot)$	Proximal mapping w.r.t. f and parameter α, $\mathrm{Prox}_{\alpha f}(\mathbf{y}) = \mathrm{argmin}_{\mathbf{x}} \left(\alpha f(\mathbf{x}) + \frac{1}{2} \|\mathbf{x} - \mathbf{y}\|_2^2 \right)$

A.2   Algebra and Probability

Proposition A.1 (Cauchy-Schwartz Inequality)

$$\langle \mathbf{x}, \mathbf{y} \rangle \leq \|\mathbf{x}\| \|\mathbf{y}\|.$$

Lemma A.1 *For any* $\mathbf{x}, \mathbf{y}, \mathbf{z}$, *and* $\mathbf{w} \in \mathbb{R}^n$, *we have the following three identities:*

$$2 \langle \mathbf{x}, \mathbf{y} \rangle = \|\mathbf{x}\|^2 + \|\mathbf{y}\|^2 - \|\mathbf{x} - \mathbf{y}\|^2, \tag{A.1}$$

$$2 \langle \mathbf{x}, \mathbf{y} \rangle = \|\mathbf{x} + \mathbf{y}\|^2 - \|\mathbf{x}\|^2 - \|\mathbf{y}\|^2, \tag{A.2}$$

$$2 \langle \mathbf{x} - \mathbf{z}, \mathbf{y} - \mathbf{w} \rangle = \|\mathbf{x} - \mathbf{w}\|^2 - \|\mathbf{z} - \mathbf{w}\|^2 - \|\mathbf{x} - \mathbf{y}\|^2 + \|\mathbf{z} - \mathbf{y}\|^2. \tag{A.3}$$

Definition A.1 (Singular Value Decomposition (SVD)) Suppose that $\mathbf{A} \in \mathbb{R}^{m \times n}$ with rank$\mathbf{A} = r$. Then $\mathbf{A}$ can be factorized as

$$\mathbf{A} = \mathbf{U} \mathbf{\Sigma} \mathbf{V}^\top,$$

where $\mathbf{U} \in \mathbb{R}^{m \times r}$ satisfies $\mathbf{U}^\top \mathbf{U} = \mathbf{I}$, $\mathbf{V} \in \mathbb{R}^{n \times r}$ satisfies $\mathbf{V}^\top \mathbf{V} = \mathbf{I}$, and $\mathbf{\Sigma} = $ Diag$(\sigma_1, \cdots, \sigma_r)$, with

$$\sigma_1 \geq \sigma_2 \geq \cdots \geq \sigma_r > 0.$$

The factorization (1) is called the *economic singular value decomposition* (SVD) of $\mathbf{A}$. The columns of $\mathbf{U}$ are called *left singular vectors* of $\mathbf{A}$, the columns of $\mathbf{V}$ are *right singular vectors*, and the numbers σ_i are the *singular values*.

Definition A.2 (Laplacian Matrix of a Graph) Denote a graph as $\mathfrak{g} = \{V, E\}$, where V and E are the node and the edge sets, respectively. $e_{ij} = (i, j) \in E$ indicates that nodes i and j are connected. Define $V_i = \{j \in V | (i, j) \in E\}$ to be the index set of the nodes that are connected to node i. The Laplacian matrix of the graph $\mathfrak{g} = \{V, E\}$ is defined as:

$$\mathbf{L}_{ij} = \begin{cases} |V_i|, & \text{if } i = j, \\ -1, & \text{if } i \neq j \text{ and } (i, j) \in E, \\ 0, & \text{otherwise.} \end{cases}$$

Definition A.3 (Dual Norm) Let $\| \cdot \|$ be a norm of vectors in $\mathbb{R}^n$, then its dual norm $\| \cdot \|^*$ is defined as:

$$\|\mathbf{y}\|^* = \max\{\langle \mathbf{x}, \mathbf{y} \rangle \,|\, \|\mathbf{x}\| \leq 1\}.$$

Proposition A.2 *Given random vector* $\boldsymbol{\xi}$, *we have*

$$\mathbb{E}\|\boldsymbol{\xi} - \mathbb{E}\boldsymbol{\xi}\|^2 \leq \mathbb{E}\|\boldsymbol{\xi}\|^2.$$

Proposition A.3 (Jensen's Inequality: Continuous Case) *If $f : C \subseteq \mathbb{R}^n \to \mathbb{R}$ is convex and $\boldsymbol{\xi}$ is a random vector over C, then*

$$f(\mathbb{E}\boldsymbol{\xi}) \leq \mathbb{E}f(\boldsymbol{\xi}).$$

Definition A.4 (Discrete-Time Martingale) A sequence of random variables (or random vectors) $X_1, X_2, \cdots$ is called a martingale if it satisfies for any time n,

$$\mathbb{E}|X_n| < \infty,$$

$$\mathbb{E}(X_{n+1}|X_1, \cdots, X_n) = X_n.$$

That is, the conditional expected value of the next observation, given all the past observations, is equal to the most recent observation.

Proposition A.4 (Iterated Law of Expectation) *For two random variables X and Y, we have*

$$\mathbb{E}Y = \mathbb{E}_X(\mathbb{E}(Y|X)).$$

A.3   Convex Analysis

The descriptions for the basic concepts of convex sets and convex functions can be found in [4].

Definition A.5 (Convex Set) A set $C \subseteq \mathbb{R}^n$ is called convex if for all $\mathbf{x}, \mathbf{y} \in C$ and $\alpha \in [0, 1]$ we have $\alpha\mathbf{x} + (1 - \alpha)\mathbf{y} \in C$.

Definition A.6 (Extreme Point) Given a nonempty convex set C, a vector $\mathbf{x} \in C$ is said to be an extreme point of C if it does not lie strictly between the endpoints of any line segment contained in C. Namely, there do not exist vectors $\mathbf{y}, \mathbf{z} \in C$, with $\mathbf{y} \neq \mathbf{x}$ and $\mathbf{z} \neq \mathbf{x}$, and a scalar $\alpha \in (0, 1)$ such that $\mathbf{x} = \alpha\mathbf{y} + (1 - \alpha)\mathbf{z}$.

Definition A.7 (Convex Function) A function $f : C \subseteq \mathbb{R}^n \to \mathbb{R}$ is called convex if C is a convex set and for all $\mathbf{x}, \mathbf{y} \in C$ and $\alpha \in [0, 1]$ we have

$$f(\alpha\mathbf{x} + (1 - \alpha)\mathbf{y}) \leq \alpha f(\mathbf{x}) + (1 - \alpha)f(\mathbf{y}).$$

C is called the domain of f.

Definition A.8 (Concave Function) A function $f : C \subseteq \mathbb{R}^n \to \mathbb{R}$ is called concave if $-f$ is convex.

Definition A.9 (Strictly Convex Function) A function $f : C \subseteq \mathbb{R}^n \to \mathbb{R}$ is called strictly convex if C is a convex set and for all $\mathbf{x} \neq \mathbf{y} \in C$ and $\alpha \in (0, 1)$ we have

$$f(\alpha \mathbf{x} + (1 - \alpha)\mathbf{y}) < \alpha f(\mathbf{x}) + (1 - \alpha) f(\mathbf{y}).$$

Definition A.10 (Strongly Convex Function and Generally Convex Function) A function $f : C \subseteq \mathbb{R}^n \to \mathbb{R}$ is called strongly convex if C is a convex set and there exists a constant $\mu > 0$ such that for all $\mathbf{x}, \mathbf{y} \in C$ and $\alpha \in [0, 1]$ we have

$$f(\alpha \mathbf{x} + (1 - \alpha)\mathbf{y}) \leq \alpha f(\mathbf{x}) + (1 - \alpha) f(\mathbf{y}) - \frac{\mu \alpha (1 - \alpha)}{2} \|\mathbf{y} - \mathbf{x}\|^2.$$

μ is called the *strong convexity modulus* of f. For brevity, a strongly convex function with a strong convexity modulus μ is called a *μ-strongly convex function*. If a convex function is not strongly convex, we also call it a *generally convex function*.

Proposition A.5 (Jensen's Inequality: Discrete Case) *If $f : C \subseteq \mathbb{R}^n \to \mathbb{R}$ is convex, $\mathbf{x}_i \in C$, $\alpha_i \geq 0$, $i = 1, \cdots, m$, and $\sum_{i=1}^{m} \alpha_i = 1$, then*

$$f\left(\sum_{i=1}^{m} \alpha_i \mathbf{x}_i\right) \leq \sum_{i=1}^{m} \alpha_i f(\mathbf{x}_i).$$

Definition A.11 (Smooth Function) A function is (informally) called smooth if it is continuously differentiable.

Definition A.12 (Function with Lipschitz Continuous Gradients) A differentiable function $f : C \subseteq \mathbb{R}^n \to \mathbb{R}$ is called to have Lipschitz continuous gradients if there exists $L > 0$ such that

$$\|\nabla f(\mathbf{x}) - \nabla f(\mathbf{y})\| \leq L \|\mathbf{y} - \mathbf{x}\|, \quad \forall \mathbf{x}, \mathbf{y} \in C.$$

For simplicity, if the constant L is explicitly specified we also call such a function an *L-smooth function*.

Definition A.13 (Function with Coordinate Lipschitz Continuous Gradients) $f(\mathbf{x})$ is said to have L_c-coordinate Lipschitz continuous gradients if:

$$|\nabla_i f(\mathbf{x}) - \nabla_i f(\mathbf{y})| \leq L_c |\mathbf{x}_i - \mathbf{y}_i|, \quad \text{all } i \in [n]. \tag{A.4}$$

Definition A.14 (Function with Lipschitz Continuous Hessians) A twice differentiable function $f : C \subseteq \mathbb{R}^n \to \mathbb{R}$ is called to have Lipschitz continuous Hessians if there exists $L > 0$ such that

$$\|\nabla^2 f(\mathbf{x}) - \nabla^2 f(\mathbf{y})\| \leq L \|\mathbf{y} - \mathbf{x}\|, \quad \forall \mathbf{x}, \mathbf{y} \in C.$$

If the constant L is explicitly specified such a function is also called to have L-Lipschitz continuous Hessians.

Proposition A.6 ([6]) *If $f : C \subseteq \mathbb{R}^n \to \mathbb{R}$ is L-smooth, then*

$$|f(\mathbf{y}) - f(\mathbf{x}) - \langle \nabla f(\mathbf{x}), \mathbf{y} - \mathbf{x} \rangle| \le \frac{L}{2} \|\mathbf{y} - \mathbf{x}\|^2, \quad \forall \mathbf{x}, \mathbf{y} \in C. \tag{A.5}$$

In particular, if $\mathbf{y} = \mathbf{x} - \frac{1}{L} \nabla f(\mathbf{x})$, then

$$f(\mathbf{y}) \le f(\mathbf{x}) - \frac{1}{2L} \|\nabla f(\mathbf{x})\|^2. \tag{A.6}$$

If f is further convex, then

$$f(\mathbf{y}) \ge f(\mathbf{x}) + \langle \nabla f(\mathbf{x}), \mathbf{y} - \mathbf{x} \rangle + \frac{1}{2L} \|\nabla f(\mathbf{y}) - \nabla f(\mathbf{x})\|^2. \tag{A.7}$$

Proposition A.7 *If $f : C \subseteq \mathbb{R}^n \to \mathbb{R}$ has L_c-coordinate Lipschitz continuous gradients on coordinate i and $\mathbf{x}$ and $\mathbf{y}$ only differ at the i-th entry, then we have*

$$|f(\mathbf{x}) - f(\mathbf{y}) - \langle \nabla_i f(\mathbf{y}), \mathbf{x}_i - \mathbf{y}_i \rangle| \le \frac{L_c}{2} (\mathbf{x}_i - \mathbf{y}_i)^2.$$

Proposition A.8 *If $f : C \subseteq \mathbb{R}^n \to \mathbb{R}$ has L-Lipschitz continuous Hessians, then* [6]

$$\left| f(\mathbf{y}) - f(\mathbf{x}) - \langle \nabla f(\mathbf{x}), \mathbf{y} - \mathbf{x} \rangle - \frac{1}{2}(\mathbf{y} - \mathbf{x})^T \nabla^2 f(\mathbf{x})(\mathbf{y} - \mathbf{x}) \right| \le \frac{L}{6} \|\mathbf{y} - \mathbf{x}\|^2,$$

$$\forall \mathbf{x}, \mathbf{y} \in C. \tag{A.8}$$

Definition A.15 (Subgradient of a Convex Function) A vector $\mathbf{g}$ is called a subgradient of a convex function $f : C \subseteq \mathbb{R}^n \to \mathbb{R}$ at $\mathbf{x} \in C$ if

$$f(\mathbf{y}) \ge f(\mathbf{x}) + \langle \mathbf{g}, \mathbf{y} - \mathbf{x} \rangle, \forall \mathbf{y} \in C.$$

The set of subgradients at $\mathbf{x}$ is denoted as $\partial f(\mathbf{x})$.

Proposition A.9 *For convex function $f : C \subseteq \mathbb{R}^n \to \mathbb{R}$, its subgradient exists at every interior point of C. It is differentiable at $\mathbf{x}$ iff (aka if and only if) $\partial f(\mathbf{x})$ is a singleton.*

Proposition A.10 *If $f : \mathbb{R}^n \to \mathbb{R}$ is μ-strongly convex, then*

$$f(\mathbf{y}) \ge f(\mathbf{x}) + \langle \mathbf{g}, \mathbf{y} - \mathbf{x} \rangle + \frac{\mu}{2} \|\mathbf{y} - \mathbf{x}\|^2, \quad \forall \mathbf{g} \in \partial f(\mathbf{x}). \tag{A.9}$$

In particular, if f is differentiable and μ-strongly convex and $\mathbf{x}^* = \mathrm{argmin}_{\mathbf{x}} f(\mathbf{x})$, then

$$f(\mathbf{x}) - f(\mathbf{x}^*) \geq \frac{\mu}{2} \|\mathbf{x} - \mathbf{x}^*\|^2. \tag{A.10}$$

On the other hand, we can have

$$f(\mathbf{x}^*) \geq f(\mathbf{x}) - \frac{1}{2\mu} \|\nabla f(\mathbf{x})\|^2. \tag{A.11}$$

Definition A.16 (Epigraph) The epigraph of $f : C \subseteq \mathbb{R}^n \to \mathbb{R}$ is defined as

$$\mathrm{epi}\, f = \{(\mathbf{x}, t) | \mathbf{x} \in C, t \geq f(\mathbf{x})\}.$$

Definition A.17 (Closed Function) If epi f is a closed set, then f is called a closed function.

Definition A.18 (Monotone Mapping and Monotone Function) A set valued function $f : C \subseteq \mathbb{R}^n \to 2^{\mathbb{R}^n}$ is called a monotone mapping if

$$\langle \mathbf{x} - \mathbf{y}, \mathbf{u} - \mathbf{v} \rangle \geq 0, \quad \forall \mathbf{x}, \mathbf{y} \in C \text{ and } \mathbf{u} \in f(\mathbf{x}), \mathbf{v} \in f(\mathbf{y}).$$

In particular, if f is a single valued function and

$$\langle \mathbf{x} - \mathbf{y}, f(\mathbf{x}) - f(\mathbf{y}) \rangle \geq 0, \quad \forall \mathbf{x}, \mathbf{y} \in C.$$

then it is called a monotone function.

Proposition A.11 (Monotonicity of Subgradient) *If $f : C \subseteq \mathbb{R}^n \to \mathbb{R}$ is convex, then $\partial f(\mathbf{x})$ is a monotone mapping. If f is further μ-strongly convex, then*

$$\langle \mathbf{x}_1 - \mathbf{x}_2, \mathbf{g}_1 - \mathbf{g}_2 \rangle \geq \mu \|\mathbf{x}_1 - \mathbf{x}_2\|^2, \quad \forall \mathbf{x}_i \in C \text{ and } \mathbf{g}_i \in \partial f(\mathbf{x}_i), i = 1, 2.$$

Definition A.19 (Envelope Function and Proximal Mapping) Given a function $f : C \subseteq \mathbb{R}^n \to \mathbb{R}$ and $a > 0$,

$$\mathrm{Env}_{af}(\mathbf{x}) = \min_{\mathbf{y} \in C} \left(f(\mathbf{y}) + \frac{1}{2a} \|\mathbf{y} - \mathbf{x}\|^2 \right)$$

is called the envelope function of $f(\mathbf{x})$, and

$$\mathrm{Prox}_{af}(\mathbf{x}) = \mathrm{argmin}_{\mathbf{y} \in C} \left(f(\mathbf{y}) + \frac{1}{2a} \|\mathbf{y} - \mathbf{x}\|^2 \right)$$

is called the proximal mapping of $f(\mathbf{x})$. $\mathrm{Prox}_{af}(\mathbf{x})$ may be set-valued if f is not convex.

Further descriptions of proximal mapping can be found in [7].

Definition A.20 (Bregman Distance) Given a differentiable strongly convex function h, the Bregman distance is defined as

$$D_h(\mathbf{y}, \mathbf{x}) = h(\mathbf{y}) - h(\mathbf{x}) - \langle \nabla h(\mathbf{x}), \mathbf{y} - \mathbf{x} \rangle .$$

The Euclidean distance is obtained when $h(\mathbf{x}) = \frac{1}{2}\|\mathbf{x}\|^2$, in which case $D_h(\mathbf{y}, \mathbf{x}) = \frac{1}{2}\|\mathbf{x} - \mathbf{y}\|^2$. The generalization to a nondifferentiable h was discussed in [5].

Definition A.21 (Conjugate Function) Given $f : C \subseteq \mathbb{R}^n \to \mathbb{R}$, its conjugate function is defined as

$$f^*(\mathbf{u}) = \sup_{\mathbf{z} \in C} (\langle \mathbf{z}, \mathbf{u} \rangle - f(\mathbf{z})) .$$

The domain of f^* is

$$\operatorname{dom} f^* = \{\mathbf{u} | f^*(\mathbf{u}) < +\infty\}.$$

Proposition A.12 (Properties of Conjugate Function) *Given $f : C \subseteq \mathbb{R}^n \to \mathbb{R}$, its conjugate function has the following properties:*

1. *f^* is always a convex function;*
2. *$f^{**}(\mathbf{x}) \leq f(\mathbf{x})$, $\forall \mathbf{x} \in C$;*
3. *If f is a proper, closed and convex function, then $f^{**}(\mathbf{x}) = f(\mathbf{x})$, $\forall \mathbf{x} \in C$.*
4. *If f is L-smooth, then f^* is L^{-1}-strongly convex on $\operatorname{dom} f^*$. Conversely, if f is μ-strongly convex, then f^* is μ^{-1}-smooth on $\operatorname{dom} f^*$.*
5. *If f is closed and convex, then $\mathbf{y} \in \partial f(\mathbf{x})$ if and only if $\mathbf{x} \in \partial f^*(\mathbf{y})$.*

Proposition A.13 (Fenchel-Young Inequality) *Let f^* be the conjugate function of f, then*

$$f(\mathbf{x}) + f^*(\mathbf{y}) \geq \langle \mathbf{x}, \mathbf{y} \rangle .$$

Definition A.22 (Lagrangian Function) Given a constrained problem:

$$\min_{\mathbf{x} \in \mathbb{R}^n} \ f(\mathbf{x}), \tag{A.12}$$

$$s.t. \ \mathbf{A}\mathbf{x} = \mathbf{b},$$

$$\mathbf{g}(\mathbf{x}) \leq \mathbf{0},$$

where $\mathbf{A} \in \mathbb{R}^{m \times n}$ and $\mathbf{g}(\mathbf{x}) = (g_1(\mathbf{x}), \cdots, g_p(\mathbf{x}))^T$, the Lagrangian function is

$$L(\mathbf{x}, \mathbf{u}, \mathbf{v}) = f(\mathbf{x}) + \langle \mathbf{u}, \mathbf{A}\mathbf{x} - \mathbf{b} \rangle + \langle \mathbf{v}, \mathbf{g}(\mathbf{x}) \rangle ,$$

where $\mathbf{v} \geq \mathbf{0}$.

Definition A.23 (Lagrange Dual Function) Given a constrained problem (A.12), the Lagrange dual function is

$$d(\mathbf{u}, \mathbf{v}) = \min_{\mathbf{x} \in C} L(\mathbf{x}, \mathbf{u}, \mathbf{v}), \tag{A.13}$$

where C is the domain of f. The domain of the dual function is $\mathcal{D} = \{(\mathbf{u}, \mathbf{v}) | d(\mathbf{u}, \mathbf{v}) > -\infty\}$.

Definition A.24 (Dual Problem) Given a constrained problem (A.12), the dual problem is

$$\max_{\mathbf{u}, \mathbf{v}} \ d(\mathbf{u}, \mathbf{v}),$$

$$s.t. \ \mathbf{v} \geq \mathbf{0}.$$

Accordingly, problem (A.12) is called the primal problem.

Definition A.25 (Slater's Condition) For convex primal problem (A.12), if there exists an $\mathbf{x}_0$ such that $\mathbf{A}\mathbf{x}_0 = \mathbf{b}$, $g_i(\mathbf{x}_0) \leq 0, i \in \mathcal{I}_1$, and $g_i(\mathbf{x}_0) < 0, i \in \mathcal{I}_2$, where $\mathcal{I}_1$ and $\mathcal{I}_2$ are the sets of indices of linear and nonlinear inequality constraints, respectively, then the Slater's condition holds.

Proposition A.14 (Properties of Dual Problem)

1. $d(\mathbf{u}, \mathbf{v})$ *is always a concave function even if the primal problem (A.12) is not convex.*
2. *The primal and the dual optimal values, f^* and d^*, always satisfy the weak duality: $f^* \geq d^*$.*
3. *When the Slater's condition holds, the strong duality holds: $f^* = d^*$.*

Definition A.26 (KKT Point and KKT Condition) $(\mathbf{x}, \mathbf{u}, \mathbf{v})$ is called a Karush–Kuhn–Tucker (KKT) point of problem (A.12) if

1. Stationarity: $\mathbf{0} \in \partial f(\mathbf{x}) + \mathbf{A}^T \mathbf{u} + \sum_{i=1}^{p} \mathbf{v}_i \partial g_i(\mathbf{x})$.
2. Primal feasibility: $\mathbf{A}\mathbf{x} = \mathbf{b}$, $g_i(\mathbf{x}) \leq 0, i = 1, \cdots, p$.
3. Complementary slackness: $\mathbf{v}_i g_i(\mathbf{x}) = 0$.
4. Dual feasibility: $\mathbf{v}_i \geq \mathbf{0}, i = 1, \cdots, p$.

The above conditions are called the KKT condition of problem (A.12). They are the optimality condition of problem (A.12) when $f(\mathbf{x})$ and $g_i(\mathbf{x}), i = 1, \cdots, p$, are all convex.

Proposition A.15 *When $f(\mathbf{x})$ and $g_i(\mathbf{x}), i = 1, \cdots, p$, are all convex, $(\mathbf{x}^*, \mathbf{u}^*, \mathbf{v}^*)$ is a pair of the primal and the dual solutions with zero dual gap if and only if it satisfies the KKT condition.*

Definition A.27 (Compact Set) A subset S of $\mathbb{R}^n$ is called compact if it is both bounded and closed.

Definition A.28 (Convex Hull) The convex hull of a set X, denoted as $\text{conv}(X)$, is the set of all convex combinations of points in X:

$$\text{conv}(X) = \left\{ \sum_{i=1}^{k} \alpha_i \mathbf{x}_i \,\middle|\, \mathbf{x}_i \in X, \alpha_i \geq 0, i = 1, \cdots, k, \sum_{i=1}^{k} \alpha_i = 1 \right\}.$$

Theorem A.1 (Danskin's Theorem) *Let $\mathcal{Z}$ be a compact subset of $\mathbb{R}^m$, and let $\phi : \mathbb{R}^n \times \mathcal{Z} \to \mathbb{R}$ be continuous and such that $\phi(\cdot, \mathbf{z}) : \mathbb{R}^n \to \mathbb{R}$ is convex for each $\mathbf{z} \in \mathcal{Z}$. Define $f : \mathbb{R}^n \to \mathbb{R}$ by $f(\mathbf{x}) = \max_{\mathbf{z} \in \mathcal{Z}} \phi(\mathbf{x}, \mathbf{z})$ and*

$$\mathcal{Z}(\mathbf{x}) = \left\{ \bar{\mathbf{z}} \,\middle|\, \phi(\mathbf{x}, \bar{\mathbf{z}}) = \max_{\mathbf{z} \in \mathcal{Z}} \phi(\mathbf{x}, \mathbf{z}) \right\}.$$

If $\phi(\cdot, \mathbf{z})$ is differentiable for all $\mathbf{z} \in \mathcal{Z}$ and $\nabla_x \phi(\mathbf{x}, \cdot)$ is continuous on $\mathcal{Z}$ for each $\mathbf{x}$, then

$$\partial f(\mathbf{x}) = \text{conv} \left\{ \nabla_x \phi(\mathbf{x}, \mathbf{z}) | \mathbf{z} \in \mathcal{Z}(\mathbf{x}) \right\}, \quad \forall \mathbf{x} \in \mathbb{R}^n.$$

Definition A.29 (Saddle Point) $(\mathbf{x}^*, \boldsymbol{\lambda}^*)$ is called a saddle point of function $f(\mathbf{x}, \boldsymbol{\lambda}) : C \times D \to \mathbb{R}$ if it satisfies the following inequalities:

$$f(\mathbf{x}^*, \boldsymbol{\lambda}) \leq f(\mathbf{x}^*, \boldsymbol{\lambda}^*) \leq f(\mathbf{x}, \boldsymbol{\lambda}^*), \quad \forall \mathbf{x} \in C, \boldsymbol{\lambda} \in D.$$

A.4   Nonconvex Analysis

Definition A.30 (Proper Function) A function $g : \mathbb{R}^n \to (-\infty, +\infty]$ is said to be proper if $\text{dom } g \neq \emptyset$, where $\text{dom } g = \{\mathbf{x} \in \mathbb{R} : g(\mathbf{x}) < +\infty\}$.

Definition A.31 (Lower Semicontinuous Function) A function $g : \mathbb{R}^n \to (-\infty, +\infty]$ is said to be lower semicontinuous at point $\mathbf{x}_0$ if

$$\liminf_{\mathbf{x} \to \mathbf{x}_0} g(\mathbf{x}) \geq g(\mathbf{x}_0).$$

Definition A.32 (Coercive Function) $F(\mathbf{x})$ is called coercive if $\inf_{\mathbf{x}} F(\mathbf{x}) > -\infty$ and $\{\mathbf{x} | F(\mathbf{x}) \leq a\}$ is bounded for all a.

Definition A.33 (Subdifferential) Let f be a proper and lower semicontinuous function.

1. For a given $\mathbf{x} \in$ dom f, the Frechel subdifferential of f at $\mathbf{x}$, written as $\hat{\partial} f(\mathbf{x})$, is the set of all vectors $\mathbf{u} \in \mathbb{R}^n$ which satisfies

$$\liminf_{\mathbf{y} \neq \mathbf{x}, \mathbf{y} \to \mathbf{x}} \frac{f(\mathbf{y}) - f(\mathbf{x}) - \langle \mathbf{u}, \mathbf{y} - \mathbf{x} \rangle}{\|\mathbf{y} - \mathbf{x}\|} \geq 0.$$

2. The limiting subdifferential, or simply the subdifferential, of f at $\mathbf{x} \in \mathbb{R}^n$, written as $\partial f(\mathbf{x})$, is defined through the following closure process:

$$\partial f(\mathbf{x}) := \{\mathbf{u} \in \mathbb{R}^n : \exists \mathbf{x}_k \to \mathbf{x}, f(\mathbf{x}_k) \to f(\mathbf{x}), \mathbf{u}_k \in \hat{\partial} f(\mathbf{x}_k) \to \mathbf{u}, k \to \infty\}.$$

Definition A.34 (Critical Point) A point $\mathbf{x}$ is called a critical point of function f if $\mathbf{0} \in \partial f(\mathbf{x})$.

The following lemma describes the properties of subdifferential.

Lemma A.2

1. *In the nonconvex context, Fermat's rule remains unchanged: If $\mathbf{x} \in \mathbb{R}^n$ is a local minimizer of g, then $\mathbf{0} \in \partial g(\mathbf{x})$.*
2. *Let $(\mathbf{x}_k, \mathbf{u}_k)$ be a sequence such that $\mathbf{x}_k \to \mathbf{x}$, $\mathbf{u}_k \to \mathbf{u}$, $g(\mathbf{x}_k) \to g(\mathbf{x})$, and $\mathbf{u}_k \in \partial g(\mathbf{x}_k)$, then $\mathbf{u} \in \partial g(\mathbf{x})$.*
3. *If f is a continuously differentiable function, then $\partial(f+g)(\mathbf{x}) = \nabla f(\mathbf{x}) + \partial g(\mathbf{x})$.*

Definition A.35 (Desingularizing Function) A function $\varphi : [0, \eta) \to \mathbb{R}^+$ satisfying the following conditions is called a desingularizing function:

(1) φ is concave and continuously differentiable on $(0, \eta)$;
(2) φ is continuous at 0, $\varphi(0) = 0$; and
(3) $\varphi'(\mathbf{x}) > 0, \forall \mathbf{x} \in (0, \eta)$.

Φ_η is the set of desingularizing functions defined on $[0, \eta)$.

Now we define the KŁ function. More introductions and applications can be found in [1–3].

Definition A.36 (Kurdyka–Łojasiewicz (KŁ) Property) A function $f : \mathbb{R}^n \to (-\infty, +\infty]$ is said to have the Kurdyka-Łojasiewicz (KŁ) property at $\bar{\mathbf{u}} \in$ dom$\partial f := \{\mathbf{x} \in \mathbb{R}^n : \partial f(\mathbf{u}) \neq \emptyset\}$ if there exists $\eta \in (0, +\infty]$, a neighborhood U of $\bar{\mathbf{u}}$, and a desingularizing function $\varphi \in \Phi_\eta$, such that for all

$$\mathbf{u} \in U \cap \{\mathbf{u} \in \mathbb{R}^n : f(\bar{\mathbf{u}}) < f(\mathbf{u}) < f(\bar{\mathbf{u}}) + \eta\},$$

the following inequality holds:

$$\varphi'(f(\mathbf{u}) - f(\bar{\mathbf{u}}))\text{dist}(\mathbf{0}, \partial f(\mathbf{u})) > 1.$$

Lemma A.3 (Uniform Kurdyka-Łojasiewicz Property) *Let Ω be a compact set and let $f : \mathbb{R}^n \to (-\infty, +\infty]$ be a proper and lower semicontinuous function. Assume that f is constant on Ω and satisfies the KŁ property at each point of Ω. Then there exists $\epsilon > 0$, $\eta > 0$, and $\varphi \in \Phi_\eta$, such that for all $\overline{\mathbf{u}}$ in Ω and all $\mathbf{u}$ in the following intersection*

$$\{\mathbf{u} \in \mathbb{R}^n : \text{dist}(\mathbf{u}, \Omega) < \epsilon\} \cap \{\mathbf{u} \in \mathbb{R}^n : f(\overline{\mathbf{u}}) < f(\mathbf{u}) < f(\overline{\mathbf{u}}) + \eta\},$$

the following inequality holds:

$$\varphi'(f(\mathbf{u}) - f(\overline{\mathbf{u}}))\text{dist}(\mathbf{0}, \partial f(\mathbf{u})) > 1.$$

Functions satisfying the KŁ property are general enough. Typical examples include: real polynomial functions, logistic loss function $\log(1+e^{-t})$, $\|\mathbf{x}\|_p$ ($p \geq 0$), $\|\mathbf{x}\|_\infty$, and indicator functions of the positive semidefinite (PSD) cone, the Stiefel manifolds, and the set of constant rank matrices.

References

1. H. Attouch, J. Bolte, P. Redont, A. Soubeyran, Proximal alternating minimization and projection methods for nonconvex problems: an approach based on the Kurdyka-Łojasiewicz inequality. Math. Oper. Res. **35**(2), 438–457 (2010)
2. H. Attouch, J. Bolte, B.F. Svaiter, Convergence of descent methods for semi-algebraic and tame problems: proximal algorithms, forward-backward splitting, and regularized Gauss-Seidel methods. Math. Program. **137**(1–2), 91–129 (2013)
3. J. Bolte, S. Sabach, M. Teboulle, Proximal alternating linearized minimization for nonconvex and nonsmooth problems. Math. Program. **146**(1–2), 459–494 (2014)
4. S. Boyd, L. Vandenberghe, *Convex Optimization* (Cambridge University Press, Cambridge, 2004)
5. K.C. Kiwiel, Proximal minimization methods with generalized Bregman functions. SIAM J. Control. Optim. **35**(4), 1142–1168 (1997)
6. Y. Nesterov, *Introductory Lectures on Convex Optimization: A Basic Course* (Springer, New York, 2004)
7. N. Parikh, S. Boyd, Proximal algorithms. Found. Trends Optim. **1**(3), 127–239 (2014)

Index

© Springer Nature Singapore Pte Ltd. 2020
Z. Lin et al., *Accelerated Optimization for Machine Learning*,
https://doi.org/10.1007/978-981-15-2910-8